Trends in Logic

Volume 46

TRENDS IN LOGIC
Studia Logica Library

VOLUME 46

SCOPE OF THE SERIES

The book series Trends in Logic covers essentially the same areas as the journal Studia Logica, that is, contemporary formal logic and its applications and relations to other disciplines. The series aims at publishing monographs and thematically coherent volumes dealing with important developments in logic and presenting significant contributions to logical research.

The series is open to contributions devoted to topics ranging from algebraic logic, model theory, proof theory, philosophical logic, non-classical logic, and logic in computer science to mathematical linguistics and formal epistemology. However, this list is not exhaustive, moreover, the range of applications, comparisons and sources of inspiration is open and evolves over time.

More information about this series at http://www.springer.com/series/6645

Sven Ove Hansson

Descriptor Revision

Belief Change Through Direct Choice

Springer

Sven Ove Hansson
Division of Philosophy
Royal Institute of Technology
Stockholm
Sweden

ISSN 1572-6126 ISSN 2212-7313 (electronic)
Trends in Logic
ISBN 978-3-319-53060-4 ISBN 978-3-319-53061-1 (eBook)
DOI 10.1007/978-3-319-53061-1

Library of Congress Control Number: 2017933665

Printed on acid-free paper

This Springer imprint is published by Springer Nature
The registered company is Springer International Publishing AG
The registered company address is: Gewerbestrasse 11, 6330 Cham, Switzerland

Preface

The use of logical modelling to study belief change is a well-established research area since more than three decades. The focus is mostly on operations that take us from a belief state in which a particular sentence is believed to one in which it is not believed, or the other way around. In the major models, these operations are based on a mechanism that can be constructed as a choice among maximally consistent sets of sentences, "possible worlds." The development of these models has been accompanied by a critical discussion in which the resulting operations of change have been shown to have implausible properties.

This book proposes a retake for the logic of belief change. From a formal point of view, two innovations are combined to arrive at a new type of model, descriptor revision, in which the implausible properties are avoided. One of these innovations consists in basing the operation on a direct choice among potential outcomes (belief states) rather than indirectly on a choice among possible worlds or their equivalents. The other innovation is a new formal device, belief descriptors, that is used for describing the goals of a belief change. They allow us to represent a much wider array of belief changes than the incorporation or rejection of a single sentence. In combination with the new choice mechanism, they also allow us to perform all these changes in a unified way.

All formal proofs are deferred to an appendix. A list of symbols and a general index are included in order to facilitate the reader's orientation in the book.

This book has grown out of several years of research. References are given throughout the text to previous publications where this work was reported at earlier stages. Readers wishing to compare the book to its precursors may have use for the following references to the Bibliography: Sections 4.1–4.2 [135], Sections 4.3–4.5 and 5.1–5.2 [124], Section 5.4 [130], Section 5.5 [126], Sections 6.1–6.3 and 7.2–7.3 [132], Section 7.4 [129], Section 7.5 [131], Section 8.1 [124], Section 8.2 [124, 128], Section 8.3 [132], Section 8.4 [126], Section 9.1 [124], Section 9.3 [126], Section 10.3 [134], Section 10.5 [120], and Section 10.6 [117].

I have benefitted greatly in this work from critical discussions on seminars and workshops in Stockholm, Lund, Copenhagen, Amsterdam, Schloss Dagstuhl, Munich, Nancy, A Caruña, Madeira, Pittsburgh, São Paulo, Campinas, Beijing, and several other places. I am grateful to everyone who took part in these discussions. Special thanks go to Johan van Benthem, Gregor Betz, John Cantwell, Hans van Ditmarsch, Eduardo Fermé, Valentin Goranko, David Makinson, and Zhang Li for their comments on an earlier version of the manuscript.

Stockholm, Sweden
December 2016

Sven Ove Hansson

Contents

Symbols xi

Preview xv

Part I Why a New Approach?

1 The State of the Art 3
1.1 A Sentential, Input-Assimilating Approach 3
1.2 The AGM Construction 6
1.3 The AGM Postulates 7
1.4 Possible World Models 9
1.5 Epistemic Entrenchment 15
1.6 Conclusion 16

2 Inside the Black Box 17
2.1 Epistemic Choice 18
2.2 The Select-and-Intersect Method 19
2.3 Is the Intersection as Good as Its Origins? 21
2.4 Do All Success Conditions Withstand Intersection? 22
2.5 Do We Select Among the Right Objects? 23

3 Questionable Patterns of Change 27
3.1 The Postulates of Finite-Based Outcome 27
3.2 The Success Postulates 29
3.3 The Recovery Postulate 30
3.4 Pure Contraction and the Inclusion Postulate 31
3.5 The Expansion Property 33
3.6 Extending the Language 34
3.7 Iterated Change 37
3.8 Ten Desiderata for an Alternative Framework 42

Part II Descriptor Revision

4 Putting the Building-Blocks Together 45
4.1 Beginning Without Sentences 45
4.2 Support Functions 49
4.3 Belief Descriptors 53
4.4 Properties of Descriptors 55
4.5 Descriptor Revision Introduced 58
4.6 Conclusion 60

5 Local Descriptor Revision 63
5.1 Monoselective Descriptor Revision 64
5.2 Linear and Centrolinear Revision 67
5.3 Indeterministic Descriptor Revision 70
5.4 Blockage Revision 73
5.5 Relations of Epistemic Proximity 78

6 Global Descriptor Revision 85
6.1 Global Monoselective and Centrolinear Revision 86
6.2 Distance-Based Global Revision 87
6.3 Global Blockage Revision 91

7 Dynamic Descriptors 95
7.1 Representing Autoepistemic Beliefs 96
7.2 Ramsey Descriptors 98
7.3 The Logic of Ramsey Descriptors 100
7.4 Alternative Approaches to Conditionals 105
7.5 Non-Monotonic Inference 107
7.6 Modalities of Belief 111

Part III Sentential Change Revisited

8 Sentential Revision 117
8.1 Sentential Monoselective Revision 118
8.2 Sentential Centrolinear Revision 120
8.3 Global Sentential Revision 126
8.4 Believability Relations 127
8.5 Multiple Revision and Making up One's Mind 130

9 Revocation 133
9.1 Monoselective Revocation 134
9.2 Centrolinear Revocation 135
9.3 Entrenchment 139
9.4 Multiple Revocation and Entrenchment 143
9.5 A Revisionary Account of Giving up Beliefs 146

10 Contraction 151
10.1 How to Construct Contraction 151
10.2 Centrolinear and Maximal Contraction 152
10.3 Two Impossibility Theorems 154
10.4 Monomaximal Contraction 155
10.5 Perimaximal Contraction. 156
10.6 Bootstrap Contraction 159

Part IV Epilogue

11 Looking Back – and Ahead 165
11.1 Looking Back 165
11.2 Looking Ahead 166

Appendices

Proofs 171

Bibliography 219

Index 231

Symbols

Set Theory

$\in$	Element
$\subseteq$	Subset
$\subset$	Proper subset
$\cup, \bigcup$	Union (p. 56)
$\cap, \bigcap$	Intersection
$\setminus$	Difference
$\varnothing$	Empty set
$\wp$	Power set
$\times$	Cartesian product
$\aleph_0$	The smallest infinite cardinal number (p. 175)
ω	The smallest infinite ordinal number (p. 120)

The Object Language

$\mathcal{L}$	The object language (p. 9)
$p, q, \ldots$	Sentences in the object language
$A, B, \ldots$	Sets of sentences in the object language
$\neg$	Negation
$\vee$	Disjunction
$\&$	Conjunction
$\rightarrow$	(Material) implication
$\leftrightarrow$	Equivalence
$\bot$	Falsum, an arbitrary contradiction (p. 9)
$\top$	An arbitrary tautology (p. 38)
$\mathcal{\&}$	Set conjunction (p. 198)
$\stackrel{\&}{\times}$	Conjunctive product (p. 145)
$\rightarrowtail$	(Sentential) conditional (p. 34)
$\twoheadrightarrow$	Elicited conditional (p. 106)
$\Diamond$	Possibility (pp. 36, 112)

$\diamondsuit$	Serious possibility (p. 111)
$\square$	Necessity (p. 112)
$\Diamond$	Iterative possibility (p. 112)
$\boxdot$	Iterative necessity (p. 112)
[ ]	Descriptor-based necessity (p. 113)
$\langle\ \rangle$	Descriptor-based possibility (p. 113)

Logic of the Object Language

Cn	Consequence relation (p. 4)
C	Inference relation (p. 110)
$\vdash$	Logical implication (p. 15)
$\vdash_{\exists}$	Choice implication (p. 160)
$\mid\!\sim$	Non-monotonic inference (p. 108)
$\perp$	Remainder (pp. 6, 157)

Descriptors

$\mathfrak{B}$	Belief operator (pp. 37, 53, 54)
$\alpha, \beta, \ldots$	Molecular (belief) descriptor (p. 54)
$\Psi, \Xi, \ldots$	(Belief) descriptor (p. 53)
$\neg$	Descriptor negation (p. 56)
$\underline{\vee}$	Descriptor disjunction (p. 56)
$\underline{\perp}$	Descriptor falsum (p. 56)
Π	Maxispecified descriptor (pp. 57, 175)
$\widehat{\Psi}$	Maximal amplification (p. 81)
$\Rightarrow$	Ramsey descriptor (p. 100)

Logic of Descriptors

$\Vdash$	Descriptor implication (p. 55)
$\dashv\Vdash$	Descriptor equivalence (p. 55)
$\twoheadleftarrow$	Dynamic implication (p. 101)

Belief States and Belief Sets

$\mathbb{K}$	Set of belief states (p. 47)
$\mathcal{K}$	Belief state (p. 47)
$\mathbb{X}, \mathbb{Y}, \ldots$	Set of belief sets (p. 57)
K	Belief set (p. 4)
$\mathbb{I}$	Set of inputs (p. 47)
$\imath$	Input (p. 47)
$\mathfrak{s}$	Support function (p. 49)
$[\![\,]\!]$	Set of satisfying belief sets (p. 64)

Possible World Models

$\mathcal{W}$	Set of possible worlds (p. 10)
W	Possible world (p. 9)
$[\,]$	Set of surrounding possible worlds (pp. 10, 200)
$\mathfrak{S}$	System of spheres (p. 201)
$\mathcal{S}$	Sphere (p. 201)
$\mathcal{R}$	Ring (in sphere system) (p. 201)

Operations of Belief Change

$\circledcirc$	Input assimilation (p. 47)
$\breve{\circledcirc}$	Indeterministic input assimilation (pp. 47, 70)
$\circ$	Descriptor revision (p. 58)
$\breve{\circ}$	Indeterministic descriptor revision (pp. 58, 71)
$\div$	Contraction (p. 4)
$\sim$	Full meet contraction (p. 6)
$\dot{-}$	Revocation (p. 133)
$\dot{-}_{\forall}$	Package revocation (p. 143)
$\dot{-}_{\exists}$	Choice revocation (p. 144)
$+$	Expansion (p. 4)
$*$	Sentential revision (p. 5)
$\breve{*}$	Indeterministic sentential revision (p. 71)
$*_{\forall}$	Package revision (p. 130)
$*_{\exists}$	Choice revision (p. 130)
$\underset{\neg}{*}$	Resolution, making up one's mind (p. 130)

Choice Mechanisms

γ	Selection function (p. 6)
C	Choice function (p. 19)
$\widehat{C}$	Monoselective choice function (p. 59)
δ	(Pseudo)distance measure (p. 88)
$ł$	Location function (p. 126)
l	Accessibility delimiter (p. 90)
ℓ	Delimiter for non-monotonic inference (p. 109)
a	Accessibility function (p. 60)
$\mathbb{K}_{\mathcal{K}}$	Directly reachable range (p. 48)
$\mathbb{K}_{\mathcal{K}}^{+}$	Indirectly reachable range (p. 48)
M	Minimal satisfying element (p. 186)
f	Sentential selector (p. 160)
$\mathfrak{r}$	Reject function (p. 148)
$\rightharpoondown$	Blockage relation (pp. 73, 91)
$\overset{=}{\rightharpoondown}$	Weak blockage relation (p. 191)
$t^{db}, t^{d\circ}$	Transformation function (p. 82)

Binary Relations on Belief Sets

(The primary relation is followed by its strict part.)

$\leqq, <$	Belief set ordering (pp. 68, 86, 110)
$\overset{\star}{\leqq}, \overset{\star}{<}$	Additive restriction of belief set ordering (p. 128)
$\overset{-}{\leqq}, \overset{-}{<}$	Subtractive restriction of belief set ordering (p. 140)
$\overset{=}{\leqq}, \overset{=}{<}$	Multiply subtractive restriction of belief set ordering (p. 144)

Binary Relations on Descriptors and Sentences

(The primary relation is followed by its strict and symmetric parts.)

$\succeq, \succ, \simeq$	Proximity (p. 79)
$\overset{\star}{\succeq}, \overset{\star}{\succ}, \overset{\star}{\simeq}$	Believability (p. 127)
$\dot{\leq}, \dot{<}, \dot{\simeq}$	Entrenchment (pp. 15, 81)
$\overset{=}{\dot{\leq}}, \overset{=}{\dot{<}}$	Multiple entrenchment (p. 144)

Negations

Overcrossed symbols abbreviate negations. For instance, $(X \not\subset Y)$ abbreviates $\neg(X \subset Y)$, $X \nrightarrow Y$ abbreviates $\neg(X \rightarrow Y)$, and $(p \,\dot{\not\simeq}\, q)$ abbreviates $\neg(p \,\dot{\simeq}\, q)$.

Preview

The first three chapters provide the justification for developing a new logical model of belief change. Chapter 1, *The State of the Art*, provides the reader with the essential background in previous research. It begins by introducing the basic features of models of belief change: Belief states are represented by logically closed sets of sentences, commonly called belief sets. Changes take the form of either introducing a new sentence into the belief set (revision) or removing a sentence from it without adding anything else (contraction). The standard model for performing such changes, the so-called AGM model, is introduced along with the axioms used to characterize it. Two of the most important equivalent formulations of the AGM model, possible world models and epistemic entrenchment, are also introduced.

Chapter 2, *Inside the Black Box*, is devoted to a critical examination of how the choice what to believe is represented in the standard model. The choice is assumed to be performed on logically infinite objects of choice (remainders or possible worlds) that are not themselves plausible representations of belief states. This is followed by a second step in which the outcome is obtained by intersecting the chosen objects. However, epistemic choiceworthiness does not seem to be preserved under intersection. It is argued that the choice what to believe is better represented as a choice among potential outcomes of the operation of change.

In Chapter 3, *Questionable Patterns of Change*, this is followed up with a discussion of the properties of the standard (AGM) operations of change. Both contraction and revision violate the highly plausible postulate of finite-based outcome. Contraction has further problems relating to the postulates of success, recovery, and inclusion, and revision turns out to have implausible monotonicity-related properties. The chapter concludes with a list of ten desiderata for an alternative approach to belief change that eschews the major problems exhibited in Chapters 2 and 3.

The four chapters that follow form the central part of the book, in which a new approach, "descriptor revision," is introduced and developed. In Chapter 4, *Putting the Building-Blocks Together*, the new model is constructed from its basic components. We begin with a skeletal input–output model that contains no sentences but only primitive (i.e. unstructured) belief states and inputs, together with a revision function $\circledcirc$ that takes us from any belief state $\mathcal{K}$ and input $\imath$ to a new belief state $\mathcal{K} \circledcirc \imath$. Sentences are added to the framework with the help of a support function that takes us from any belief state $\mathcal{K}$ to the set of sentences representing the beliefs that it supports. After that, the two major components of the new framework are introduced. The first is belief descriptors, a versatile construct for describing belief states. The metalinguistic expression $\mathfrak{B}p$ denotes that p is believed (i.e. that it is an element of the supported set). Truth-functional combinations are also used, thus $\neg\mathfrak{B}p$ denotes that p is not believed and $\mathfrak{B}p \vee \mathfrak{B}\neg p$ that either p or $\neg p$ is believed. Sets of these expressions are used to denote combined properties, hence $\{\neg\mathfrak{B}p, \neg\mathfrak{B}q\}$ denotes that neither p nor q is believed. These expressions are used as inputs for belief revision (replacing primitive inputs such as $\imath$). Due to their versatility, all changes can be performed with a single, uniform change operation $\circ$. In order to revise by a sentence p we use the input (success condition) $\mathfrak{B}p$, in order to remove the sentence q the input $\neg\mathfrak{B}q$, etc. In order to perform these operations, the second major component of the new framework is introduced, namely a selection mechanism (a monoselective choice function) that directly selects the output among a given set of belief sets that are potential outcomes (the "outcome set"). Notably, no use is made of constructs such as possible worlds that have a central role in the traditional framework.

In Chapter 5, *Local Descriptor Revision*, the properties of this construction are explored. Several variants of the selection mechanism are presented, and their properties are investigated. It turns out that the different variants of descriptor revision can be axiomatically characterized with plausible postulates. Controversial axioms such as those discussed in Chapter 3 are avoided in the new framework. The chapter also introduces relations of epistemic proximity that are relations among descriptors. We write $\mathfrak{B}p \succeq \neg\mathfrak{B}q$ to denote that it is closer at hand for the epistemic agent to believe in p than not to believe in q. This is a generalization of the relations of epistemic entrenchment of the AGM framework. (p is less entrenched than q if and only if $\neg\mathfrak{B}p$ is more proximate than $\neg\mathfrak{B}q$.)

In Chapter 6, *Global Descriptor Revision*, the perspective is widened to global (iterated) belief change. This means that the formal framework is extended to cover successive changes, such as $K \circ \mathfrak{B}p \circ \neg\mathfrak{B}q$. Again, several constructions are presented and axiomatically characterized. The most orderly of these constructions is based on pseudodistances (distance measures that allow the distance from X to Y to be different from the distance from Y to X). For any elements X and Y of the outcome set, i.e. the set of belief sets that are eligible as outcomes, there is a number $\delta(X, Y)$ denoting how far away Y is from X. When revising a belief set K by some descriptor Ψ, the outcome $K \circ \Psi$ is the belief set satisfying Ψ that is closest to K, as measured with δ. If we revise $K \circ \Psi$ by Ξ, then the outcome $K \circ \Psi \circ \Xi$ is the belief set δ-closest to $K \circ \Psi$ that satisfies Ξ, etc.

Chapter 7, *Dynamic Descriptors*, is devoted to descriptors that convey how an agent's beliefs are disposed to be changed in response to different inputs. We introduce Ramsey descriptors that have the form $\Psi \Rightarrow \Xi$ where Ψ and Ξ are (static) descriptors of the types introduced in Chapter 4. For example, $\mathfrak{B}(p \vee q) \Rightarrow \neg\mathfrak{B}r$

denotes that if the agent changes her beliefs to believe that $p \vee q$, then she will not believe in r. The special case represented by $\mathfrak{B}p \Rightarrow \mathfrak{B}q$ corresponds to standard Ramsey test conditionals. Ramsey descriptors are axiomatically characterized with a set of plausible postulates that are generalizations of postulates commonly used in the logic of conditional sentences. It is also shown that Ramsey descriptors can unproblematically be inserted into belief sets. Revision by Ramsey descriptors can be allowed, without the problems associated with Ramsey test conditionals in the AGM framework and related models of belief change. Standard (sentential) conditionals are also investigated, and two alternatives to the Ramsey test are introduced. Criticism is offered against the common view that the logic of non-monotonic inference is a fragment of the logic of conditional sentences. Finally, various methods to introduce modalities into the belief change framework are explored.

In the next three chapters, we return to the major operations of the traditional approach, namely (sentential) revision and contraction, and investigate how they can be developed as special cases of descriptor revision. Chapter 8, *Sentential Revision*, shows how local and global operations of revision by a sentence p can be derived as descriptor revision by $\mathfrak{B}p$, using the operations developed in Chapters 5 and 6. Operations of multiple revision and making up one's mind are also investigated, and so are the relations of believability on sentences that can be derived from relations of epistemic proximity. The most important results in this chapter are two theorems showing that the major revision operations of the AGM framework, namely partial meet revision and its transitively relational variant, are both reconstructible as subcases of sentential revision in the descriptor framework.

Chapter 9, *Revocation*, is devoted to operations of revocation, by which is meant a generalization of contraction in which a specified sentence is removed in a process that may possibly also involve the addition of some new information to the belief set. A couple of such operations are constructed as special cases of descriptor revision, and they are characterized in terms of their properties. Entrenchment relations are derived from operations of epistemic proximity, and they are subjected to a fairly close examination. Entrenchment relations are also constructed for sets of sentences. These extended entrenchment relations are studied in connection with operations of multiple revocations, in which a set of sentences rather than a single sentence is removed. Finally, the alternative approach of "revisionary revocation" is introduced. Its basic idea is that revocation of a sentence p takes the form of revision by some sentence $\mathfrak{r}(p)$ that can be interpreted as saying that there are sufficient reasons not to believe in p.

In Chapter 10, *Contraction*, we investigate operations of contraction in the classical sense, i.e. operations that remove a specified sentence from the belief set without adding anything else. The chapter begins with a study of the contraction operations that can be obtained as special cases of descriptor revision. This is followed by the main results of this chapter: Two impossibility theorems show that AGM contractions cannot be straightforwardly reconstructed as descriptor operations. After that, three constructions of belief contraction are investigated that are based on descriptor revision but deviate in different ways from its central feature of using just a single, one-step selection among the set of potential outcomes.

In the final Chapter 11, *Looking Back – and Ahead*, we return to the ten desiderata for the new framework that were presented in Chapter 3 and summarize how they have been satisfied in the chapters that followed. The chapter concludes with a list of remaining problems and areas for future research.

Part I
Why a New Approach?

Chapter 1
The State of the Art

This chapter is a brief and fairly elementary introduction to the theory of belief change, covering belief sets and the operations of contraction, expansion and revision (Section 1.1), the AGM construction with selection functions (Section 1.2), the AGM postulates for contraction and revision (Section 1.3), possible world models (Section 1.4), and epistemic entrenchment (Section 1.5). A couple of other important approaches are briefly mentioned in the final Section 1.6. Readers not in need of such an elementary introduction can pass on directly to the next chapter.

1.1 A Sentential, Input-Assimilating Approach

Belief change (belief revision) arose as a subject of its own in the 1980s. It grew out of two converging research traditions. One of these was studies by computer scientists of procedures for updating databases. Truth maintenance systems [44] and the notion of database priorities [46] were particularly important contributions. The other research tradition was endeavours by philosophers to give precise formal accounts of changes in complex systems such as scientific theories and legal systems. Isaac Levi contributed many of the ideas in this tradition that made fruitful formal investigations possible [160, 162].

One important feature of this framework is its emphasis on rationality in belief change. The theory of belief change is normative in the same sense as decision theory (that deals with rational decision making) and epistemology (that deals with rational belief). These, by the way, are two disciplines that have many connections with belief change.

Another important feature of the framework is its exclusive focus on beliefs that can be represented by sentences in a given formal language. This is obviously an idealization since we have many beliefs that are difficult or impossible to represent by sentences, in particular beliefs connected to complex sensory impressions. You may for instance have beliefs (and knowledge) about how a musical piece sounds, without being able to express those beliefs in words. The reason for this focus on

S.O. Hansson, *Descriptor Revision*, Trends in Logic 46,
DOI 10.1007/978-3-319-53061-1_1

sentences is of course that sentences are eminently suitable for formal treatment and, as we will see, also for precisely formulated rationality requirements.

The epistemic agent's beliefs at a particular point in time are conventionally represented by a set of sentences, called a "belief set". The belief set is usually denoted K. (This choice of a letter is due to a philosophically defunct interpretation as a representation of *k*nowledge.) The belief set is assumed to be logically closed (closed under logical consequence), which means that all sentences that follow logically from it are already among its elements. Logical closure is obviously unrealistic since we are not aware of all the logical consequences of our beliefs.[1] From a formal point of view it is a highly useful idealization since it simplifies the logical treatment. Isaac Levi has provided the best justification of this idealization: We can interpret the belief set as consisting of the sentences that the epistemic agent is committed to believe in, rather than her actual beliefs. [164, pp. 6–9] According to Levi, we are doxastically committed to believe in all the logical consequences of our beliefs, even though our performance does not live up to this commitment.

Logical consequence can be expressed with a consequence relation Cn, such that for any set X of sentences, $\mathrm{Cn}(X)$ is the set of its logical consequences. Our requirement that K is logically closed can then be expressed with the simple formula $K = \mathrm{Cn}(K)$.

The theory of belief change is concerned with changes in belief induced by external influence, such as evidence. The basic format for change is that an information-carrying input is received and gives rise to a new belief set in which the information contained in the input has been accepted. The input usually takes the form of a sentence and an instruction saying what to do with it. Operations with such inputs can be called sentential operations. Standardly, they come in three types, corresponding to the instructions "remove this sentence", "add this sentence", and "add this sentence and retain consistency". The central problem in belief change is to find a new belief set that follows these instructions without unduly changing other beliefs.

The instruction "remove this sentence" is performed with an operation of *contraction*, usually denoted $\div$. The instruction "add this sentence" is performed with the operation of *expansion*, denoted $+$. It is usually interpreted as a simple set-theoretical operation. We first add p to the belief set, obtaining $K \cup \{p\}$. This is typically not a logically closed set, so we have to close it under logical consequence, which gives rise to the standard expansion operation:

$$K + p = \mathrm{Cn}(K \cup \{p\})$$

Expansion has the virtue of simplicity, but it also has the damaging property of leading to inconsistency whenever we assimilate some information that contradicts what we believed before. (If $\neg p \in K$ then $K + p$ is inconsistent.) Therefore we need

[1] Note, however, that logical closure does not imply that the reasoner is able to draw all valid logical conclusions, only those that are covered by the consequence operation used in the formal system. The consequence operation employed in belief change theory is standardly assumed to include classical truth-functional propositional logic. It may or may not contain something more. Logical closure is often described as requiring "logical omniscience", but that is a misleading way to describe an albeit perfect ability to draw classical truth-functional inferences in a particular object language.

the more sophisticated operation of *revision* that corresponds to the instruction "add this sentence and retain consistency". The outcome of revising K by the sentence p is written $K * p$.

Whereas expansion is a purely logical operation, definable in terms of logical consequence (Cn), both contraction and revision involve choices that are not determined by logic alone. Much of the discussion on belief change is devoted to various rationality requirements on the latter two operations. Such requirements can be of two types, substantial and structural. The difference is perhaps best seen from a couple of simple examples. First, suppose that Emily gives up her previous belief that William Shakespeare wrote Hamlet and instead begins to believe that Miguel de Cervantes wrote this and all the other allegedly Shakespearean plays. We would consider her belief change to be irrational for substantial reasons that are related to the evidence that is available to her. If she starts to believe that Russell Oberlin is a countertenor but not a falsettist, while at the same time retaining her previous belief that all countertenors are falsettists, then her beliefs are irrational for structural rather than substantial reasons. The crucial criterion why the latter is a structural rather than a substantial failure of rationality is that we can ascertain its irrationality without even knowing the meaning of the two key terms "falsettist" and "countertenor". Structural (formal) rationality falls within the purview of logic and will therefore be at our focus here.

We have already touched upon one rationality requirement, namely that the outcome of an operation on a belief set should be a new belief set, which means that it has to be logically closed. In formal language this is written $K \div p = \text{Cn}(K \div p)$ for contraction and $K * p = \text{Cn}(K * p)$ for revision. Another important requirement is that the operation should be successful. For revision the success condition is $p \in K * p$. For contraction the corresponding simple condition $p \notin K \div p$ does not quite work since it cannot be satisfied if p is a tautology and $K \div p$ is a belief set. (All tautologies are elements of all logically closed sets.) Therefore the common success condition for contraction contains an exception clause:

If $p \notin \text{Cn}(\varnothing)$, then $p \notin K \div p$.

($\text{Cn}(\varnothing)$, the logical closure of the empty set, is equal to the set of tautologies.)

In addition to the formulation of such rationality requirements, or postulates as they are usually called, research in belief change is largely devoted to the construction of various operations that satisfy some of these postulates. Much effort has been spent on making these two approaches meet. Often we can show that an operation satisfies a certain set of postulates if and only if it can be constructed in a particular way. A formal result showing this is called a representation theorem. Such theorems are important since they connect two types of intuitions that we can have about a formal structure. Let us now turn to the model usually considered to be the gold standard of belief change, not least due to the wealth of elegant representation theorems that it has produced.

1.2 The AGM Construction

Many research papers have been called seminal, but few deserve that designation as much as the article in the Journal of Symbolic Logic in 1985 by Carlos Alchourrón (1931–1996), Peter Gärdenfors, and David Makinson, "On the Logic of Theory Change: Partial Meet Contraction and Revision Functions" [1]. The immediate prehistory of this article has been told both by Makinson and Gärdenfors [70, 180]. It is the most quoted and no doubt the most influential paper in the literature on belief change. Its centrepiece is the three authors' construction of an operation of contraction.

When we contract a belief set K by a sentence p, the outcome should be a logically closed subset of K that does not imply p. The problem is that there are many such subsets, and somehow we must choose one of them. In their treatment of this problem, the AGM trio began by noting that among the many subsets of K not implying p, some are inclusion-maximal, i.e. they are as large as they can be without implying p. These sets are called p-remainders, and the set of p-remainders of K is denoted $K \perp p$.

Intuitively, when contracting K by p we want to keep as much of K as we can while still removing p. This could lead us to take one of the elements of $K \perp p$ as the contraction outcome. However, it may be impossible to single out one of these as epistemically preferable to all the others. If several p-remainders share the top position, then our post-contraction beliefs will be those that are held in all the top-ranked p-remainders. Formally, this is achieved by introducing a selection function (choice function) γ that takes us from $K \perp p$ to a subset $\gamma(K \perp p)$ that consists, intuitively speaking, of the best or most choiceworthy elements of $K \perp p$. The outcome consists of that which is common to all of them, i.e. their intersection:

$$K \div p = \bigcap \gamma(K \perp p)$$

In the limiting case when p is a tautology, $\gamma(K \perp p)$ is defined to be equal to $\{K\}$. This yields $K \div p = K$ when p is a tautology, according to the simple principle that an attempt to remove what cannot be removed results in no change at all. The operation $\bigcap \gamma(K \perp p)$ is called *partial meet contraction* since its outcome is the meet (intersection) of a part of the remainder set $K \perp p$.

Partial meet contraction has two limiting cases that need to be mentioned. If $\gamma(K \perp p) = K \perp p$ for all p, then $\div$ is *full meet contraction*, also denoted $\sim$.[2] If $\gamma(K \perp p)$ is always a singleton (a set with exactly one element) when $K \perp p$ is non-empty, then $\div$ is a *maxichoice contraction*. Both these operations have quite implausible properties, but they are important limiting cases in formal investigations of the model [99, pp. 74–77].

A much more interesting special case is that in which γ is based on a transitive relation covering all remainders of K, i.e. all sets X such that $X \in K \perp p$ for some p. (A relation $\trianglelefteq$ is transitive if it holds for all x, y, and z that if $x \trianglelefteq y$ and $y \trianglelefteq z$ then $x \trianglelefteq z$. Betterness is usually held to be transitive.) If a selection function γ for K

[2]On full meet contraction, see [106, 113].

always selects the elements of $K \perp p$ that are top-ranked according to such a relation, then the resulting contraction is a *transitively relational partial meet contraction.*

The AGM approach to revision is based on the simple observation that if p cannot be consistently added to K, then that is because $\neg p$ is in K. ($K + p$ is inconsistent if and only if K implies $\neg p$.) Therefore, all we have to do to make p consistently addable is to first remove $\neg p$. This line of reasoning can also be found in earlier work by Isaac Levi [160, p. 426–427]. It gives rise to the following construction of revision in terms of contraction and expansion:

$$K * p = (K \div \neg p) + p \qquad \text{(the Levi identity)}$$

It turns out that if revision is defined in this way, then the operation of contraction on which the operation $*$ of revision was based can be regained as follows:

$$K \div p = K \cap (K * \neg p) \qquad \text{(the Harper identity)}$$

An operation is called a *partial meet revision* if and only if it is obtainable via the Levi identity from a partial meet contraction, and it is a *transitively relational partial meet revision* if and only if it is obtainable in that way from a transitively relational partial meet contraction.

Belief changes in real life are often much more complex than (single) contractions, expansions, or revisions. The most general form of change is of course the replacement of one belief set K_1 by another belief set K_2. According to Isaac Levi, if K_2 contradicts K_1 then it may be difficult to justify a direct shift from K_1 to K_2. But he continues to point out that we do not have to go directly from K_1 containing p to K_2 containing $\neg p$. Instead, we can perform a sequence of operations such that we never add a sentence without first removing its negation. [162, pp. 63–64] We can take this to be the general format for operations that replace one belief set by another, i.e. for belief change in general:

The decomposition principle [61, pp. 130-131] (Cf. [99, p. 8].)
Every legitimate belief change is decomposable into a sequence of contractions and expansions.

The decomposition principle has not been much discussed, but it is an important presupposition in the AGM model and related approaches. It is highly simplifying since it drastically limits the need for diversity in belief change operations to only two operations: the trivial operation of expansion and the (not so trivial) operation of contraction. The principle may not be as ungainsayable as it appears to be at first glance, but we will return to that in Chapter 3.

1.3 The AGM Postulates

The AGM operations have all been characterized with elegant sets of postulates. Beginning with partial meet contraction (the general case), one of the central results in the 1985 paper [1] was that a sentential operation $\div$ on a belief set K is a partial meet contraction if and only if it satisfies the following six postulates, generally known as the six basic Gärdenfors postulates (or AGM postulates) of contraction:

$K \div p = \mathrm{Cn}(K \div p)$ (closure)
$K \div p \subseteq K$ (inclusion)
If $p \notin K$, then $K \div p = K$. (vacuity)
$p \notin (K \div p) \setminus \mathrm{Cn}(\varnothing)$ (success[3])
If $p \leftrightarrow q$ is a logical truth, then $K \div p = K \div q$. (extensionality)
$K \subseteq (K \div p) + p$ (recovery)

Closure, which we have already discussed, says that the outcome of revising a belief set should be a new belief set. According to *inclusion*, an operation whose purpose is to remove something, adds nothing new. *Vacuity* tells us that removing what is not there is no change at all. The *success* criterion says that the operation succeeds in its purpose whenever that is at all possible. *Extensionality* says, essentially, that the syntactical form of sentences has no impact. Finally, the *recovery* postulate puts a limit on how much we may remove in order to get rid of a sentence. The other five postulates would allow operations of contraction that make excessive removals, for instance an operation such that $K \div p = \mathrm{Cn}(\varnothing)$ for all non-tautological elements p of K. This would of course be a highly implausible (and arguably irrational) operation. It requires that if you give up one of your beliefs on empirical subject-matter, then you let go of them all. The recovery postulate rules out such drastic contractions. It does so by making sure that when removing p we retain so much of K that all our previous beliefs will be regained if we immediately reinstate p. (The recovery postulate has been the subject of considerable controversy, but we will leave this as well as other controversies over postulates to Chapter 3.)

The 1985 paper also contained an equally elegant representation theorem for the transitively relational variant of partial meet contraction. A sentential operation is a transitively relational partial meet contraction if and only if it satisfies the six basic postulates just referred to and in addition the following two that are usually called the supplementary contraction postulates:

$(K \div p) \cap (K \div q) \subseteq K \div (p\&q)$ (conjunctive overlap)
If $p \notin K \div (p\&q)$, then $K \div (p\&q) \subseteq K \div p$. (conjunctive inclusion)

Both these postulates refer to contraction by conjunctions, i.e. by sentences of the form $p\&q$. To exemplify *conjunctive overlap*, suppose that if I were to give up my belief that Pamela is a law-abiding citizen (p), then I would retain my belief that her daughter Rebecca is a law-abiding citizen (r). Furthermore suppose that if I were instead to give up my belief that Rebecca's father Quentin is a law-abiding citizen (q), then I would likewise retain my belief that Rebecca is so (r). Given all this, it would seem reasonable to assume that if I were to give up my belief that both Pamela and Quentin are law-abiding citizens ($p\&q$), then I would retain my belief r in that case as well. More generally speaking, if a belief r can withstand both contraction

[3]More commonly written: If $p \notin \mathrm{Cn}(\varnothing)$, then $p \notin K \div p$.

by p and contraction by q, then it can also withstand contraction by $p\&q$ (which can of course be performed by removing only p or by removing only q). This is exactly what the postulate tells us.

For *conjunctive inclusion*, suppose that contracting by $p\&q$ leads to loss of the belief in p, i.e., $p \notin K \div (p\&q)$. We have then removed p in order to rid ourselves of $p\&q$. This manoeuvre can be expected to lead to the loss of all beliefs that we would throw out in order to get rid of p, i.e. everything that we remove from K to obtain $K \div p$ is also removed to obtain $K \div (p\&q)$. Another way to express this is that in this case, everything that is retained in $K \div (p\&q)$ is also retained in $K \div p$, which is what the postulate says.

The 1985 paper also contains postulates for the corresponding operations of revision [1]. The AGM trio showed that partial meet revision is exactly characterized by the following six postulates, several of which are closely analogous to the basic contraction postulates:

$K * p = \text{Cn}(K * p)$ (closure)
$K * p \subseteq K + p$ (inclusion)
If $\neg p \notin K$, then $K + p \subseteq K * p$. (vacuity)
$p \in K * p$ (success)
If $p \leftrightarrow q$ is a logical truth, then $K * p = K * q$. (extensionality)
If p is consistent, then so is $K * p$. (consistency)

It follows from inclusion and vacuity that if $\neg p \notin K$, then $K * p = K + p$. Thus, if the outcome $K + p$ of expanding K by p is consistent, then it is identical to the revision outcome $K * p$.

In order to characterize transitively relational partial meet revision, the following two postulates have to be added:

$K * (p\&q) \subseteq (K * p) + q$ (superexpansion)
If $\neg q \notin K * p$, then $(K * p) + q \subseteq K * (p\&q)$. (subexpansion)

There are several other, equivalent presentations of AGM theory. Two of them will be presented in the rest of this chapter since they are needed as reference points in the chapters that follow.

1.4 Possible World Models

By a possible world, in the logical sense, is meant a maximally consistent set of sentences, in other words a logically consistent set so large that nothing can be added to it without making it inconsistent. Let $\mathcal{L}$ denote the (set of all sentences in the) language and let $\perp$ be a logically inconsistent sentence. Then $\mathcal{L}\perp\perp$ is the set of possible worlds. Possible worlds have properties that make them highly convenient for the logician. Perhaps most importantly, if W is a possible world and p is a sentence, then either p or its negation is included in W, i.e. either $p \in W$ or $\neg p \in W$. This is what makes possible worlds (in the logical sense) plausible descriptions of complete states of the world.

Possible world models have a central role in modal logic and many other areas of logic. A close connection between possible worlds and the AGM model was discovered by Adam Grove [80]. The following notation can be used to express the relationship: For each belief set K, let $[K]$ be the set of possible worlds that surround it:

$$[K] = \{W \in \mathcal{L}\perp_{\perp} \mid K \subseteq W\}$$

It can be shown fairly easily that every belief set K is identical with the intersection of all the possible worlds that surround it [99, p. 52]:

$$K = \bigcap[K]$$

If K is inconsistent, then $[K] = \varnothing$. It is quite reasonable that an inconsistent belief set should correspond to an empty set of worlds, since there is no world that contains it.

Beginning at the other end, let $\mathcal{W}$ be a set of possible worlds, i.e. $\mathcal{W} \subseteq \mathcal{L}\perp_{\perp}$. Its intersection $\bigcap \mathcal{W}$ is a belief set. In this way, belief sets and sets of possible worlds are completely exchangeable. Without any loss or gain, we can talk about sets of possible worlds instead of talking about belief sets. The informal interpretation should also be clear: Your belief state can be represented by the set of all possible worlds that it is compatible with, and the intersection of all those possible worlds is your belief set.[4]

Sentences can be represented in the same way:

$$[p] = \{W \in \mathcal{L}\perp_{\perp} \mid p \in W\}$$

However, although each sentence can be represented by a set of possible worlds, the reverse relationship does not hold. There are sets $\mathcal{W}$ of possible worlds with no p such that $\mathcal{W} = [p]$.[5]

Possible worlds are often represented geometrically, and such representations are useful in belief change. A surface (usually a rectangle) corresponds to the set of all possible worlds. Each point on the surface represents a possible world. Both belief sets and sentences are represented by subareas of the rectangle. Importantly, a larger area corresponds to a smaller belief set. For any two belief sets X and Y, $X \subset Y$ holds if and only if $[Y] \subset [X]$.

Figure 1.1 illustrates (sentential) revision in this framework. The circle in the middle contains exactly those possible worlds that are compatible with the current belief set K, i.e. this area represents $[K]$. The area covered by the parabola represents those possible worlds in which p holds. We are going to perform the operation $K * p$. In this case, $[K] \cap [p] \neq \varnothing$. This means that there are worlds W such that both $W \in [K]$ and $W \in [p]$, or equivalently, both $K \subseteq W$ and $p \in W$. Thus $K \cup \{p\}$ is consistent. As we saw above, $K * p = \mathrm{Cn}(K \cup \{p\})$ in this case, and since $[\mathrm{Cn}(K \cup \{p\})] = [K \cup \{p\}] = [K] \cap [p]$ it follows that $[K * p]$ is represented by the intersection of the circle and the parabola in the diagram.

---

[4]Note that a belief set K is a subset of a possible world W if and only their union $K \cup W$ is consistent.

[5]A less cursory introduction to possible worlds and their relations to belief sets can be found in [99, pp. 51–57, 287–304].

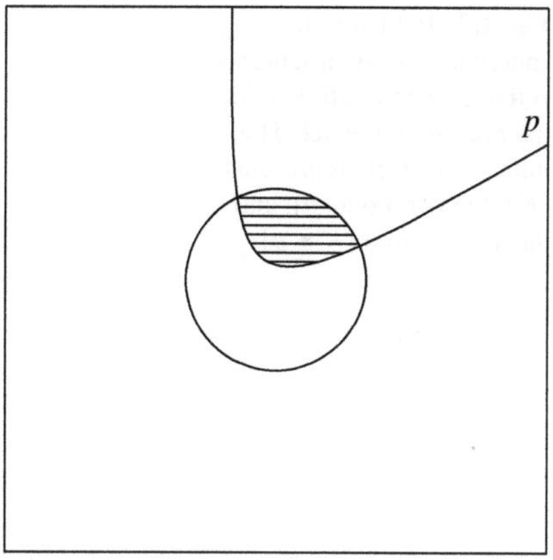

Fig. 1.1 Revision by a sentence p that is compatible with the present belief set. The circle represents $[K]$, the parabola $[p]$, and the shaded area $[K * p]$.

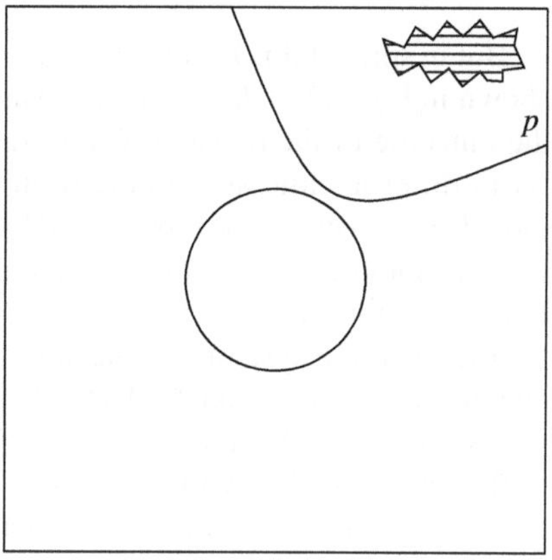

Fig. 1.2 Revision by a sentence p that is incompatible with the present belief set. The circle represents $[K]$, the parabola $[p]$, and the shaded area $[K * p]$.

But there is another case of belief revision, namely that in which $K \cup \{p\}$ is inconsistent or, equivalently, $[K] \cap [p] = \varnothing$. This case illustrated in Fig. 1.2. The revision should satisfy the success condition $p \in K * p$, or equivalently $[K * p] \subseteq [p]$. The shaded area in the diagram is one of many that satisfy this criterion and it can therefore be chosen as the revision outcome.

The procedures for revision that we have followed in these two cases can be summarized in the following recipe:

Revision recipe 1
$[K * p]$ is a subset of $[p]$ that is
(1) non-empty if $[p] \neq \varnothing$, and
(2) equal to $[K] \cap [p]$ if $[K] \cap [p] \neq \varnothing$.

It is fairly easy to show that this recipe is equivalent with partial meet revision. In other words, an operation $*$ of revision on a belief set K can be obtained in this way if and only if it is a partial meet revision. (See [80] or [99, pp. 290–291].)

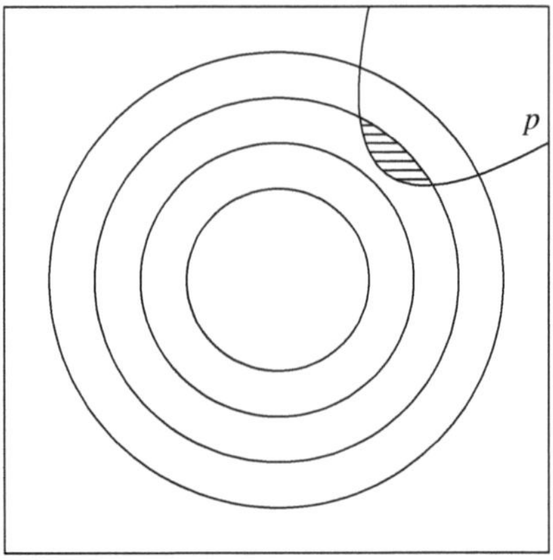

Fig. 1.3 Revision in a sphere system by a sentence p that is incompatible with the present belief set. The innermost circle represents $[K]$, the parabola $[p]$, and the shaded area $[K * p]$.

But in spite of this beautiful result, the recipe appears to be overly permissive. As shown in Fig. 1.2, if $K + p$ is inconsistent, then it allows any set of p-worlds to be the outcome of the revision $K * p$. Intuitively, we would expect the revised belief set to differ as little as possible from the original belief set K. In the geometrical model, similarity can be represented by closeness. Then the outcome of revising $[K]$ by $[p]$ should consist of those elements of $[p]$ that are as close as possible to $[K]$. For that purpose, Adam Grove proposed that the p-worlds outside of $[K]$ should be arranged in a system of concentric spheres (just as in David Lewis's account of counterfactual conditionals [167]). Each sphere represents a degree of closeness or similarity to $[K]$. Such a system of spheres corresponds, of course, to an ordering[6] of the set of possible worlds.[7]

Revision in a sphere model is shown in Fig. 1.3. The revision recipe that we apply here is a specified variant of the previous one:

Revision recipe 2
$[K * p]$ is a subset of $[p]$. If $[p] \neq \varnothing$, then $[K * p]$ is the intersection of $[p]$ with the innermost sphere that it has a non-empty intersection with.

In his 1988 paper Adam Grove showed that this recipe is equivalent with transitively relational partial meet revision. In other words, an operation $*$ of revision on a belief

[6]An *ordering* (weak ordering) is a binary relation $\leq$ that satisfies completeness ($X \leq Y$ or $Y \leq X$) and transitivity (if $X \leq Y$ and $Y \leq Z$ then $X \leq Z$). An ordering is a *linear ordering* if it also satisfies antisymmetry (if $X \leq Y$ and $Y \leq X$ then $X = Y$). An *equivalence relation* is a binary relation $\leq$ that satisfies reflexivity ($X \leq X$), symmetry (if $X \leq Y$ then $Y \leq X$), and transitivity. A *partial ordering* is a binary relation that satisfies reflexivity, antisymmetry, and transitivity. A *strict ordering* is a binary relation $<$ that satisfies transitivity and trichotomy (exactly one of $X < Y$, $Y < X$, and $X = Y$).

[7]Readers interested in a more precise definition of a system of spheres will find one in Definition A.4 (p. 201). In that definition, ($\mathfrak{S}1$) tells us that spheres are concentric (totally ordered by set inclusion). ($\mathfrak{S}2$) says that $\mathcal{W}$ itself is the minimal sphere. We assume that $\mathcal{W} = [K]$ for some belief set K. ($\mathfrak{S}3$) says that the set of all possible worlds is itself the maximal sphere. ($\mathfrak{S}4$) ensures that for all sentences p there is a minimal sphere containing some p-world.

set K is obtainable from a system of spheres with this recipe if and only if it is a transitively relational partial meet revision. (See [80] or [99, pp. 296–299].)

To represent contraction in possible world models, we should note that due to the inclusion postulate for contraction, $K \div p \subseteq K$ or, equivalently, $[K] \subseteq [K \div p]$. Furthermore, due to the success postulate, whenever p is non-tautologous, then $p \notin K \div p$ or, equivalently, $[K \div p] \cap [\neg p] \neq \varnothing$. We should therefore construct $[K \div p]$ so that it includes the whole of $[K]$ and also contains some $\neg p$-world(s). There are two cases.

In the first case, $p \notin K$. This means that there are already some $\neg p$-worlds in $[K]$. This is illustrated in Fig. 1.4. The obvious solution is to leave K unchanged after contraction by p. This solution complies with the vacuity postulate that requires $K \div p = K$ in this case.

The principal case, when $p \in K$ or, equivalently, $[K] \cap [\neg p] = \varnothing$, is illustrated in Fig. 1.5. Here we need to add some $\neg p$-worlds, as exemplified with the shaded part of the parabola. The procedures followed in these two diagrams are summarized in the following recipe:

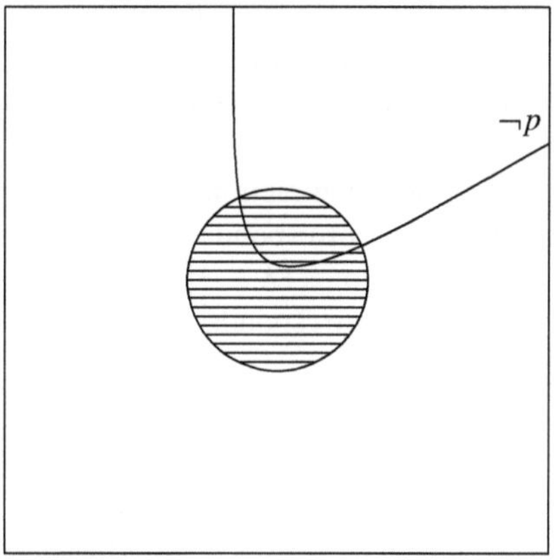

Fig. 1.4 Contraction by a sentence p that is not implied by the present belief set. The circle represents $[K]$, the parabola $[\neg p]$, and the shaded area $[K \div p]$.

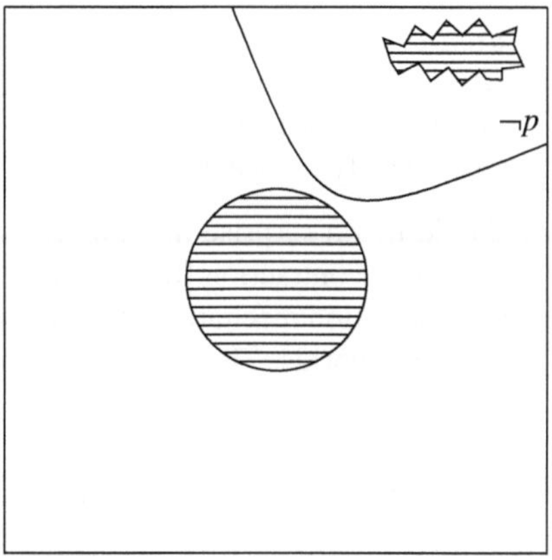

Fig. 1.5 Contraction by a sentence p that is implied by the present belief set. The circle represents $[K]$, the parabola $[\neg p]$, and the shaded areas $[K \div p]$.

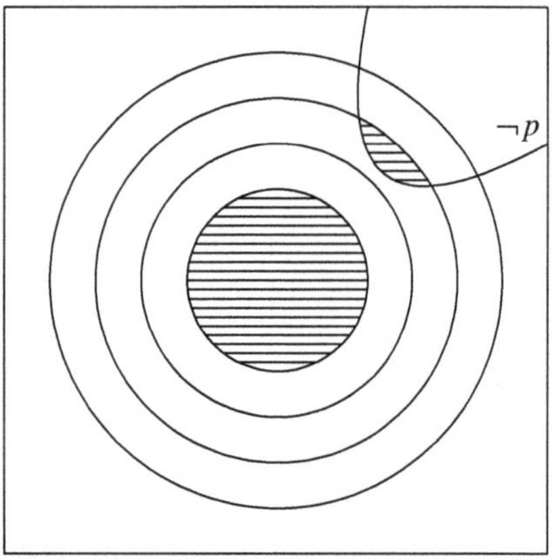

Fig. 1.6 Contraction in a sphere system by a sentence p that is implied by the present belief set. The innermost circle represents $[K]$, the parabola $[\neg p]$, and the shaded areas $[K \div p]$.

Contraction recipe 1
$[K \div p]$ is the union of $[K]$ and a subset of $[\neg p]$ that is
(1) non-empty if $[\neg p]$ is non-empty, and
(2) equal to $[K] \cap [\neg p]$ if $[K] \cap [\neg p]$ is non-empty.

Contraction performed according to this recipe corresponds exactly to partial meet contraction, i.e. an operation $\div$ of contraction on a belief set K can be obtained in this way if and only if it is a partial meet contraction. The two limiting cases of partial meet contraction have straightforward counterparts in this model. Full meet contraction corresponds to adding the whole of $[\neg p]$ to $[K]$ in the case shown in Fig. 1.5, i.e. to let $[K \div p] = [K] \cup [\neg p]$ whenever $[K] \cap [\neg p]$ is empty. Maxichoice contraction corresponds to adding only one element of $[\neg p]$ (a "point" on the surface) in the same case.

But this recipe is just as unspecified and over-permissive as the first revision recipe. We can solve this problem in the same way as for revision, namely with a sphere system. As shown in Fig. 1.6 we should then add those elements of $[\neg p]$ that belong to the innermost sphere containing any $\neg p$-worlds. This corresponds to the following recipe:

Contraction recipe 2
$[K \div p]$ is the union of $[K]$ and a subset of $[\neg p]$. If $[\neg p] \neq \varnothing$ then $[K \div p]$ is the union of $[K]$ and the intersection of $[\neg p]$ with the innermost sphere that it has a non-empty intersection with.

It should be no surprise that this construction coincides with transitively relational partial meet contraction. In summary, it makes no difference for the outcome if we apply a selection mechanism to remainder sets as in partial meet operations or to sets of possible worlds. The same operations of contraction and revision are obtainable with both constructions.[8]

[8]These results are based on a one-to-one correspondence called "Grove's bijection" between the remainder set $K \perp p$ and the set of possible worlds not containing p. Two crucial facts show

1.5 Epistemic Entrenchment

A seemingly quite different approach to belief change classifies the sentences in K according to how easily retractible they are. This is expressed with a relation of *epistemic entrenchment*; $p \leq q$ ("p is at most as entrenched as q") means that the epistemic agent is at least as willing to give up p as to give up q. It is often more instructive to consider the corresponding strict relation $p < q$ ("p is less entrenched than q"), defined such that $p < q$ if and only if $p \leq q$ holds but $q \leq p$ does *not* hold. Epistemic entrenchment was introduced and analyzed by Gärdenfors and Makinson [69, 71]. The standard entrenchment relation satisfies the following five properties: ($p \vdash q$ is an alternative notation for $q \in \text{Cn}(\{p\})$ and $\vdash q$ for $q \in \text{Cn}(\varnothing)$.)

If $p \leq q$ and $q \leq r$, then $p \leq r$. (transitivity)
If $p \vdash q$, then $p \leq q$. (dominance)
Either $p \leq p\&q$ or $q \leq p\&q$. (conjunctiveness)
$p \notin K$ if and only if $p \leq q$ for all q. (minimality)
If $p \leq q$ for all p, then $\vdash q$. (maximality)

Entrenchment-based contraction is constructed such that for all non-tautologous sentences p:

$$K \div p = \{q \in K \mid p < p \vee q \text{ or } p \in \text{Cn}(\varnothing)\}$$

Gärdenfors and Makinson showed that entrenchment-based contraction coincides with transitively relational partial meet contraction. ([71], cf. [99, pp. 188–192].) Therefore it also coincides with sphere-based contraction, as introduced in the previous section. The connection between entrenchment relations and spheres is geometrically very simple. Given a sphere system centered on $[K]$ we can define an entrenchment relation $\leq$ such that:

$p \leq q$ if and only if either $[\neg q] = \varnothing$ or the innermost sphere containing some $\neg p$-world is included in the innermost sphere containing some $\neg q$-world.

Figure 1.7 provides an illustration of this connection between a sphere system and an entrenchment relation. It is easy to show that a relation $\leq$ defined from spheres in this way satisfies all the five standard properties of entrenchment relations, and that the entrenchment-based contraction obtained from $\leq$ coincides with the sphere-based contraction obtainable from the same sphere system. [99, pp. 300–301].

(Footnote 8 continued)
why this can be so. First, if K is a belief set, $p \in K$, and $X \in K \perp p$, then $\text{Cn}(X \cup \{\neg p\})$ is a maximal consistent subset of the language. Secondly, if K is a belief set, $p \in K$, and Y is an inclusion-maximal subset of the language not implying p, then $Y \cap K \in K \perp p$. See [80] or [99, pp. 53–55] for proofs of these connections.

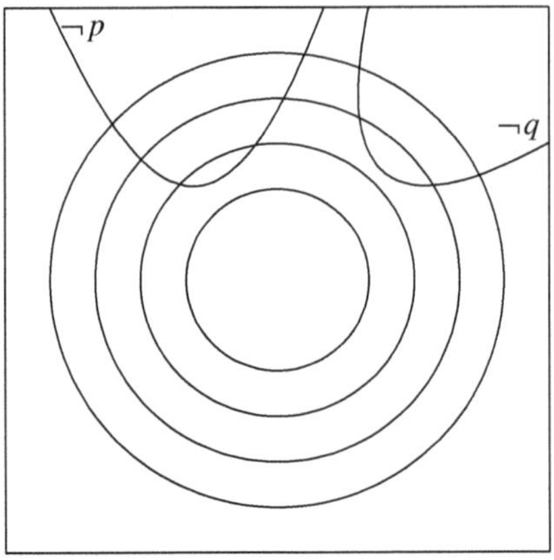

Fig. 1.7 Epistemic entrenchment in a sphere system. The innermost circle represents $[K]$ and the two parabolas $[\neg p]$ respectively $[\neg q]$. In this case, $[\neg p]$ comes closer to the centre than $[\neg q]$, which means that p is less entrenched than q.

1.6 Conclusion

This was on purpose a brief, almost rudimentary, summary of the classical approach to belief change. The AGM framework is much richer than has been shown here. Additional equivalent presentations are available; perhaps the most important of these are safe and kernel contraction in which at least one sentence is removed from each minimal p-implying subset of the belief set K [4, 90, 224]. Other important developments are operations that add or remove sets of sentences rather than single sentences (usually called "multiple operations") and operations that are defined for all belief sets, not only for a single belief set K (usually studied in the form of "iterated operations" such as $K * p * q$ and $K \div p \div q$). Both multiple and iterated operations will be discussed in some of the following chapters. Another faithful extension of the framework is its application to sets that are not closed under logical consequence; such sets are usually called "belief bases". The reader is referred to [99] for a fairly thorough introduction to the standard framework and some of its extensions. For an overview of more recent research in the area, see [52].

In summary, the AGM model of belief change is an unusually simple, elegant, and fecund theory. Not surprisingly for such a highly simplified theory, there are many features of actual belief systems that it does not capture. Let us now turn our attention to its limitations.

Chapter 2
Inside the Black Box

As we saw in Chapter 1, there are two major formal approaches to rationality in belief change. One is the *constructive approach.* We can design various mechanisms for operations of belief change, and it is then a topic for discussion whether these constructions are built on plausible principles. In this approach we may ask for instance whether selection functions and relations of epistemic entrenchment correspond to credible ways for a cognitive agent to change its beliefs. The other alternative is the *axiomatic approach* in which we consider various properties that belief change operations can have, expressed with the AGM postulates and others of the same sort. The tenability of each of these postulates can be scrutinized with the help of examples of reasonable changes in belief.

Both these approaches have weaknesses. A major problem with the constructive approach is that if a mechanism is implausible or difficult to explain, it may nevertheless yield the right results. An unconvincing construction can be defended as a "black box", a gear that we should be happy with because it does what it is supposed to do, even if we do not fathom how it does so. On the other hand, postulates only provide a partial description of how beliefs are changed. That a change operation satisfies a set of plausible postulates does not prevent it from also satisfying other, quite implausible ones.

The best solution to this problem is to combine the constructive and the axiomatic approach, in other words to specify mechanisms that we consider to be plausible and characterize them completely in terms of axioms. That was the route taken by the AGM authors, and in doing so they set a standard for subsequent researchers in the field. The purpose of this and the following chapter is to uncover problems in the AGM framework that can justify the development of alternative frameworks for belief change. In the present chapter we will follow the constructive approach, and investigate the use of selection mechanisms in both partial meet contraction and sphere systems. In Section 2.1 the notion of epistemic choice is discussed, and choice functions are introduced. In Section 2.2 it is clarified how in both partial meet contraction and sphere models, the application of a selection function is followed by

S.O. Hansson, *Descriptor Revision*, Trends in Logic 46,
DOI 10.1007/978-3-319-53061-1_2

the intersection of the selected sets. The plausibility of this sequence of operations is scrutinized in Section 2.3, and the formal limits to its applicability are pointed out in Section 2.4. Finally, in Section 2.5 we discuss the crucial question whether the AGM selection mechanisms are applied to the right objects.

2.1 Epistemic Choice

Choice has a central role in the theory of belief change. Operations of change take the form of replacing a belief set by another that satisfies a given success condition (such as $p \in K * p$ for revision and $p \notin (K \div p) \setminus \mathrm{Cn}(\varnothing)$ for contraction). In typical cases, there are many belief sets satisfying this condition, and exactly one of them is the outcome. The process of identifying one of the alternatives as the outcome is usually conceived as a choice. However, it must be recognized that the notion of choice is far from unproblematic in an epistemic context. We do not normally choose what to believe in the same way that we choose between dishes in a restaurant. Most belief changes seem to be uncontrollable effects of external influences rather than the results of voluntary choices made by the subject ([119, pp. 143–145]. See also: [8, 22, 141, 142, 192, 197, 199, 223, 232, 251, 252].)

> Svetlana has two sisters, Olga and Aleksandra. Olga is severely ill. One day Svetlana came to Pavel and said, sobbing: "Now I have only one sister."
>
> "How terrible", he said. "I knew that Olga was approaching the inevitable but I had hoped that she would live to see her grandchild."

Logically speaking, what Svetlana said only gave Pavel reason to believe that either Olga or Aleksandra had died. Nevertheless, his belief that Olga had died came to him immediately, unpreceded by any choice or other premeditation. But presumably, if he had carefully compared the alternatives and chosen which of them to believe in, the outcome would have been the same. We can take such reconstructibility in terms of premeditated choice as a criterion of rational belief change. Spontaneous behaviour can be rational, but only if it coincides with what one could have done if guided by rational reflection.

This "as if" approach to rationality (that is also common in decision theory [140, pp. 381–382]) has important implications for the formal representation of belief change. If a process of belief change takes place *as if* it was an actual choice, then that provides us with a reasonable justification for representing it as a choice.

Choices have been extensively studied in economics and in particular in social choice theory [237]. The standard formal representation of choice used in these disciplines, namely choice functions, has been taken over by belief change theory.[1]

[1] Arrow introduced choice functions in economics. He said: "We do not want to prescribe that $C(S)$ contains only a single element; for example, S may contain two elements between which the chooser is indifferent." [7, p. 4]. At that time choice functions were already used in logic, but the standard definition in logic was different. A choice function for a set $\mathfrak{X}$ of non-empty sets was defined as a

A choice function is defined over a set $\mathcal{A}$ of alternatives. It can be used to make a selection among any subset of $\mathcal{A}$:

Definition 2.1 *C is a* choice function *for a set $\mathcal{A}$ if and only if for each subset $\mathcal{B}$ of $\mathcal{A}$:*

(1) $C(\mathcal{B}) \subseteq \mathcal{B}$, and

(2) $C(\mathcal{B}) \neq \varnothing$ if $\mathcal{B} \neq \varnothing$.

A choice function C for $\mathcal{A}$ is based on *a relation $\multimap$ if and only if for all $\mathcal{B} \subseteq \mathcal{A}$: $X \in C(\mathcal{B})$ if and only if $X \in \mathcal{B}$ and $X \multimap Y$ for all $Y \in \mathcal{B}$.*

According to this definition, $C(\mathcal{B})$ can have more than one element. In everyday talk about choice, choices sometimes have this property, sometimes not:

Example 1:
"I am going to throw away these old LP records unless you want some of them. Choose those you want, and then I will throw away the rest."

Example 2:
"Since you have done so much for me I want to give you an LP record from my collection. You are free to choose whichever you like."

Choice functions, as defined above, represent the type of choice instantiated in the first of these examples.

2.2 The Select-and-Intersect Method

In social choice theory, when a choice function delivers an outcome with more than one element, this means that all those elements are (considered to be) equally choiceworthy. It is then left to the decision-maker to further narrow down the choice to one single object. Which element of $C(\mathcal{B})$ she ends up with is presumed to be arbitrary from the viewpoint of rationality. Hence, if Alex, Bailey, and Casey are three willing candidates for marriage, then $C(\{\text{Alex, Bailey, Casey}\}) = \{\text{Alex, Bailey}\}$ does not indicate a wish for bigamy but rather vacillation between Alex and Bailey.

Therefore, strictly speaking, choice functions in social choice theory only cover the first of two stages in a choice process. The second stage that slims down the outcome to a single element is often described as a matter of picking rather than choosing [245]. We can call this the *select-and-pick* method.

(Footnote 1 continued)
function C such that $C(X) \in X$ for all $X \in \mathfrak{X}$. [147] – On the use of choice functions in logic, see also [138].

In belief change as well, choice functions with multiple outputs leave us with a need for a further process that takes us from several objects to a single one.[2] In belief change, however, the second stage is different.[3] It consists in forming the intersection of the sets chosen in the first stage. This intersection is taken to be the outcome of the operation. This has been called the *select-and-intersect method* [135]. It comes in two major versions, both of which were introduced in Chapter 1. In partial meet contraction, the first stage is a selection among remainders, and in sphere-based revision it is a selection among possible worlds. The second stage, intersection, is the same in both cases.

At first glance, the select-and-intersect method may seem to be an almost impeccable way to deal with ties. When we hesitate between two or more potential outcomes, then it would seem natural to use their intersection, i.e. what they all have in common, as the output. But closer inspection will reveal that the select-and-intersect method can be questioned on at least three accounts. First, we can dispute the *preservation of optimality under intersection*. In the first step of the select-and-intersect process, options are chosen that are in some sense optimal. In partial meet contraction the first step passes on the best or most choiceworthy remainders that satisfy the success condition to the second stage for intersection. But is that optimality retained after intersection? Or would perhaps the intersection of some other set of remainders be more choiceworthy, while still satisfying the success condition? If the latter is true, then the achievement of the first stage was lost in the second.

Secondly, the *preservation of success under intersection* cannot always be taken for granted. In partial meet contraction, the success condition is the elimination of some input sentence p. All the sets chosen in the first stage satisfy that condition (since elements of $K \perp p$ do not contain p). It follows that their intersection, the final outcome of the operation, does not contain p either. In other words, this success condition is preserved under intersection. But does that apply to all success conditions that we may wish to apply? If not, then that is a constraint on the applicability of the select-and-intersect method.

Thirdly and perhaps most importantly, the *adequacy of the options selected for intersection* is contestable. In the AGM approach, the primary selection is made among remainders or (in the sphere model) possible worlds. Are these plausible outcomes? As noted above, the use of intersection can be justified as a means to adjudicate between equally plausible outcomes. It would seem more difficult to justify the select-and-intersect method if the objects chosen for intersection are not plausible outcomes of the operation.

In the next three sections we are going to look more closely at each of these problems for the select-and-intersect method.

---

[2] A few studies have been devoted to indeterministic belief change operations. These are operations that deliver, for each input, a set that may contain more than one possible outcome [66, 169].

[3] This difference would seem to have implications for the view that the use of choice functions in both areas reveals an underlying unity between practical and theoretical reasoning. On that view, see [205, 215, 217].

2.3 Is the Intersection as Good as Its Origins?

Although the use of choice functions in belief change is largely modelled after social choice theory, the use of intersection among options is unknown in social choice. The reason for this is obvious: in a social choice context optimality is not preserved under intersection [118].

> GAME SHOW HOST: Congratulations! You have won the first prize. This means that you now have a choice between two options. One is a Porsche 991 and 50 litres of petrol. The other is a Lamborghini Huracán and 50 litres of petrol. Which of them do you choose?
> CONTESTANT: I am unable to choose between them. The two alternatives are exactly equally good.
> GAME SHOW HOST: Thanks for telling us. We will now follow our standard procedure for such cases of indecision, and give you the intersection between the two sets you could not choose between. One of the sets contains a Porsche 991 and 50 litres of petrol, and the other a Lamborghini Huracán and 50 litres of petrol. Let me congratulate you once more. You are now the happy owner of the intersection of those two sets, namely 50 litres of petrol, of the highest quality.

This absurdity would have no relevance for belief change if it could be shown that contrary to other collections of objects, logically closed sets of sentences do not lose in choiceworthiness by being intersected with other equally choiceworthy objects. However, no such argument seems to be forthcoming. This problem was first pointed out by Tor Sandqvist. He proposed that we consider two collections of beliefs sets, $\mathcal{A}$ and $\mathcal{B}$. Suppose that each belief set in $\mathcal{A}$ is preferable to each belief set in $\mathcal{B}$. From this, he says, it does not follow that the belief set $\bigcap \mathcal{A}$ is preferable to the belief set $\bigcap \mathcal{B}$. The reason for this is that the elements of $\mathcal{A}$ may be "each very valuable but such that their intersection is practically worthless – namely, if whatever makes each of them so valuable fails to be that which they all have in common." [227, p. 292].

This argument is in need of a supporting example, but the construction of such an example is made difficult by the fact that we may have different standards of choiceworthiness for belief sets. Belief change theory is in general neutral between such standards, but if we wish to illustrate how choiceworthiness can be lost in intersection, the standard of choiceworthiness has to be made explicit. The following example has been chosen because its standard of choiceworthiness is particularly susceptible to deterioration through intersection:

> Ibraham chooses between five sets of religious beliefs, namely the full set of Roman Catholic beliefs (C_1), that of Lutheran beliefs (C_2), that of Sunni beliefs (I_1), that of Shia beliefs (I_2), and finally the beliefs of Spinozan pantheism (P). Judging these belief systems according to their ability to give him guidance and peace of mind, he considers each of C_1, C_2, I_1, and I_2 to be equally choiceworthy, and each of them to be more choiceworthy than P. However, $C_1 \cap C_2 \cap I_1 \cap I_2$, the state of hesitation between the four belief systems he ranks highest, is much

worse than P. It gives him no peace of mind, and the guidance it provides on how to conduct his life is tantalizingly incomplete. For instance it tells him that there is only one road to salvation, but leaves him ignorant of which that road is.

This problem also has a reverse form that comes out most clearly in sphere-based possible world models of revision. In these models, the original belief set K is assumed to be the intersection of all the possible worlds that have maximal plausibility. As explained in Section 1.4, it follows that the possible worlds that have K as a subset are all equally plausible. To see why this is problematic, note that my present belief set K neither contains the statement that Proxima Centauri b, the closest known exoplanet, has intelligent life (p) nor the statement that it does not ($\neg p$). Consequently, there are possible worlds containing $K \cup \{\neg p\}$ and also possible worlds that contain $K \cup \{p\}$. It follows from the sphere-based construction that these worlds are all equally plausible. This is counter-intuitive since I hold $\neg p$ to be more plausible than p. On a more basic level, it is difficult to see what it means to apply a concept of plausibility – or any other property that correlates in the intended way with epistemic choiceworthiness – to a single possible world.

2.4 Do All Success Conditions Withstand Intersection?

Up to now we have only discussed two success conditions, namely those of contraction (absence of the input sentence) and revision (presence of the input sentence). Both these success conditions have the following characteristic:

A property on sets is *preserved under intersection* if and only if the following holds for all non-empty collections $\mathfrak{X}$ of sets:

If each element of $\mathfrak{X}$ has the property, then so does $\bigcap \mathfrak{X}$ [126].

It is the preservation under intersection of the respective success conditions that makes the select-and-intersect method operable for contraction and revision. If p is absent from all elements of $\mathfrak{X}$, then it is also absent from $\bigcap \mathfrak{X}$. Similarly, if p is present in all elements of $\mathfrak{X}$, then it is also present in $\bigcap \mathfrak{X}$. The following example shows that we may sometimes wish to perform an operation of change with a success condition that is not preserved under intersection.

According to the public prosecutor's indictment, the accused has committed either murder or voluntary manslaughter. Susan is the judge assigned to the case. According to procedural law, she has three options. She can find the accused guilty of murder, find him guilty of voluntary manslaughter, or acquit him. She is convinced that he has killed the victim, but finds it difficult to adjudicate whether it was murder or not. Although the procedural law admits disjunctive indictments, it does not allow disjunctive verdicts. She therefore has to make up her mind so that she can either conclude that the accused committed murder or that he is guilty of voluntary manslaughter.

The success condition for the shift in her beliefs that the situation requires can best be described as a requirement that she either comes to believe that the accused committed murder (m) or that he committed voluntary manslaughter (v). This is of course different from believing that he committed either murder or voluntary manslaughter ($m \vee v$), which she already does. The belief change called for can be formalized as an operation of *choice revision*, in which the input is a set of sentences rather than a single sentence [60, 64]. The success condition of choice revision by a set A of sentences is that the output should contain at least one element of A. To see that this condition is not preserved under intersection, we can use our example $A = \{m, v\}$ and consider the two potential outcomes $X_1 = \text{Cn}(\{m\})$ and $X_2 = \text{Cn}(\{v\})$. The success condition is satisfied by both X_1 and X_2 but not by their intersection $X_1 \cap X_2$.[4] Choice revision also defies the decomposition principle discussed in the previous chapter, i.e. it does not seem to be reconstructible in terms of expansion and contraction.

2.5 Do We Select Among the Right Objects?

When choice functions are used in social choice theory, they operate on sets containing objects available for choice, such as physical objects or social states of affairs. The standard properties of choice functions have been developed from our intuitions about their application to objects we can choose between. In belief change, we use choice functions to obtain a new belief set. To choose a belief set means to choose among potential belief sets, just as choosing a dessert means to choose among desserts. Therefore, we should expect the choice functions (selection functions) of belief change theory to be applied to potential belief sets.

However, as we saw in Chapter 1, selection functions are standardly applied to sets of remainders and possible worlds. It is not difficult to show that neither of these are plausible belief sets. Beginning with possible worlds, we have already noted that if W is a possible world, then it holds for each sentence q in the language that either $q \in W$ or $\neg q \in W$. The absurdity of belief sets with this property was noted by two of the AGM authors already in 1982 [3, p. 21]. An example of how the sphere model works can serve to illustrate the point: On one occasion I had a belief set K containing the sentence "There is milk in my fridge" (p). When opening my fridge I found this to be wrong and revised my belief set by $\neg p$. The sphere model (as in Fig. 1.3, substituting $\neg p$ for p) depicts this change as a process in which I first selected the most plausible possible worlds in which $\neg p$ is true, and then adopted the intersection of all those worlds as my new belief set. In each of the options selected in the first stage I would be a full-fledged *Besserwisser*, willing to assign a confident "true" or "false" to every statement that can be made in the language. Needless to say, my experience of coming to believe that I had no milk did not involve an intermediate

[4] See Section 4.4 for a formal characterization of the preservation of success conditions under intersection of belief sets.

stage in which I vacillated between different forms of purported omniscience. A reconstruction of the process in such terms seems far-fetched.

Remainders do not have this property, but they have another problematic property:

Observation 2.2 ([3, p. 20]) *Let $p \in K$ and $X \in K \perp p$, and let q be any sentence. Then either $p \vee q \in X$ or $p \vee \neg q \in X$.*

As Alchourrón and Makinson noted, this is a "rather counterintuitive" property, in particular when q is (intuitively speaking) content-wise unconnected with both p and the rest of K [3]. To see that, we can again consider my belief change when I found no milk in the fridge. Let q denote that Socrates had the hiccups on his sixth birthday. According to the partial meet account of belief revision, my adoption of the belief $\neg p$ began with retraction of p from my original belief set. The retraction followed the select-and-intersect pattern. Therefore, in the initial selection phase I chose among a collection of belief sets, in each of which I believed in one of the two statements "either there is milk in the fridge or Socrates had the hiccups on his sixth birthday" ($p \vee q$) and "either there is milk in the fridge or Socrates did not have the hiccups on his sixth birthday" ($p \vee \neg q$). Both of these are strange beliefs for someone to hold who has no idea what happened to Socrates on his sixth birthday. I should be able to give up my belief that I have milk in the fridge without passing through an intermediate stage in which I vacillate between such outlandish belief states.[5]

Furthermore, a plausible belief state should be one that a human mind can harbour. Since we are finite beings, we cannot have belief states that require infinite representations. If a belief state can only be represented by infinite sets, then it is not a belief state that human beings can have or entertain having. (Nor is it representable in a computer.) We should therefore expect all belief sets that are considered in a belief change process to satisfy the following condition:

Definition 2.3 *A logically closed set X of sentences is* finite-based *if and only if there is some finite set X' such that $X = \mathrm{Cn}(X')$.*

A simple way to achieve this would be to use a logically finite language, i.e. a language that contains only a finite number of (pairwise) non-equivalent sentences. This would mean that the language has only a finite number of atoms.[6] However, such a language is bound to have gratuitous limits to its expressive power [108, 109]. Consider the following list of sentences:

[5] On the implausibility of maxichoice contraction of belief sets, see also [1], [99, pp. 76–77], and [109, p. 33]. Maxichoice contraction is less implausible for belief bases (that are not logically closed) than for belief sets, see [175] and [99, p. 77].

[6] A language is syntactically finite if it has only a finite number of non-identical sentences. All syntactically finite languages are logically finite, but the converse does not hold. For instance, a language that contains the atom a and the conjunction sign is syntactically infinite since it contains the infinite set of sentences $\{a, a\&a, a\&a\&a, \dots\}$. Contrary to logical finiteness, syntactic finiteness is a property of the language itself (rather than a property of the logic).

v_{50} = Less than 50 paintings by Johannes Vermeer are extant.
v_{51} = Less than 51 paintings by Johannes Vermeer are extant.
v_{52} = Less than 52 paintings by Johannes Vermeer are extant.
...
$v_{1.000.000}$ = Less than 1.000.000 paintings by Johannes Vermeer are extant.
...

I believe in each of the sentences on this list, and therefore my set of beliefs contains infinitely many logically non-equivalent sentences. A logically finite language cannot treat all pairs of sentences on the list as non-equivalent. This is a serious restriction on its expressive power. However, my belief in all of these sentences can be represented by a finite-based belief set. The reason for this is that all the sentences on this infinite list follow logically from the first of them, viz. v_{50}. Therefore a belief set that contains v_{50} implies all of the others. This example shows that the requirement of finite-basedness allows for much more expressive power than that of a logically finite language.

Let us now apply the criterion of finite-basedness to the two types of intermediates used in the AGM approach, namely remainders and possible worlds. It is fairly easy to show that neither of them can be finite-based if the language is logically infinite. In addition we can show that both of them will come in infinite numbers:

Observation 2.4 ([109]) *Let the language $\mathcal{L}$ consist of infinitely many logically independent atoms and their truth-functional combinations. Let K be a belief set and let $p \in K \setminus \text{Cn}(\varnothing)$. Then:*

(1) If W is a possible world (i.e. $W \in \mathcal{L} \perp \perp$), then W is not finite-based.
(2) There are infinitely many W such that $p \in W \in \mathcal{L} \perp \perp$.
(3) If $X \in K \perp p$, then X is not finite-based, and
(4) $K \perp p$ is infinite.

Hence, if the language is logically infinite, then all remainders and all possible worlds lack a finite representation. Furthermore, the remainders or possible worlds that we have to select among in a partial meet contraction or a sphere-based revision are always infinite in number.

Thus, even if both the original belief set (K) and the outcome of an operation ($K \div p$ or $K * p$) are finite-based, the transition from the former to the latter requires that we create an infinite set of irreducibly infinite entities, which are then eliminated (through intersection). In other words, the road from a finite-based belief set to another finite-based belief set takes a detour into Cantor's paradise. For those of us who are in favour of cognitive realism and linguistic representability, this is not a desirable deviation.

Someone might wish to argue that this excursion into infinity is useful and perhaps even necessary since we are trying to model the doxastic behaviour of rational agents rather than that of actual agents. Supposedly, results obtained for ideal rational agents with transfinite reasoning powers have normative force as ideals for actual agents. However, the best use of limited cognitive resources may require that one follows principles and processes that would not be useful for logically omniscient beings. Therefore, normative guidance is best obtained from studies of another type of ideal agents, namely agents that have limited cognitive capacity of which they make rational use.

Chapter 3
Questionable Patterns of Change

In Chapter 2 we saw that the standard use of the select-and-intersect method in belief change has several features that are difficult to justify. However, these features all concern the mechanisms used to construct the operations, not the properties of the actual operations. If we adopt the black box approach, then we do not need to worry about the realism of the mechanisms used to construct belief change operations. Instead, our focus should be on the properties of the operations that these mechanisms give rise to. In this chapter we will consider a series of postulates from that point of view. First we look at finite-based outcome, an eminently plausible property that does not hold in the AGM framework (Section 3.1). After that we will consider three postulates for contraction that have the opposite problem, i.e. they hold for AGM operations but their plausibility is questionable. The three are: success (Section 3.2), recovery (Section 3.3), and inclusion (Section 3.4). We will then consider the expansion property for revision (Section 3.5) and some properties of extensions of the framework that include conditional sentences (Section 3.6) and iterated change (Section 3.7). The chapter ends in Section 3.8 with a list of ten desiderata for an alternative framework that eschews the major problems of the standard framework pointed out in this and the previous chapter.

3.1 The Postulates of Finite-Based Outcome

According to the black box approach it does not matter if intermediate constructs in belief change are logically infinite or otherwise cognitively unrealistic. However, even in this approach, the cognitive realism of outcomes cannot be taken lightly. We should expect the following property to hold:

> An operation on belief sets satisfies *finite-based outcome* if and only if it yields a finite-based belief set as outcome whenever the original belief set is finite-based [89, p. 604].

S.O. Hansson, *Descriptor Revision*, Trends in Logic 46,
DOI 10.1007/978-3-319-53061-1_3

Finite-based outcome can be specified for contraction and revision as follows:

If K is finite-based, then so is $K \div p$. (finite-based contraction)
If K is finite-based, then so is $K * p$. (finite-based revision)

Both these postulates can be supported with arguments that refer specifically to the respective operation. Revision incorporates a single sentence into the belief set. Since a single sentence only contains a finite amount of information, it cannot take us from a finite-based belief set to one that is logically infinite. Similarly, contraction means loss of information and therefore, when we contract a finite-based belief set by some sentence, the outcome should be finite-based.

Although finite-based outcome is not one of the AGM postulates (and does not follow from them), all three AGM authors have gone on record endorsing what that postulate requires.

> We suggest, finally, that the intuitive processes themselves, contrary to casual impressions, are never really applied to theories as a whole, but rather to more or less clearly identified bases for them. For a theory is an infinite object, having as it does an infinite number of elements, and it is only by working on some finite generator or representative of the theory that the outcome of a process such as contraction can ever in practice be determined. (Alchourrón and Makinson [3, pp. 21-22])

> In all applications, knowledge sets [belief sets] will be *finite* in the sense that the consequence relation $\vdash$ partitions the elements of K into a finite number of equivalence classes. (Gärdenfors and Makinson [71, p. 90])

Makinson has also written that belief change operations on sets that are closed under logical consequence "form an idealized mathematical exercise, and that in real life the operations are always applied to bases for theories" [175, p. 384].

However, finite-based outcome does not hold for either partial meet contraction or partial meet revision:

Observation 3.1 *Let K be a belief set, p a sentence, $\div$ a partial meet contraction on K, and $*$ a partial meet revision on K. Then:*

(1) Even if K is finite-based, it does not hold in general that $K \div p$ is finite-based.

(2) Even if K is finite-based, it does not hold in general that $K * p$ is finite-based.

(3) If K is not finite-based, then neither is $K \div p$.[1]

Obviously, finite-based outcome can be added as a postulate, but there does not seem to be any plausible property of AGM selection functions that corresponds to it.[2]

---

[1]However, it is possible for $K * p$ to be finite-based even if K is not. Let p be an atom, let S be an infinite set of atoms such that $p \notin S$, and let $K = \mathrm{Cn}(\{p \vee s \mid s \in S\})$.

[2]There seems to be only one construction on record by which finite-based outcome can be achieved while retaining all the AGM postulates. This construction replaces the selection function γ by a sentential selector f that takes us from one sentence in the language to another, so that $K \div p$ is

Therefore, Observation 3.1 reveals a problem for the standard approach to belief revision.

Several other postulates are problematic for the standard approach. For finite-based outcome the difficulty is that the postulate does not hold for the AGM operations although intuitively, it should. For most of the other contested postulates it is the other way around: they hold in spite of weighty arguments why they should not.

3.2 The Success Postulates

The success postulate for revision ($p \in K * p$) requires that the input sentence is always accepted. In real life, however, new information is often rejected if it contradicts strongly held previous beliefs.

> It follows from the postulates for revision that the system is totally trusting at each stage about the input information; it is willing to give up whatever elements of the background theory must be abandoned to render it consistent with the new information. Once this information has been incorporated, however, it is at once as susceptible to revision as anything else in the current theory. Such a rule of revision seems to place an inordinate value on novelty, and its behavior towards what it learns seems capricious. [34, p. 251]

This property of operations of revision has also been called "primacy of new information" [35], "recency-prejudice" [206, p. 14], and "dictatorship" of the most recent evidence [77, p. 40]. The need to relax the success condition for revision has been pointed out by many authors, for instance by David Makinson who observed that "we may not want to give top priority to new information. If it conflicts with the old information in K, we may wish to weigh it against the old material, and if it is really just too far-fetched or incredible, we may not wish to accept it" [179, p. 14].

The success condition for belief contraction is also remarkably strict; the input sentence is always removed unless it is a tautology. Hans Rott noted that it is desirable to "allow a reasoner to refuse the withdrawal of [a sentence] not only in the case where

---

(Footnote 2 continued)
identified with $\bigcap(K \perp f(p))$ (where $f(p)$ is a sentence) rather than with $\bigcap \gamma(K \perp p)$. On this construction, see [108, 109, 111, 114, 115].

A highly interesting construction was proposed to me by David Makinson in correspondence. Let K be finite-based. As previously shown in [152] and [182, pp. 244–247] there is then a unique smallest set of atomic sentences S that are present in at least one sentence of every set A of sentences such that $K = \text{Cn}(A)$. Let $\mathcal{L}_S$ be the set of sentences in the language that contain no other atoms than those in S. Then $K = \text{Cn}(K \cap \mathcal{L}_S)$. Let $\div$ be an operation on K such that for all $p \in \mathcal{L}$: $K \div p = \text{Cn}(\gamma((K \cap \mathcal{L}_S) \perp p))$ for a selection function γ satisfying the standard AGM conditions. Makinson has shown that $\div$ satisfies five of the six basic AGM contraction postulates and the sixth, namely recovery, in the following weakened form:

If $p \in \mathcal{L}_S$, then $K \subseteq \text{Cn}((K \div p) \cup \{p\})$. (limited recovery)

He has also shown that if γ is transitively relational, then $\div$ satisfies both conjunctive overlap and conjunctive inclusion.

Both of these two constructions satisfy the above-mentioned postulate finite-based contraction. However, neither of them is immune against most of the other problems with the traditional approach that are discussed in this and the previous chapter.

[that sentence] is a logical truth. There may well be other sentences ('necessary truths') which are of topmost importance for him" [213, p. 54].

The strictness of the two success conditions should be listed among the problematic features of the AGM model. However, this is a feature that is relatively easy to modify within the framework itself. It is for instance possible to divide inputs into two categories, acceptable and non-acceptable, and treat the former in accordance with the AGM model while the latter induce no change at all. Such non-prioritized belief revision has been the subject of quite extensive studies. Many of these are faithful to the original framework, both in the sense of modifying the construction only moderately and in the sense of satisfying most if not all of the other postulates.[3] It is an important desideratum for an alternative framework for belief change that it should be fully workable without the requirement that all inputs are accepted.

3.3 The Recovery Postulate

We noted in Section 1.3 that partial meet contraction satisfies the following postulate:

$K \subseteq (K \div p) + p$ (recovery)

In combination with the inclusion postulate, recovery implies the following:

If $p \in K$, then $K = (K \div p) + p$.

Recovery says that nothing is lost if we first remove and then reintroduce some belief [67, pp. 93–94]. It is easy to find examples that seem to validate recovery. A person who first loses and then regains her belief that she has a dollar in her pocket can be described as returning to her previous state of belief. However, other examples can also be presented, in which recovery is more questionable.

For many years, Derek was confident that his wife was faithful to him (f). But one day a neighbour told him stories that convinced him she was cheating on him. When he confronted her, she could explain everything, and he regained his previous belief in her faithfulness. But something strange happened. He never regained his belief that she loved him (l). He could not explain why. All misunderstandings had been straightened up, and everything else was as before, but still he was unable to believe in her love any more. Hence, $l \in K$ but $l \notin (K \div f) + f$.

I believed that Cleopatra had a son (s). Therefore I also believed that Cleopatra had a child (c or equivalently $s \vee d$ where d denotes that Cleopatra had a daughter). Then I received information that made me give up my belief in c, and I contracted my belief set accordingly, forming $K \div c$. Soon afterwards I learned from a reliable source that Cleopatra had a child. It seems perfectly reasonable for me to then add c (i.e., $s \vee d$) to my set of beliefs without also reintroducing s [82].

[3] For constructions of non-prioritized belief change operations, see: [10, 15, 32, 47, 50, 51, 54, 75, 95, 96, 98, 137, 179, 186, 204, 229].

I previously entertained the two beliefs "George is a criminal" (c) and "George is a mass murderer" (m). When I received information that induced me to give up the first of these beliefs (c), the second (m) had to go as well (since c would otherwise follow from m).

I then received new information that made me believe that George is a shoplifter (s). The resulting new belief set is the expansion of $K \div c$ by s, $(K \div c) + s$. Since c follows from s, $(K \div c) + c$ is a subset of $(K \div c) + s$. By recovery, $(K \div c) + c$ includes m. It follows from this that $(K \div c) + s$ also includes m.

Thus, since I previously believed George to be a mass murderer, I cannot now believe him to be a shoplifter without believing him to be a mass murderer [87].

Derek's pattern of belief change may be irrational (but we should not judge him too harshly). The other two examples would not seem to be easily demoted to that category.[4] It is therefore reasonable to conclude that the recovery postulate is too demanding. But on the other hand, recovery is unavoidable if $K \div p$ is the intersection of some elements of $K \perp p$.[5] It is in other words a direct consequence of applying the select-and-intersect method to remainders, and therefore the counterexamples to recovery speak directly against that method.

3.4 Pure Contraction and the Inclusion Postulate

Contraction is defined as an operation in which a specified previous belief is removed but no new beliefs are added. Therefore, it satisfies the inclusion postulate:

$$K \div p \subseteq K$$

But in spite of the central role of contraction in the belief change literature, it is difficult to find examples of "pure" contraction in which no new belief is added. Of course there are belief changes in real life that are driven by a need to give up a certain belief. However, such changes tend to be caused by the acquisition of some new information that is added to the belief set [103]. The following is a typical example:

---

[4] In a defence of recovery, Makinson argued that the theories considered in these and similar examples are implicitly "clothed" with additional justificational structure. In his view, recovery remains appropriate for "naked", logically closed theories without such additional structure [178]. In a somewhat similar vein, Glaister argued that in the Cleopatra example, the loss of my belief that Cleopatra had a child is better represented by a multiple contraction than by a contraction by $s \vee d$ [78]. On the recovery postulate, see also [30, 67, 82, 97, 165, 175].

[5] To see this, let X be a p-remainder of the belief set K, i.e. $X \in K \perp p$, and let q be any sentence in K. Now consider the sentence $p \rightarrow q$. Since it follows logically from K it is also one of its elements. We are going to prove that $p \rightarrow q \in X$. Suppose to the contrary that $p \rightarrow q \notin X$. Then, since X is a p-remainder of K, the addition of $p \rightarrow q$ to X will have to result in a set that implies p. Thus $X \cup \{p \rightarrow q\} \vdash p$, thus $X \vdash (p \rightarrow q) \rightarrow p$, thus $X \vdash p$, contrary to our assumption that X is a p-remainder. This contradiction shows that our supposition $p \rightarrow q \notin X$ was wrong, thus $p \rightarrow q \in X$. This holds for all $X \in K \perp p$. Since the partial meet contraction $\bigcap \gamma(K \perp p)$ is the intersection of a collection of p-remainders of K, it follows that $p \rightarrow q \in \bigcap \gamma(K \perp p)$ for all $q \in K$.

> Not long ago a friend told me he was quite sure that the Vatican City State is a full member of the United Nations, which I believed it was not. This made me uncertain and induced me to enter a state of hesitation concerning the issue in question. I therefore removed the sentence "The Vatican City State is not a full member of the United Nations" (v) from my set of beliefs, without adding its negation.

In the belief revision literature, this would be treated as a contraction, and an operation $\div$ would take us from the initial belief set K to a new belief set $K \div v$ such that $v \notin K \div v \subseteq K$. However, this is not what happened. The reason why I gave up v was that I acquired the new belief that my friend is convinced that the Vatican City State is a full member of the United Nations (c). We would then have $c \notin K$ but $c \in K \div v$, contrary to the inclusion postulate. Importantly, this is not due to some particular feature of the example, but to a general pattern in how we change our beliefs: Rational rejection must have a basis in some new information that forces the old belief out, and since this new information is accepted it will be retained after the operation of change has been completed.[6]

The most plausible candidates for pure contraction are hypothetical contractions performed for the sake of argument, i.e. in order to give a contradictory belief a hearing [63, 64, 164]:

> AHMED: This painting must be by Johannes Vermeer. Look at the characteristic visible brushstrokes in the folds of the mantle to the right in the picture.
> FATMA: I totally disagree. In my view Vermeer never left such visible brushstrokes in his paintings. But we can leave that open and look instead at the man with the blue hat to the left. The blue colour has turned dark and greenish. It is probably azurite, at any rate not lapis lazuli that Vermeer always used. And then look at the supposedly parallel lines on the walls. They do not meet in the same point, which is absolutely incompatible with Vermeer's perfect linear perspective.

With the phrase "we can leave that open" the second interlocutor withdraws (or offers to withdraw) her belief that Vermeer never left visible brushstrokes on his canvases. She does so without accepting any new belief that would expel the belief in question. Therefore, this can be represented as a pure contraction, an operation that satisfies the inclusion postulate. However, the relevance of such hypothetical contractions in belief change is questionable since these contractions are not seriously undertaken by the agent. In particular, the use of contraction as a suboperation in belief revision, as encoded in the Levi identity ($K * p = (K \div \neg p) + p$), requires that the contraction is real and not just hypothetical. Therefore, the failure of the inclusion postulate and the consequent unrealism of (pure) contraction create problems for the decomposition principle that we discussed in Section 1.2. If we do not perform contractions in the AGM sense, but instead revisions by sentences that drive out other sentences, then how can contractions be obligatory suboperations of all belief-contravening revisions?[7]

---

[6]This refers to the modelling of human beliefs. Pure contraction of databases is unproblematic.

[7]On the use of revision to remove sentences, see Section 9.5.

3.5 The Expansion Property

Let us now turn to the operation of revision. Two of the basic AGM postulates for revision, namely inclusion ($K * p \subseteq K + p$) and vacuity (If $\neg p \notin K$ then $K + p \subseteq K * p$) imply in combination the following property:

If $\neg p \notin K$, then $K * p = K + p$. (expansion property of revision)

The expansion property can be described as a principle of deductivism. It says that if the agent adopts a new belief that does not contradict her previous beliefs, then she comes to believe in everything that follows logically from the combination of her old beliefs and the new one, but nothing beyond that. The expansion property can also be derived from the Levi identity ($K * p = \text{Cn}((K \div \neg p) \cup \{p\})$) that is closely connected with the decomposition principle, in combination with the vacuity postulate for contraction (If $p \notin K$ then $K \div p = K$).[8]

However, plausible counter-examples to the expansion property are not difficult to find:

> John is a neighbour about whom I initially know next to nothing.
>
> *Case 1:* I am told that he goes home from work by taxi every day (t). This makes me believe that he is a rich man (r).
>
> *Case 2:* When told t, I am also told that John is a driver by profession (d). In this case I am not made to believe that he is a rich man (r) [85].

In case 1 we have $r \in K * t$, and due to the expansion property $K * t = \text{Cn}(K \cup \{t\})$. Since K is logically closed it follows that $t \rightarrow r \in K$. In case 2, the expansion property yields $K * (t\&d) = \text{Cn}(K \cup \{t\&d\})$. Combining this with $t \rightarrow r \in K$ we obtain $r \in K * (t\&d)$, contrary to the description of case 2.

This example shows that the expansion property is at variance with a common (and fully rational) pattern of belief change: When we acquire a new belief that does not contradict our previous beliefs (such as t in the example), then we often complement it with additional beliefs (such as r in the example) that "round off" the belief set and make it more coherent, but do not follow deductively.

The expansion property can also go wrong in the opposite direction. Sometimes when we revise by a sentence p that does not logically contradict the belief set K, this leads to the exclusion of some sentence that the new information makes implausible but does not exclude logically. In such cases we can have $\neg p \notin K$ but $K + p \not\subseteq K * p$:

> I believed that one of the three heirs, Amelia, Barbara, and Carol, murdered the rich eccentric ($a \vee b \vee c$). Then I received information convincing me that both Amelia and Barbara are innocent ($\neg a\&\neg b$). However, since I had no specific information binding Carol to the crime this did not make me believe that Carol was the murderer.

[8] Let $\neg p \notin K$. Vacuity yields $K \div \neg p = K$, and then the Levi identity yields $K * p = K + p$.

Valentina was uncertain whether or not her husband is faithful to her (f), but she still believed that he loves her (l). However, when she learnt that he is unfaithful to her, she lost her belief that he loves her [135].

In the first example we have $K \nvdash \neg(\neg a \& \neg b)$ but it can be seen from $c \notin K * (\neg a \& \neg b)$ that $\text{Cn}(K \cup \{\neg a \& \neg b\}) \nsubseteq K * (\neg a \& \neg b)$. In the second example we similarly have $K \nvdash f$ but $l \in \text{Cn}(K \cup \{\neg f\})$ and $l \notin K * \neg f$, thus $\text{Cn}(K \cup \{\neg f\}) \nsubseteq K * \neg f$.

The expansion property is strongly associated with the decomposition principle of the AGM framework. Therefore the counterexamples that call this property into question bring out fundamental problems in the framework. Some of the repercussions will be seen in the next section.

3.6 Extending the Language

The AGM framework is based on a supraclassical sentential logic, i.e. one that satisfies all the rules of the classical truth-functional logic of sentences. Usually, no other logical principles than the truth-functional ones have been assumed.[9] The reason for this is that additions to the language of sentences with non-truth-functional properties have turned out to be more difficult to implement than what one would perhaps expect.

Most of the discussion on such extensions has referred to Ramsey test conditionals, i.e. conditional sentences that satisfy the so-called Ramsey test. The test is based on a suggestion by Frank Ramsey [210, pp. 247–248] that has been further developed by Roderick M. Chisholm [31], Robert Stalnaker [240, pp. 101–105], and others. The basic idea is very simple: "If p then q" is taken to be maintained by the epistemic agent if and only if she would believe in q after revising her present belief state by p. To express this in formal language, let $p \rightarrowtail q$ denote "if p then q", or more precisely: "if p were the case, then q would be the case". One version of the Ramsey test says:

$p \rightarrowtail q$ is maintained by the agent in K if and only if $q \in K * p$.

(Ramsey test, version 1)

Attempts have been made to include sentences of the form $p \rightarrowtail q$ in the object language of the AGM framework. This means that these sentences will be included in the belief set when the agent assents to them, thus:

$p \rightarrowtail q \in K$ if and only if $q \in K * p$. (Ramsey test, version 2)

[9] It has sometimes been proposed that the object language should be extended to first-order predicate logic (with representations of "all" and "some"). However, no such extension seems to have been carried out. Presumably, this is because not much would be gained by it. When referring to a finite number of objects, "all" and "some" can be rendered with repeated uses of "and" and "or", respectively.

The step from to the first to the second version of the Ramsey test, i.e. the inclusion of these conditionals in belief sets, may seem undramatic but it has far-reaching consequences for the formal framework.[10] This is because the second version implies the following property:

If $K \subseteq K'$, then $K * p \subseteq K' * p$. (revision monotonicity)

The proof that revision monotonicity follows from the second version of the Ramsey test is quite simple: Let $K \subseteq K'$ and $q \in K * p$. The Ramsey test yields $p \rightarrowtail q \in K$, then $K \subseteq K'$ yields $p \rightarrowtail q \in K'$, and finally one more application of the Ramsey test yields $q \in K' * p$ [68].

Revision monotonicity holds in all belief change systems that include a Ramsey test conditional, but its impact differs between such systems. It has no impact at all in systems where no available belief set can be a proper subset of another available belief set. However, the AGM framework is not such a system. To the contrary, it contains two guarantees for the presence of belief sets that are proper subsets of other belief sets. One of these guarantees is based on the inclusion and success postulates for contraction. Together they provide us, for each belief set K and each sentence $p \in K \setminus \text{Cn}(\varnothing)$, with a belief set $K \div p$ that is a proper subset of K. The other guarantee follows from the expansion property that provides us, for each belief set K and sentence p such that $p \notin K$ and $\neg p \notin K$, with a belief set $K * p$ that is a proper superset of K. In both these cases, revision monotonicity has implausible implications. The following example refers to the contractive case:

> In my original belief set K, I believed that Vasily's brother Boris is a cross-dresser (transvestite) (c).
>
> *Case i:* I realize that my belief that Boris is a cross-dresser was based on a confusion, and contract that belief. After that I am told that Boris will come to Vasily's party next Friday (p). I now believe that Boris will come to the party in trousers (t).
>
> *Case ii:* I do not retract my belief in c. But just as in the previous case I am told that Boris will come to Vasily's party next Friday (p). I do not believe that he will come in trousers.

From Case i we learn that $t \in ((K \div c) * p)$. It follows from the inclusion property of contraction that $K \div c \subseteq K$, and then revision monotonicity yields $t \in K * p$, contrary to the reasonable pattern described in Case ii of the example. Note that this implausible result derives only from the Ramsey test (that alone yields revision monotonicity) and the inclusion property of contraction.

A similarly implausible result can be based on revision, provided that it satisfies the expansion property.

[10] Due to these consequences, Isaac Levi accepts only the first version of the test. ([163], see also [65].).

In my original belief set K all I know about Ellen is that she is the sister of my new friend Karen.

Case i: Karen says that Ellen is married (m). This makes me believe that Ellen has a husband (h).

Case ii: Karen tells me that Ellen is a lesbian (l). Then she tells me that Ellen is married (m). This does not make me believe that Ellen has a husband.

Since $\neg l \notin K$, the expansion property yields $K * l = K + l$, thus $K \subseteq K * l$. We know from Case i that $h \in K * m$. Revision monotonicity yields $h \in (K * l) * m$, which is of course implausible, as recognized in Case ii. The only formal properties that we need to derive this result are the Ramsey test (yielding revision monotonicity) and the expansion property of revision.

The problems that Ramsey test conditionals bring with them into the AGM framework were first pointed out in a famous impossibility theorem by Peter Gärdenfors [68].[11] The theorem shows that the following three conditions cannot all be satisfied:

1. Version 2 of the Ramsey test.
2. The six basic AGM postulates (or a weakened version of them).
3. There are three sentences p, q, and r such that each of $p\&q$, $q\&r$, and $p\&r$ is inconsistent and that $\neg p \notin K$, $\neg q \notin K$, and $\neg r \notin K$.

André Fuhrmann has shown similar problems for modal sentences of the form $\Diamond p$, denoting that p is possible [61]. He employs the following definition of possibility:

$\Diamond p$ holds at K if and only if $\neg p \notin K$. (Fuhrmann's test, version 1)

This is one of several plausible explications of possibility in a belief change context.[12] It corresponds to a relatively weak notion of possibility:

PHILOSOPHER'S HUSBAND: I can't find my keys.
PHILOSOPHER: Perhaps they are on the car seat, where you left them last week.
PHILOSOPHER'S HUSBAND: No that's not possible.
PHILOSOPHER: What do you mean by "not possible"?
PHILOSOPHER'S HUSBAND: Please do not quibble about words. I just mean that I know they are not there.

If we wish to include possibility statements, thus interpreted, into the object language, then we can modify Fuhrmann's test as follows:

$\Diamond p \in K$ if and only if $\neg p \notin K$. (Fuhrmann's test, version 2)

---

[11] Segerberg [233] generalized this result to a wider range of underlying logics. On the Ramsey test in belief revision, see also [23, 24, 85, 144, 212, 221].

[12] Alternatively we could assume that "p is possible" holds whenever there is something (some belief change) that would bring the agent to believe in p. Such a definition cannot be meaningfully introduced into the original AGM framework since the standard operation of revision satisfies the success postulate ($p \in K * p$) and consequently $\Diamond p \in K$ would be true for all p and K. However, if $*$ is an operation of non-prioritized revision, then such a definition can provide a useful account of possibility. See [91] and Section 7.6.

But just as for Ramsey test conditionals, this step creates problems. The second version of Fuhrmann's test implies the following property:

There are no belief sets K and K' such that $K \subset K'$. (non-inclusion)

The proof is quite simple.[13] Suppose that $K \subset K'$. Then there is some q such that $q \in K'$ and $q \notin K$. It follows from $q \notin K$ and the test that $\Diamond\neg q \in K$, and $K \subset K'$ yields $\Diamond\neg q \in K'$. But it also follows from $q \in K'$ and the test that $\Diamond\neg q \notin K'$. This contradiction concludes the proof.

Due to the vacuity postulate, the non-inclusion property is incompatible with the AGM framework. Therefore it should be no surprise that the introduction of this type of possibility representation into that framework gives rise to inconsistencies, as shown in detail by Fuhrmann [61].

Yet another plausible addition to the belief change framework is a functional predicate $\mathfrak{B}$ denoting what the agent believes. Arguably, a rational agent should be aware of her own beliefs. Let $\mathfrak{B}$ denote "the agent believes that". Then autoepistemic beliefs (beliefs about one's own beliefs) can be represented by $\mathfrak{B}$-containing expressions in the object language, and general patterns of self-knowledge can be expressed in the following two principles:

$\mathfrak{B}p \in K$ if and only if $p \in K$. (positive self-knowledge)
$\neg\mathfrak{B}p \in K$ if and only if $p \notin K$. (negative self-knowledge)

Positive self-knowledge causes no problem in the formal system, but just like the second version of Fuhrmann's test, negative self-knowledge implies non-inclusion.[14]

We can sum up this section as follows: Most results in the AGM framework do not go beyond a simple sentential language with a truth-functional logic. Some of the more plausible extensions of that language contain composite expressions whose truth conditions refer to the presence or absence in the belief set of some of their component sentences. Important examples are conditional, possibility, and epistemic expressions. Such expressions are difficult to accommodate in the AGM framework, as attested by several impossibility theorems. Moreover, the problems we have identified are all connected with the presence in the AGM framework of belief sets that are proper subsets of other belief sets.

3.7 Iterated Change

As originally defined, the AGM operations are *local* in the sense that they only provide means to change one particular belief set. This can be seen clearly from the definition:

---

[13]This is a strengthened version of a result from [99, p. 363].

[14]The proof is similar. Suppose that $K \subset K'$. Then there is some q with $q \in K'$ and $q \notin K$. It follows from $q \notin K$ and negative self-knowledge that $\neg\mathfrak{B}q \in K$, and $K \subset K'$ yields $\neg\mathfrak{B}q \in K'$. But it follows from $q \in K'$ and negative self-knowledge that $\neg\mathfrak{B}q \notin K'$. Contradiction. – On autoepistemic beliefs in belief change representations, see [58, 171, 172, 249, 250].

Original definition of partial meet contraction: [1]
(1) If $K \perp p \neq \varnothing$, then $\varnothing \neq \gamma(K \perp p) \subseteq K \perp p$.
(2) If $K \perp p = \varnothing$, then $\gamma(K \perp p) = \{K\}$.
(3) $K \div p = \bigcap \gamma(K \perp p)$

Suppose that we wish to use the same selection function γ for two different belief sets K_1 and K_2. Let $\top$ be a tautology. It follows from clause (2) that $\gamma(K_1 \perp \top) = \{K_1\}$ and $\gamma(K_2 \perp \top) = \{K_2\}$, thus $\gamma(K_1 \perp \top) \neq \gamma(K_2 \perp \top)$. But we also have $K_1 \perp \top = K_2 \perp \top = \varnothing$, so for γ to be a function it must be the case that $\gamma(K_1 \perp \top) = \gamma(K_2 \perp \top)$. We can conclude from this contradiction that in this framework, each selection function can only be used for one belief set.

By a *global* operation of contraction is meant one that can contract any belief set by any sentence. Similarly, a global operation of revision is one that allows us to revise any belief set by any sentence. Contrary to local operations, global operations can be used to perform whole series of changes, such as $K \div p \div q$, $K * p * q$, and $K * p \div q$. Therefore, the extension from local change to global change has usually been described as the introduction of *iterated* change. How to achieve it is one of the most discussed issues in the literature on belief change.[15]

It is important to distinguish between two problems of global (iterated) change, namely the *construction problem* and the *properties problem.* The construction problem consists in finding a mechanism that differs from selection functions as used above in being applicable to all belief sets. The properties problem consists in identifying plausible postulates for global (iterated) change. Obviously, the two problems are closely connected to each other; one would hope for the construction to be axiomatically characterized by a set of plausible postulates.

The construction problem has a surprisingly simple solution: We can replace the original definition of partial meet contraction by the following:

Alternative definition of partial meet contraction:
(1') $\gamma(K \perp p) \subseteq K \perp p$, and if $K \perp p \neq \varnothing$ then $\gamma(K \perp p) \neq \varnothing$.
(2') $K \div p = \bigcap \gamma(K \perp p)$, unless $\gamma(K \perp p) = \varnothing$ in which case $K \div p = K$.

Note that this definition, as applied to a single belief set K, is equivalent with the original definition. The only difference concerns the limiting case of contracting by a tautology. The two definitions both yield $K \div p = K$ whenever p is a tautology, but they achieve this result in different ways. The original definition deviates from the standard definition of a selection function in order to deal with this case (clause (2)). The alternative definition follows the usual definition of a selection function, and instead makes an exception in the derivation of the operation from the selection function (clause (2')).

This reorganization of the definition allows us to use one and the same selection function for all belief sets. This solves one of the two iterated change problems, namely the construction problem. It has often been claimed that after contracting K by

[15] For a brief summary, see [52, pp. 307–309]. See also [5, 14, 17, 18, 20, 21, 25, 28, 29, 33, 36, 41, 48, 49, 111, 115, 143, 146, 151, 154, 193–196, 219, 220, 244].

p with the selection function γ we have a new belief set, but no new selection function to be used in further operations on this new belief set. As the alternative construction shows, that need not be the case. Partial meet contraction can be constructed as a global operation. Since partial meet revision is definable from partial meet contraction via the Levi identity, this means that we have global operators for both contraction and revision.

But what implications does this reconstruction of the definition have for the properties problem? What properties does global partial meet contraction have, in addition to those of the local operation? The answer to that question may be surprising:

Observation 3.2 ([115, p. 160])[16] *Let $\mathcal{L}$ be infinite, let K_1 and K_2 be logically closed, and let $p_1 \in K_1 \setminus \mathrm{Cn}(\varnothing)$ and $p_2 \in K_2 \setminus \mathrm{Cn}(\varnothing)$. If $K_1 \perp p_1 = K_2 \perp p_2$ then $K_1 = K_2$ and $\vdash p_1 \leftrightarrow p_2$.*

Corollary to Observation 3.2 *Let $\mathfrak{X}$ be a set of belief sets, and for each $K \in \mathfrak{X}$ let γ_K be a selection function for the set of all K-remainders.*[17] *Then there is a global partial meet contraction γ for the set of remainders of elements of $\mathfrak{X}$,*[18] *such that $\gamma_K(K \perp p) = \gamma(K \perp p)$ for all $K \in \mathfrak{X}$ and all sentences p.*

In other words, any combination of selection functions for each belief set K can be unified into a single, global selection function. The corollary also answers our question what properties global partial meet contraction has in addition to those of the local operation. The answer is: none.

This refutes the common criticism against the AGM model that it does not provide us with means for iterated (global) change. If we use the alternative definition of the operation, then one and the same partial meet contraction (and one and the same selection function) can be used for contractions of all belief sets.

But do we not have another problem here? The corollary shows that with this construction, the operations of change on different belief sets will be completely independent of each other, and no specifically global postulates can be obtained. Isn't that implausible? Not necessarily, since expectations on general logical principles for iterated (global) operations of revision and contraction may be too high. To see why, let us consider the four most discussed such principles, namely the so-called Darwiche-Pearl postulates for revision: [36][19]

If $q \vdash p$, then $(X * p) * q = X * q$. (DP1)
If $q \vdash \neg p$, then $(X * p) * q = X * q$. (DP2)
If $X * q \vdash p$, then $(X * p) * q \vdash p$. (DP3)
If $X * q \nvdash \neg p$, then $(X * p) * q \nvdash \neg p$ (DP4)

---

[16]This does not hold for belief bases [88].

[17]This is the set $\{X \mid (\exists p \in \mathcal{L})(X \in K \perp p)\}$.

[18]This is the set $\{X \mid (\exists K \in \mathfrak{X})(\exists p \in \mathcal{L})(X \in K \perp p)\}$.

[19]The Darwiche-Pearl postulates express an intuition about the epistemic ordering of possible worlds, namely that when we revise by a sentence p, then the ordering among p-worlds should be unchanged, and so should the ordering among $\neg p$-worlds. The change takes the form of a shift of the relative positions of these two parts of the original ordering of worlds.

The following are counter-examples to each of them:

The three switches (Counterexample to DP1) (Stalnaker [242, pp. 205-206]) There are three electric switches, each of which can be either up or down. a denotes that the first switch is up, b that the second switch is up, and c that the third switch is up. They are connected to two lamps, a plasma lamp and a quartz lamp. The plasma lamp is lit (p) if and only if either $a\&b$ or $\neg a\&\neg b$. The quartz lamp is lit (q) if and only if either $a\&b\&c$ or $\neg a\&\neg b\&\neg c$. Your original belief set contains the three statements a, $\neg b$, and c. Originally, you have no direct information about whether the plasma or the quartz lamp are lit.

Case 1: Your learn that the quartz lamp is on. You conclude that either b or $\neg a\&\neg c$, but you do not know which. Thus $b \notin K * q$.

Case 2: You first learn that the plasma lamp is on. This makes you conclude that either b or $\neg a$. After that you learn that the quartz lamp is on. This makes you conclude that b. Thus $b \in K * p * q$.

The adder and the multiplier (Counterexample to DP2) (Konieczny and Pino Pérez [151, p. 352])[20] A circuit contains both an adder and a multiplier. Initially we know nothing about whether they work or not.

Case 1: We first receive the message "both the adder and the multiplier are working" (p). After that we receive the message "the adder is not working" (q). In the resulting belief state $K * p * q$ we believe that the multiplier is working.

Case 2: We just receive the message "the adder is not working" (q). I In the resulting belief state $K * q$ we still have no belief about whether the multiplier is working or not.

Cracks in the petrol tank (Counterexample to DP3) [132] During my time as a clerk at the headquarters of Destination Paradise airlines, service technicians discovered cracks in the petrol tanks of some of the older planes. These airplanes were always immediately grounded until the tank had been replaced by a new one. The routine for grounding these planes was very reliable, but some of the pilots were worried that deep cracks could develop in a few days. Such an occurrence in the interval between two service inspections could potentially cause a severe accident.

Case 1: At 9 a.m. I overheard a conversation in the coffee room. One of the secretaries said: "I have been told that airplane DP3 has a crack in the petrol tank." This made me believe that DP3 had, at 9 a.m., a crack in the petrol tank (p). It also made me believe that the plane was in for repair. One hour later my boss told me: "I have been called to a meeting. There was a terrible accident just a few minutes ago. DP3 has caught fire in the air and crashed, and apparently there is not much hope that anyone has survived." This made me believe that DP3 had crashed (q). It also made me give up my previous belief that it had a crack in its tank one hour

[20]For other counterexamples to DP2, see [29, pp. 71-72] and [242, pp. 206-208]. For discussions on DP2 see also [154] and [146].

ago, since if that information were correct, then the airplane would not have been in service. Hence, $K * p * q \nvdash p$.

Case 2: I did not overhear the coffee table conversation. But at ten o'clock my boss told me that DP3 had just caught fire in the air and crashed (q). Since a crack in the petrol tank was the cause of such an accident that immediately came to my mind, I now also believed that there had been a crack in the petrol tank at 9 a.m. that day (p). Hence, $K * q \vdash p$.

The Commander-in-Chief's message (Counterexample to DP4) [132]

Case 1: After an intelligence briefing about enemy activities, the General becomes convinced that the enemy intends to attack next morning on the northern flank (p). But late in the evening he receives a message from the Commander-in-Chief, saying: "We have very reliable information that the enemy's current troop movements have the purpose to deceive us about their plans for the next few days." The General now believes that the enemy has tried to deceive him about its plans for tomorrow (q). Consequently, his belief in p is replaced by belief in its negation. Hence, $K * p * q \vdash \neg p$.

Case 2: The general has no opinion about the enemy's plans for tomorrow. He receives the same message from the Commander-in-Chief as in the previous case, but after receiving that message he still has no opinion on whether p is true or not. Hence, $K * q \nvdash \neg p$.

There is an underlying problem that prevents the construction of plausible postulates for iterated contraction and revision: In a succession of changes, each input sentence influences how the subsequent inputs are received. Furthermore, this influence is not purely logical, but in the logical language we cannot express its non-logical components. The following simple dialogue should make the point:

MY FRIEND ANN: You know Patricia and Quinn, don't you?
ME: Yes, the couple that love each other so dearly.
ANN: Not any longer.
(I contract the sentence $p\&q$, where p denotes that Patricia loves Quinn and q that Quinn loves Patricia, from my belief set K. But I have no idea whether it is Patricia, Quinn, or both, that ceased loving the other.)
ANN: I met Patricia. She stays in Richard's flat now.
(I revise my belief set $K \div (p\&q)$ by r, "Patricia stays in Richard's flat". So she has found a new lover! Thus, $\neg p \in K \div (p\&q) * r$.)
ME: Is he ...
ANN: Yes he is her brother.
(I revise my belief set $K \div (p\&q) * r$ by s, denoting that Richard is Patricia's brother. I now see her stay in his apartment in a new light, and $\neg p \notin K \div (p\&q) * r * s$.)

The subtle ways in which r influences the impact of having contracted $p\&q$, and s modifies the influence of r, are not easily captured by logical relationships. Indeed, p, q, r, and s are all logically independent of each other. In a series of belief changes like this, each step in the series involves a choice of how best to accommodate the input. How that choice is made will affect the ways in which further changes are performed. In such a series of changes, the effects of non-logical relationships among sentences tend to be multiplied. That is the reason why attempted postulates for iterated revision do not have sufficiently general validity.[21] If global change operations have properties that go beyond those of local change, then those properties may not be expressible in a purely truth-functional object language.

3.8 Ten Desiderata for an Alternative Framework

We can summarize the findings of this and the previous chapter in the form of ten desiderata for an improved formal framework for belief change.

1. Selection functions should operate directly on plausible outcomes, i.e. on potential belief sets, rather than on cognitively unmanageable objects such as remainders or possible worlds.
2. Operations of revision and contraction that do not satisfy the success postulates should be available.
3. Other types of belief change than contraction and revision should be available, including operations whose success conditions are not preserved under intersection.
4. The postulate of finite-based outcome should be satisfiable.
5. The recovery postulate for contraction should not hold in general.
6. Contraction-like operations that do not satisfy the inclusion postulate should be available.
7. The operation of revision should not be required to satisfy the expansion property.
8. Conditional sentences satisfying the Ramsey test should be includible in the belief sets.
9. Modal sentences and sentences expressing autoepistemic beliefs should be includible in belief sets.
10. Operations of change should be iterable.

These ten desiderata will be used in Chapter 11 to evaluate the constructions introduced in the intermediate chapters.

[21] See [115] for a discussion of postulates for iterated contraction, with essentially the same conclusion.

Part II
Descriptor Revision

Chapter 4
Putting the Building-Blocks Together

The previous two chapters were devoted to negative work. We have inventoried problems and implausible properties that are connected with the traditional approach to belief change. But the purpose of all this negative work was positive. In Section 3.8 we summarized our findings in the form of a list of desiderata for an alternative approach. In this chapter, the outlines of such an approach will be constructed. The rest of the book is devoted to its further development and evaluation.

We will start from scratch. Section 4.1 introduces a very general model for belief change that is based on primitive belief states and inputs, neither of which has any sentential structure. This model has the advantage of making few controversial assumptions but also the disadvantage of low expressive power. It is used as a starting-point to which more structure will successively be added in a guarded fashion, allowing us to see what assumptions are needed to obtain the resulting increase in expressive power. In Section 4.2 sentences are associated with the belief states. In Section 4.3 we introduce descriptors, a versatile tool for expressing properties of belief states, and in Section 4.4 their properties are investigated. In Section 4.5 descriptors are used as a general means for expressing success conditions of operations of change. In Section 4.6 the main features of the resulting model of belief change are summarized. At this point we will have arrived at the fundamental framework for belief change, descriptor revision, that will be further investigated in the rest of the book.

4.1 Beginning Without Sentences

It is almost universally assumed in the belief change literature that beliefs are fully representable as sentences in some language. The totality of beliefs held by an agent is represented by a belief set that is a logically closed set of sentences. Inputs specify a sentence (or sometimes a set of sentences) that has to be either added to the belief set or removed from it. The use of sentences has the immense advantage of making

S.O. Hansson, *Descriptor Revision*, Trends in Logic 46,
DOI 10.1007/978-3-319-53061-1_4

logical treatments possible. Logic operates with sentences, and it is an astoundingly efficient and versatile tool for modelling a wide array of phenomena [100]. However, like other modelling tools it puts emphasis on some aspects of the objects it models at the expense of others. One of the major characteristics of logical models is the linguistic structure that they impose on their subject matter.

Some belief changes can be adequately described in terms of sentences. When I learned that Georg Friedrich Händel wrote the *Messiah* in 1741, the resulting effect on my belief state can be summarized by saying that I started to believe in the sentence "Georg Friedrich Händel wrote the *Messiah* in 1741".[1] However, there are many belief changes that cannot easily be expressed in sentential terms. For instance, when I first heard the *Messiah* I acquired a whole set of new beliefs based on my auditory impressions, namely beliefs about how the music sounds, but I was not able to express all these beliefs in sentences. Similarly, I have beliefs about how Barack Obama's voice sounds, what Picasso's *Guernica* looks like, how my favourite brand of cheese tastes, and how hydrogen sulphide smells. In all these cases my beliefs take the form of "mental pictures" or sensory impressions that can only partially be translated into words. Such perceptually based beliefs are typically adopted "automatically", without any decision or reflection. (See [119], [192, p. 62], and [197, p. 313].) They form a large part of our beliefs. This is one of the reasons why the police use identity parades, photo-lineups, and facial composites in addition to asking witnesses to verbally describe a suspect. A witness may know what a suspect looks like without being able to express this knowledge in words.

Belief change theory is usually assumed to represent changes in the beliefs of individual persons. With this interpretation the exclusion of non-sentential beliefs is a significant limitation. The theory can also refer to belief-holders other than individual persons. In some such cases the sentential format may be less problematic. We can for instance use the theory to model database management. In that case sentential representation is at least in principle fully adequate since the contents of databases are typically representable by sentences. Another example is changes in collectively created and maintained stocks of information or knowledge, such as the corpus of scientific beliefs. Collective information processes are usually based on sentential representations since these are needed for inter-individual communication [112, 125, 136]. However, in order to cover the central case of changes in the beliefs of individual human beings, it is useful to investigate a more general approach that does not require all beliefs to be expressible in sentences. For that purpose we can use a set of primitive belief states, i.e. belief states that are not assumed to have any particular internal structure. Such a belief state may comprise both sentential and non-sentential beliefs. Changes have the effect of taking us from one such belief state to another (or vacuously keeping us in the original one).

---

[1] And in other sentences containing the same information. The joint information content of sentences with the same meaning is called a proposition. All this could alternatively be expressed in terms of propositions.

Definition 4.1 *A* (deterministic) generic belief state model *is a triple* $\langle \mathbb{K}, \mathbb{I}, \circledcirc \rangle$, *where* $\mathbb{K} = \{\mathcal{K}_1, \mathcal{K}_2, \dots\}$ *is a set of* belief states, $\mathbb{I} = \{\imath_1, \imath_2, \dots\}$ *a set of* inputs, *and* $\circledcirc$ *an* input assimilation function[2] *from* $\mathbb{K} \times \mathbb{I}$ *to* $\mathbb{K}$.

In such a model all changes are brought about by inputs, and we use the universal operation $\circledcirc$ to express their impact. For each $\mathcal{K} \in \mathbb{K}$ and $\imath \in \mathbb{I}$, $\mathcal{K} \circledcirc \imath$ is the outcome of subjecting $\mathcal{K}$ to the input $\imath$. $\mathcal{K} \circledcirc \imath$ is a new belief state on which further operations can be performed. Therefore this framework allows for iterated change such as $\mathcal{K} \circledcirc \imath_1 \circledcirc \imath_2 \dots \circledcirc \imath_n$ for arbitrary inputs $\imath_1, \imath_2, \dots, \imath_n$.

Although this is a fairly general framework it relies on a couple of assumptions that should be stated. It is *input-assimilating*, by which is meant that all changes stem from an input. Input-assimilating models highlight the causes and mechanisms of change. The inputs are usually interpreted as externally generated, which means that these models contain no representation of internally generated changes such as the loss or deterioration of information or the drawing of new inferences from old information. This can be remedied by allowing for internally generated inputs.[3] Furthermore, inputs come *consecutively*, i.e., one at a time. However, this is not a serious limitation since the set $\mathbb{I}$ of inputs can contain "combined inputs" in the same manner as the inputs of multiple contraction in AGM-style models. (We can define an operation $\div$ such that $K \div \{p, q\}$ has the success condition that neither p nor q should be an element of the outcome.) A much more important limitation is the *lack of explicit representation of time*. It does not seem possible to include a representation of time in an input assimilation model of belief change without making it inordinately complex and unmanageable for most purposes.

The belief state model in Definition 4.1 was called deterministic. That is because the income assimilation function determines for each input exactly what the new belief state will be. In other words, for each $\mathcal{K} \in \mathbb{K}$ and each $\imath \in \mathbb{I}$, we have $\mathcal{K} \circledcirc \imath \in \mathbb{K}$. Another option is to use an input assimilation function that takes us to a non-empty set of belief states. In formal terms we then have a function $\check{\circledcirc}$ such that $\varnothing \neq \mathcal{K} \check{\circledcirc} \imath \subseteq \mathbb{K}$.[4] Such an indeterministic function can be used to reflect that we do not (and perhaps cannot) know exactly what the outcome will be. Alternatively it can signify that the outcome is, in an ontological sense, undetermined. In this book, the focus will be on deterministic models of belief change, but we will return to indeterministic models in Section 5.3.

The structure introduced in Definition 4.1 can also be used to represent an agent's overall state of mind rather than the part of her state of mind that constitutes her belief state. We can for instance conceive the elements of $\mathbb{K}$ as incorporating value judgments, emotions, and desires. Such extensions will not be discussed here, but they can be useful tools for investigating the relationships among these different

---

[2]In the terminology of automata theory it is a transition function.

[3]Changes consisting of the drawing of new inferences from old information have been included in some belief change models; see [83], [91, pp. 20–21], [112, 204].

[4]The symbol $\check{}$ above the symbol representing a (deterministic) belief change operation will be used to denote the indeterministic generalization of that operation.

components of mental states, for instance the effects of changes in belief on value judgments and vice versa.

In the subsequent sections we will add structure to the generic belief state models. But before doing so we will have a look at some interesting properties of these models that can be expressed already with the structure that we have. The following notation is useful:

Definition 4.2 *Let* $\langle \mathbb{K}, \mathbb{I}, \circledcirc \rangle$ *be a generic belief state model and let* $\mathcal{K} \in \mathbb{K}$. *Then:*

(1) $\mathbb{K}_{\mathcal{K}} = \{\mathcal{K} \circledcirc \imath \mid \imath \in \mathbb{I}\}$ *is the set of* directly reachable *belief sets from* $\mathcal{K}$.

(2) $\mathbb{K}^{+}_{\mathcal{K}} = \{\mathcal{K} \circledcirc \imath_1 \circledcirc \ldots \circledcirc \imath_n \mid \{\imath_1, \ldots, \imath_n\} \subseteq \mathbb{I}\}$ *is the set of* indirectly reachable *belief sets from* $\mathcal{K}$.

The following are interesting reachability-related properties of generic belief state models:

$\mathbb{K}_{\mathcal{K}} \neq \{\mathcal{K}\}$ for some $\mathcal{K} \in \mathbb{K}$. (changeability)
$\mathcal{K} \in \mathbb{K}_{\mathcal{K}}$ for all $\mathcal{K} \in \mathbb{K}$. (retainability)
$\mathbb{K}_{\mathcal{K}} = \mathbb{K}$ for all $\mathcal{K} \in \mathbb{K}$. (direct access)
$\mathbb{K}^{+}_{\mathcal{K}} = \mathbb{K}$ for all $\mathcal{K} \in \mathbb{K}$. (successive access)

Retainability can be seen as a technical property; it ensures that the option of changing nothing is represented in the input set. Direct access says that we can go directly (through one single input) from any belief state to any other belief state. This is a problematic property since there seem to be situations where several successive inputs are needed to reach a new belief state. For instance, if $\mathcal{K}$ is a belief state in which the agent is a devout religious believer and $\mathcal{K}'$ one in which she is a staunch atheist, then there may be no single input that would take her from $\mathcal{K}$ to $\mathcal{K}'$. It is much more plausible that a series of inputs can take her there through a mechanism whereby the earlier of these inputs facilitate her assimilation of those coming later. If that is always possible, then successive access holds.

The following two properties express intuitions that run contrary to those expressed by direct access and successive access:

If $\mathcal{K} \neq \mathcal{K}'$, then $\mathcal{K} \circledcirc \imath \neq \mathcal{K}' \circledcirc \imath'$. (non-convergence)
If $\mathcal{K} \circledcirc \imath_1 \neq \mathcal{K}$, then $\mathcal{K} \circledcirc \imath_1 \circledcirc \ldots \circledcirc \imath_n \neq \mathcal{K}$. (non-reversion)

Observation 4.3 *(1) No belief state model satisfies changeability, successive access, and non-reversion.*

(2) No belief state model with at least two belief states satisfies retainability, direct access, and non-convergence.

(3) If a belief state model satisfies retainability and non-convergence, then it satisfies non-reversion.

Non-convergence and non-reversion are both plausible under the assumption that we "carry our history with us" in the sense that previous beliefs leave traces behind them, for instance in the form of beliefs about what one believed previously.

The following three finiteness properties all refer to the number of alternative belief states that are in some sense available. They are stated here in order of increasing strength.

$\mathbb{K}_{\mathcal{K}}$ is finite for all $\mathcal{K}$. (finite direct access set)
$\mathbb{K}^{+}_{\mathcal{K}}$ is finite for all $\mathcal{K}$. (finite successive access set)
$\mathbb{K}$ is finite. (finite outcome set)

The framework introduced in Definition 4.1 has the important advantage of being general enough to cover a wide range of more specified models of belief revision (and mental dynamics in general) within one and the same formal structure. It can therefore be used to compare different such models. However, no such general investigation of different frameworks will be pursued here. Instead the remainder of this book is primarily devoted to one particularly promising type of model that can be developed within this framework. A couple of comparisons with other models will be made, namely with the AGM model (Sections 8.1, 8.2, and 10.3) and dynamic epistemic logic (Section 7.6).

4.2 Support Functions

With the introduction of generic belief state models we have discarded in one fell swoop all the assumptions about relations between sentence structure and operations of change that were found to be problematic in Chapter 3. But we may have thrown out too much. Actual belief states sustain both beliefs that are expressible in sentences and beliefs that are not. By removing sentences altogether we have deprived ourselves of all means to say something interesting about the special characteristics of the former class of beliefs. In order to regain that capability we will now reintroduce sentences in a cautious manner, avoiding some of the more controversial assumptions of the traditional approach.

The first and crucial step is to assign to each belief state a set consisting of exactly those sentences (in a given language) that represent beliefs held in that state. Formally, this assignment is expressed with a *support function* $\mathfrak{s}$ that takes us from elements of $\mathbb{K}$ to sets of sentences in the object language $\mathcal{L}$.

Definition 4.4 ([85, p. 525]) *Let $\mathbb{K}$ be a set of belief states and $\mathcal{L}$ a language. A* support function *for $\mathbb{K}$ in $\mathcal{L}$ is a function $\mathfrak{s}$ such that $\mathfrak{s}(\mathcal{K}) \subseteq \mathcal{L}$ for all $\mathcal{K} \in \mathbb{K}$.*

In the intended interpretation, $\mathfrak{s}(\mathcal{K})$ is the set of sentences in $\mathcal{L}$ that are supported (believed by the epistemic agent) in the belief state $\mathcal{K}$. Importantly, a support function always refers to a specific language. One and the same belief state $\mathcal{K}$ may be associated with several support functions, $\mathfrak{s}_{\mathcal{L}_1}, \mathfrak{s}_{\mathcal{L}_2}, \ldots$, for different languages. There may also be different support functions referring to different epistemic attitudes that the agent may have to sentences in one and the same language, such as the epistemic attitudes of assuming something, taking it for granted, believing it, and being sure of it. We may for instance distinguish between the set $\mathfrak{s}^{s}_{\mathcal{L}}$ of sentences in $\mathcal{L}$ that the agent is sure of in the state $\mathcal{K}$ and the set $\mathfrak{s}^b_{\mathcal{L}}$ of sentences in the same language that she believes in. In studies comparing different epistemic attitudes it will be useful to have more than one support function. Here the focus will be on a single epistemic attitude, namely that of belief, and a single object language.

In this framework, operations of change are primarily performed on belief states, not on the sets of supported sentences that are associated with them. In other words, we do not apply the input assimilation function $\circledcirc$ to the set $\mathfrak{s}(\mathcal{K})$ of sentences. Instead we apply it to the (non-sentential) belief state $\mathcal{K}$, and then we apply $\mathfrak{s}$ to the outcome $\mathcal{K} \circledcirc \imath$ to obtain the new set of supported sentences, $\mathfrak{s}(\mathcal{K} \circledcirc \imath)$.

The introduction of support functions makes it possible to express a series of important properties of belief change models, such as:

$$\mathfrak{s}(\mathcal{K}) = \mathrm{Cn}(\mathfrak{s}(\mathcal{K})) \qquad \text{(closure)}$$
$$\perp \notin \mathfrak{s}(\mathcal{K}) \qquad \text{(consistency)}$$

As before, Cn is a consequence operation that includes classical truth-functional consequence. In what follows we will assume that closure holds, i.e. that the support function assigns a belief set to each belief state. We will also mostly assume that the assigned belief sets are consistent. However, the presence of inconsistent belief sets may not be as devastating here as it is in frameworks such as the original AGM model where belief changes take place directly on belief sets. In classical truth-functional logic, there is only one logically closed inconsistent set, namely the whole language. Therefore, if K_1 and K_2 are inconsistent belief sets, i.e. $\perp \in K_1$ and $\perp \in K_2$, then $K_1 = K_2$. Since further changes are performed on the belief sets that are now identical, no posterior change can reintroduce the lost distinction.[5] In contrast, the present framework can accommodate distinct inconsistent belief states,[6] i.e. belief states $\mathcal{K}_1$ and $\mathcal{K}_2$ such that $\perp \in \mathfrak{s}(\mathcal{K}_1)$, $\perp \in \mathfrak{s}(\mathcal{K}_2)$, $\mathfrak{s}(\mathcal{K}_1) = \mathfrak{s}(\mathcal{K}_2)$, and $\mathcal{K}_1 \neq \mathcal{K}_2$.[7] Since further changes are performed on $\mathcal{K}_1$ and $\mathcal{K}_2$, not on $\mathfrak{s}(\mathcal{K}_1)$ and $\mathfrak{s}(\mathcal{K}_2)$, distinctions can be reintroduced at a later stage, for instance through revision by some input $\imath$ such that $\mathfrak{s}(\mathcal{K}_1 \circ \imath) \neq \mathfrak{s}(\mathcal{K}_2 \circ \imath)$. This is a property that corresponds to an important feature of actual belief systems, namely that inconsistencies are reparable in a way that does not blur all distinctions.[8]

---

[5]As noted by Hans Rott [223], this problem is not present in extended versions of the AGM model where the outcome of a contraction or revision is not just a belief set but a larger object that contains information about how additional changes will be performed.

[6]More precisely: different belief states that generate inconsistencies in the language of the support function.

[7]The same is true of belief base models in which the belief state is represented by a set of sentences that is not logically closed. Different such belief bases may have the same logical closure and therefore represent belief states with the same belief set [84, 88, 89, 94].

[8]In addition, actual belief systems are capable of containing local inconsistencies that do not corrupt the entire belief system. It is "quite feasible to believe both that Jesus was a human being and that Jesus was not a human being, without believing that the moon is made of cheese" [139, p. 49]. To represent this feature we can employ a support function $\mathfrak{s}$ that does not satisfy closure under classical consequence (but possibly some weaker, paraconsistent closure condition). On local inconsistencies, see [139].

The following properties are related to direct access and successive access that were introduced in Section 4.1.[9]

If $\perp \notin \mathrm{Cn}(\{p\})$, then there is some input $\imath$ with $p \in \mathfrak{s}(\mathcal{K} \circledcirc \imath)$. (direct believability)
If $\perp \notin \mathrm{Cn}(\{p\})$, then there is a series $\imath_1, \ldots, \imath_n$ of inputs with $p \in \mathfrak{s}(\mathcal{K} \circledcirc \imath_1 \circledcirc \ldots \circledcirc \imath_n)$.
(successive believability)
If $p \notin \mathrm{Cn}(\varnothing)$, then there is some input $\imath$ with $p \notin \mathfrak{s}(\mathcal{K} \circledcirc \imath)$. (direct removability)
If $p \notin \mathrm{Cn}(\varnothing)$, then there is a series $\imath_1, \ldots, \imath_n$ of inputs with $p \notin \mathfrak{s}(\mathcal{K} \circledcirc \imath_1 \circledcirc \ldots \circledcirc \imath_n)$.
(successive removability)
There is some input $\imath$ with $s(\mathcal{K} \circledcirc \imath) = \mathrm{Cn}(\varnothing)$. (direct depletability)
There is a series $\imath_1, \ldots, \imath_n$ of inputs with $s(\mathcal{K} \circledcirc \imath_1 \circledcirc \ldots \circledcirc \imath_n) = \mathrm{Cn}(\varnothing)$.
(successive depletability)

These are all fairly strong and arguably problematic properties. As mentioned above in connection with direct access, some persons may have beliefs (such as articles of religious faith) that nothing can make them give up. There may also be potential beliefs that they will never adopt, come whatever may. However, it can be questioned whether such stubbornness is compatible with rationality. The present framework allows for different answers to that question, expressible in terms of whether or not the believability and removability postulates hold. (By way of comparison, the equivalents of direct believability and direct removability hold in the AGM framework.[10])

The following are two finiteness properties that refer to the properties of individual belief states.

$\mathfrak{s}(\mathcal{K})$ is finite-based. (finite representability)

If $\mathfrak{s}(\mathcal{K})$ is finite-based, then so is $\mathfrak{s}(\mathcal{K} \circledcirc \imath)$. (finite-based outcome)

Finite-based outcome, the weakest of the two, was discussed in Section 3.1, where we found its absence in the AGM framework to be problematic. Interestingly, it is prone to conflict with the finiteness properties introduced in Section 4.1.

Observation 4.5 *Let $\mathfrak{s}$ be a support function for the belief states of some generic belief state model, and let the language $\mathcal{L}$ to which it refers be logically infinite. Then:*

(1) Direct believability, finite-based outcome, and finite direct access set are not all satisfied.

(2) Successive believability, finite-based outcome, and finite successive access set are not all satisfied.

[9] The direct versions of these properties are discussed in [121].

[10] Since $p \in K * p$ and $p \notin (K \div p) \setminus \mathrm{Cn}(\varnothing)$.

Finally, let us introduce properties indicating how much information about the belief state $\mathcal{K}$ is contained in the supported set $\mathfrak{s}(\mathcal{K})$.

If $\mathfrak{s}(\mathcal{K}) = \mathfrak{s}(\mathcal{K}')$, then $\mathcal{K} = \mathcal{K}'$. (injectivity)

If $\mathfrak{s}(\mathcal{K}) = \mathfrak{s}(\mathcal{K}')$, then $\mathfrak{s}(\mathcal{K} \circledcirc \imath) = \mathfrak{s}(\mathcal{K}' \circledcirc \imath)$ for all $\imath \in \mathbb{I}$. (sententiality)

According to injectivity, any difference between belief states is manifested on the sentential level. For instance, suppose that the only difference between $\mathcal{K}$ and $\mathcal{K} \circledcirc \imath$ is that in the latter you have looked somewhat more closely at your neighbour's hedge, and your mental picture of it has changed in consequence. Injectivity requires that there is some sentence (presumably about the hedge) that you could utter to express your beliefs in one of $\mathcal{K}$ and $\mathcal{K} \circledcirc \imath$ but not in the other. Notably, it does not require that all the differences between the two belief states can be expressed linguistically, only that at least one of them can.

Sententiality is the weaker of the two properties. It says that if two belief states are indistinguishable in terms of what sentences they support, then no series of changes will make their successors distinguishable in that respect.[11] This excludes the existence of belief states that are statically but not dynamically equivalent on the linguistic level, i.e. such that they cannot be distinguished in terms of the beliefs they support, but their successors after some operation(s) of change can be distinguished.[12] It also excludes belief changes that weaken or strengthen beliefs without moving any of them across the belief/non-belief border.[13] But contrary to injectivity, sententiality allows for the existence of essentially non-linguistic properties of belief states that will never show up when the beliefs are expressed linguistically [93].

The plausibility of these properties depends on the language $\mathcal{L}$ that the support function $\mathfrak{s}_{\mathcal{L}}$ operates with. The more expressive power the language has, the less problematic is the assumption that two distinct belief states must have some difference that is expressible in the language.[14] However, this assumption will never be entirely unproblematic since it deprives us of the possibility of distinguishing on the linguistic level between different inconsistent belief states (given that $\mathfrak{s}_{\mathcal{L}}$ satisfies the closure property, $\mathfrak{s}_{\mathcal{L}}(\mathcal{K}) = \mathrm{Cn}(\mathfrak{s}_{\mathcal{L}}(\mathcal{K}))$.)

---

[11] It can be applied repeatedly, and can therefore equivalently be expressed as follows: If $\mathfrak{s}(\mathcal{K}) = \mathfrak{s}(\mathcal{K}')$, then $\mathfrak{s}(\mathcal{K} \circledcirc \imath_1 \circledcirc \dots \circledcirc \imath_n) = \mathfrak{s}(\mathcal{K}' \circledcirc \imath_1 \circledcirc \dots \circledcirc \imath_n)$ for all series $\imath_1, \dots, \imath_n$ of elements of $\mathbb{I}$.

[12] On the difference between static and dynamic equivalence of belief states, see [83].

[13] Suppose that an input $\imath$ (1) strengthens p in $\mathcal{K}$, but (2) does not move any sentence across the belief/non-belief border. It would seem to follow from (1) that there is some series $\imath_1, \dots, \imath_n$ of inputs such that $p \notin \mathfrak{s}(\mathcal{K} \circledcirc \imath_1 \circledcirc \dots \circledcirc \imath_n)$ and $p \in \mathfrak{s}(\mathcal{K} \circledcirc \imath \circledcirc \imath_1 \circledcirc \dots \circledcirc \imath_n)$, but it follows from (2) that $\mathfrak{s}(\mathcal{K} \circledcirc \imath) = \mathfrak{s}(\mathcal{K})$. This contradicts sententiality. On operations that strengthen or weaken beliefs, see [28].

[14] This is particularly pertinent if autoepistemic or conditional beliefs are included in the belief set. See Chapter 7.

4.3 Belief Descriptors

In order to reconstruct a sentential framework we need to represent not only belief states but also inputs in sentential terms. Standard operations of belief change are defined in terms of their success conditions, such as $p \in K * p$ for revision and $p \notin (K \div p) \setminus \mathrm{Cn}(\varnothing)$ for contraction. These are statements about what is believed in the belief state that the operation results in. We need a versatile way to express such properties of belief states.

For that purpose, we will introduce a metalinguistic belief predicate $\mathfrak{B}$. As arguments it takes sentences in the object language in which beliefs are expressed. For any sentence $p \in \mathcal{L}$, the expression $\mathfrak{B}p$ denotes that p is believed in the belief state under consideration. A belief set satisfies $\mathfrak{B}p$ if and only if it has p as an element. An expression like this, consisting of $\mathfrak{B}$ followed by a sentence in the object language, will be called an *atomic belief descriptor*. The term "atomic" signals that these sentences are the smallest building-blocks in the language of belief descriptors that we are now building. However, atomic belief descriptors are not atomic in the sense of being logically independent. For instance, from $\mathfrak{B}p$ and $\mathfrak{B}q$ we can conclude that $\mathfrak{B}(p\&q)$.[15]

Atomic belief descriptors can be combined with the usual truth-functional connectives, classically interpreted. Hence, $\mathfrak{B}p \vee \mathfrak{B}q$ denotes that either p or q is believed, and $\neg\mathfrak{B}r$ that r is not believed. The truth conditions of these expressions follow the standard pattern: $\neg\mathfrak{B}p$ is satisfied whenever $\mathfrak{B}p$ is not satisfied, $\mathfrak{B}p \;\&\; \mathfrak{B}q$ whenever both $\mathfrak{B}p$ and $\mathfrak{B}q$ are satisfied, $\mathfrak{B}p \vee \mathfrak{B}q$ whenever either $\mathfrak{B}p$ or $\mathfrak{B}q$ is satisfied. These composite expressions are called *molecular belief descriptors*.

Finally, we can form sets of (molecular) belief descriptors, such as $\{\mathfrak{B}(p \vee q), \neg\mathfrak{B}p, \neg\mathfrak{B}q\}$. Sets of molecular belief descriptors will be called *composite belief descriptors* or in short just *descriptors*. A composite belief descriptor is satisfied if and only if all its elements are satisfied. Hence the descriptor $\{\mathfrak{B}(p \vee q), \neg\mathfrak{B}p, \neg\mathfrak{B}q\}$ is satisfied by the belief set $\mathrm{Cn}(\{p \vee q\})$ but not by the belief set $\mathrm{Cn}(\{q\})$.[16]

A descriptor is (obviously) called finite if it has a finite number of elements. Strictly speaking, finite descriptors are superfluous since they can be replaced by the conjunction of their elements. For instance, $\{\mathfrak{B}(p \vee q), \neg\mathfrak{B}p, \neg\mathfrak{B}q\}$ is satisfied by exactly the same belief sets that are satisfied by the molecular descriptor $\mathfrak{B}(p \vee q) \;\&\; \neg\mathfrak{B}p \;\&\; \neg\mathfrak{B}q$. However, the set-theoretical notation is often more convenient, and it will be used freely in what follows.

Upper-case Greek letters such as $\Psi, \Xi, \ldots$ will be used to denote (composite) descriptors. Occasionally, when a notation is needed for molecular descriptors, lower-case Greek letters such as $\alpha, \beta, \ldots$ will be used for that purpose.

---

[15] Frank Ramsey noted in 1925 that "A believes p" is not a truth function of p but can instead be treated as "one of other atomic propositions". [210, p. 9n].

[16] Composite descriptors with one element will be used interchangeably with the molecular descriptor that they contain. For instance, $\{\mathfrak{B}p\}$ and $\mathfrak{B}p$ will be used interchangeably.

All this is important enough to be summarized in a formal definition:

Definition 4.6 ([124]) *An* atomic belief descriptor *is a sentence* $\mathfrak{B}p$ *with* $p \in \mathcal{L}$. *It is* satisfied *by a belief state* $\mathcal{K}$ *according to a support function* $\mathfrak{s}$ *in* $\mathcal{L}$ *if and only if* $p \in \mathfrak{s}(\mathcal{K})$.

A molecular belief descriptor *(denoted by lower-case Greek letters* α, $\beta, \dots$*) is a truth-functional combination of atomic descriptors. Conditions of satisfaction are defined inductively, such that* $\mathcal{K}$ *satisfies* $\neg\alpha$ *according to* $\mathfrak{s}$ *if and only if it does not satisfy* α, *it satisfies* $\alpha \vee \beta$ *if and only if it satisfies either* α *or* β, *etc.*

A composite belief descriptor *(in short: descriptor; denoted by upper-case Greek letters* Ψ, $\Xi, \dots$*) is a non-empty set of molecular descriptors. A belief state* $\mathcal{K}$ *satisfies a composite descriptor* Ψ *according to* $\mathfrak{s}$ *if and only if it satisfies all its elements.*

A descriptor is satisfiable within *a set of belief states if and only if it is satisfied by at least one of its elements.*

As defined here, the symbol $\mathfrak{B}$ is not part of the object language, and therefore it cannot be used to express an agent's beliefs about her own beliefs. (It is possible to include an autoepistemic belief predicate into the language. It may or may not coincide with $\mathfrak{B}$, depending on whether the agent's autoepistemic beliefs accord with her epistemic conduct. See Section 7.1.) It should also be noted that our definition does not allow $\mathfrak{B}$ to be iterated.[17] Therefore expressions such as $\mathfrak{B}\mathfrak{B}p$ or $\mathfrak{B}(\mathfrak{B}p \rightarrow \mathfrak{B}q)$ are not well-formed. The reason for this is that it is very unclear what such expressions could possibly mean, given the metalinguistic interpretation of $\mathfrak{B}$.

Descriptors are well suited to express the success conditions of different types of belief change operations. In revision, a specified sentence p should be included in the outcome, in other words the success condition has the characteristic form $\mathfrak{B}p$. In contraction, a specified sentence p is instead required not to be present in the outcome, thus a success condition of the form $\neg\mathfrak{B}p$ has to be satisfied. The success conditions of many other, less common, types of operations can be expressed analogously. Multiple revision by a set $\{p_1, \dots, p_n\}$ of sentences has two variants, package revision that requires all of them to be believed in the new belief state, and choice revision that only requires that at least one of them be believed.[18] Package revision has the success condition $\{\mathfrak{B}p_1, \dots, \mathfrak{B}p_n\}$, and choice revision the success condition $\mathfrak{B}p_1 \vee \dots \vee \mathfrak{B}p_n$. Similarly, multiple contraction by a set $\{p_1, \dots, p_n\}$ of sentences has two variants, package contraction that requires all of them to be removed and choice contraction that only requires that at least one of them be removed [64]. Package contraction has to satisfy the success condition $\{\neg\mathfrak{B}p_1, \dots, \neg\mathfrak{B}p_n\}$, and choice contraction the success condition $\neg\mathfrak{B}p_1 \vee \dots \vee \neg\mathfrak{B}p_n$. The operation of replacement is constructed to remove one specified sentence and incorporate another [110].

[17] More precisely: It does not allow the formation of expressions in which an instance of $\mathfrak{B}$ appears within the scope of another instance of $\mathfrak{B}$.

[18] This terminology is used in [107] and [239, p. 280]. It is based on the terminology for two types of multiple contraction used in [64]. Hans Rott uses the terms "bunch revision" and "pick revision" for the same concepts [217, p. 65].

Its success condition has the form $\{\neg\mathfrak{B}p, \mathfrak{B}q\}$. Finally, the operation of "making up one's mind" aims at either belief or disbelief in a specified sentence p. Its success condition is $\mathfrak{B}p \vee \mathfrak{B}\neg p$ [264]. In summary, descriptors can be used to express a wide range of success conditions in a precise and unified way. We will use this locution to construct a uniform type of belief change that covers all operations whose success conditions are expressible with descriptors. But before that we need to have a brief look at some of the formal properties of descriptors.

4.4 Properties of Descriptors

Descriptors refer to what sentences a belief state supports, i.e. to the contents of the belief set $\mathfrak{s}(\mathcal{K})$ supported by a belief state $\mathcal{K}$. We can therefore assume that if $\mathfrak{s}(\mathcal{K}) = \mathfrak{s}(\mathcal{K}')$, then $\mathcal{K}$ and $\mathcal{K}'$ satisfy the same descriptors. For simplicity, we can then refer to descriptors as satisfied by belief sets rather than by belief states. The symbol $\Vdash$ will be used for that relation of satisfaction:

Definition 4.7 ([124]) *Let K be a belief set and let Ψ and Ξ be descriptors.*

$K \Vdash \Psi$ means that K satisfies Ψ, and $\Psi \Vdash \Xi$ that all belief sets satisfying Ψ also satisfy Ξ.

The corresponding equivalence relation is written $\dashv\Vdash$; hence $\Psi \dashv\Vdash \Xi$ holds if and only if both $\Psi \Vdash \Xi$ and $\Xi \Vdash \Psi$ hold.

As can be seen from the definition, $\Vdash$ is (for simplicity) used to denote two binary relations. First, it stands for a relation between belief sets and descriptors, such that $K \Vdash \Psi$ holds if and only if K satisfies Ψ (in the sense of satisfaction specified in Definition 4.6, which means that it has to satisfy all elements of Ψ). Secondly, $\Vdash$ also represents a relation between descriptors, such that $\Psi \Vdash \Xi$ holds if and only if it holds for all belief sets K that if $K \Vdash \Psi$ then $K \Vdash \Xi$.

The following observation summarizes some elementary properties of descriptors:

Observation 4.8 *(1) Let K be a belief set and α a molecular descriptor. Then either $K \Vdash \alpha$ or $K \Vdash \neg\alpha$.*

(2) Let $\alpha_1, \ldots, \alpha_n$ be molecular descriptors. Then $\{\alpha_1, \ldots, \alpha_n\} \dashv\Vdash \{\alpha_1 \& \ldots \& \alpha_n\}$.

(3) For any descriptors Ψ and Ξ: $\Psi \Vdash \Xi$ if and only if there is some Ψ' such that $\Xi \subseteq \Psi'$ and $\Psi' \dashv\Vdash \Psi$.

Part (1) of the observation cannot be extended to formulas in which the belief set K has been replaced by a descriptor. It does not hold in general that if Ψ is a descriptor and α a molecular descriptor, then either $\Psi \Vdash \alpha$ or $\Psi \Vdash \neg\alpha$.[19]

A descriptor can be inconsistent in the sense that no belief set can satisfy it. The following notation is introduced to express such inconsistency:

[19] To see that, let p and q be logically independent elements of $\mathcal{L}$, and let $\Psi = \{\mathfrak{B}p\}$ and $\alpha = \mathfrak{B}q$.

Definition 4.9 ([124]) $\doteq$ *(descriptor falsum) denotes* $\{\mathfrak{B}p, \neg\mathfrak{B}p\}$ *for an arbitrary p.*

It is important to distinguish $\doteq$ from the falsum $\perp$ of the object language (that is introducable as $p\&\neg p$ for an arbitrary p). The inconsistent belief set $K = \mathrm{Cn}(\{\perp\})$ satisfies the condition $K \vdash \perp$, but no belief set satisfies the condition $K \Vdash \doteq$.

We can apply ordinary conjunction and disjunction to molecular descriptors, forming sentences such as $\alpha\&\beta$ and $\alpha \vee \beta$. For composite descriptors, we can use set union with essentially the same effect as conjunction. The parallel is obvious: a belief set satisfies $\alpha\&\beta$ if and only if it satisfies α and it also satisfies β. Similarly, it satisfies $\Psi \cup \Xi$ if and only if it satisfies Ψ and it also satisfies Ξ. For disjunction, the following construction can be used:

Definition 4.10 ([126]) *The* descriptor disjunction $\veebar$ *is defined by the relationship* $\Psi \veebar \Xi = \{\alpha \vee \beta \mid \alpha \in \Psi \text{ and } \beta \in \Xi\}$.

Observation 4.11 *Let K be a belief set and let* Ψ *and* Ξ *be descriptors. Then:*
$K \Vdash \Psi \veebar \Xi$ *if and only if either* $K \Vdash \Psi$ *or* $K \Vdash \Xi$.

It follows from Definition 4.6 that the negation of a molecular descriptor α is a descriptor $\neg\alpha$ such that for any belief set X: $X \Vdash \neg\alpha$ if and only if $X \nVdash \alpha$. A generalization of negation to composite descriptors should have the same property, in other words the negation of a composite descriptor Ψ would have to be another descriptor $\neg\!\!\!\neg\, \Psi$ such that for any belief set X: $X \Vdash \neg\!\!\!\neg\, \Psi$ if and only if $X \nVdash \Psi$. For any finite descriptor $\{\alpha_1, \dots, \alpha_n\}$ we can use the set:

$$\neg\!\!\!\neg\{\alpha_1, \dots, \alpha_n\} = \{\neg\alpha_1 \vee \dots \vee \neg\alpha_n\}$$

as its negation. However, as the following observation shows, there are infinite descriptors for which no construction with the desired property is possible. In other words, there are non-negatable descriptors.

Observation 4.12 *Let the object language* $\mathcal{L}$ *have infinitely many logically independent atoms. Then there are non-negatable descriptors, i.e. descriptors* Ψ *such that there is no descriptor* $\neg\!\!\!\neg\Psi$ *satisfying the condition that for any belief set X:* $X \Vdash \neg\!\!\!\neg\, \Psi$ *if and only if* $X \nVdash \Psi$.

To each descriptor Ψ we can assign a *characteristic set* of belief sets, namely the set consisting of those belief sets that satisfy the descriptor. The descriptors that characterize a single belief set are worth special attention since they are very useful in formal proofs.

Definition 4.13 ([126]) *A descriptor* Ψ *is* maxispecified *(maximally specified) if and only if there is exactly one belief set Y in* $\wp(\mathcal{L})$ *such that* $Y \Vdash \Psi$. *It is then a maxispecified descriptor* for Y.

There are many (equivalent) maxispecified descriptors for each belief set. For instance, if $X = \mathrm{Cn}(\{q\})$ then both $\{\mathfrak{B}q\} \cup \{\neg\mathfrak{B}x \mid x \notin X\}$ and $\{\mathfrak{B}x \mid x \in X\} \cup \{\neg\mathfrak{B}x \mid x \notin X\}$ are maxispecified descriptors for X. For convenience, one of the maxispecified descriptors for a belief set X will be denoted as follows:

Definition 4.14 ([126]) *Let X be a belief set. Then Π_X is the maxispecified descriptor for X such that:*

$\Pi_X = \{\mathfrak{B}x \mid x \in X\} \cup \{\neg\mathfrak{B}x \mid x \notin X\}$

Whereas all (single) belief sets can be characterized by a descriptor, there are sets of belief sets that cannot:

Definition 4.15 *A set $\mathbb{Y}$ of belief sets is* descriptor-definable *if and only if there is some descriptor Ψ such that for all belief sets Y:*

$Y \in \mathbb{Y}$ if and only if $Y \Vdash \Psi$.

Observation 4.16 ([124]) *(1) Let $\mathbb{Y}$ be a finite set of belief sets. Then $\mathbb{Y}$ is descriptor-definable.*

(2) If $\mathcal{L}$ is logically infinite[20] *then there are sets of belief sets that are not descriptor-definable.*

We now have the formal means to analyze an issue that was brought up informally in Section 2.4, namely which success conditions are preserved under intersection. We noted that if each element of a set of belief sets satisfies the success condition for revision by the sentence p (i.e. they all contain p), then their intersection also satisfies that condition (i.e. the intersection contains p). Similarly, if all belief sets in a collection satisfy the success condition for contraction by the sentence p (i.e. none of them contains p), then their intersection does the same (i.e. it does not contain p). This is what makes the select-and-intersect method viable for both revision and contraction. But not all success conditions are preserved under intersection. Since success conditions can be represented by descriptors we can now express this condition in a fully formalized way:

Definition 4.17 *A descriptor Ψ is* preserved under intersection *if and only if it holds for all sets $\mathbb{Y}$ of belief sets that if $Y \Vdash \Psi$ for all $Y \in \mathbb{Y}$, then $\bigcap \mathbb{Y} \Vdash \Psi$.*

The following observation identifies an important class of descriptors that are preserved under intersection.

Observation 4.18 ([135][21]) *A descriptor is preserved under intersection if each of its elements has one of the three forms*

(i) $\mathfrak{B}p$,

(ii) $\neg\mathfrak{B}p$, *or*

(iii) $\mathfrak{B}p_1 \vee \ldots \vee \mathfrak{B}p_n \vee \neg\mathfrak{B}q$, *with* $q \vdash p_1 \vee \ldots \vee p_n \rightarrow p_k$ *for some* p_k.

[20] A set of sentences is logically infinite if and only if it has infinitely many equivalence classes in terms of logical equivalence. Cf. Section 2.5.

[21] This observation is related to the well-known theorem that a theory is equivalent to a Horn theory if and only if the set of its models is closed under intersection. This was proved (in a generalized form) in [187]. A more accessible proof can be found in [38, pp. 254–257], and an excellent introduction to Horn clauses in [148].

4.5 Descriptor Revision Introduced

We are now ready for the final step in the construction of a new framework for belief change: the transition from primitive inputs, i.e. elements of $\mathbb{I}$, to changes based on success conditions, expressed with descriptors. We are looking for a way to revise a belief state $\mathcal{K}$ by a descriptor Ψ rather than by an element of $\mathbb{I}$. This means that we need an operation $\circ$ of belief change that takes descriptors as inputs. Such an operation will supersede and unify the traditional operations, thus $\mathcal{K} \circ \mathfrak{B}p$ takes the role of revision, $\mathcal{K} \circ \neg\mathfrak{B}p$ that of contraction, $\mathcal{K} \circ \{\neg\mathfrak{B}p_1, \dots, \neg\mathfrak{B}p_n\}$ that of multiple (package) contraction, etc.

Importantly, the use of descriptors instead of elements of $\mathbb{I}$ as inputs does not require the introduction of new outputs. We can assume that $\mathbb{I}$ is exhaustive in the sense that every new belief state that is directly reachable from $\mathcal{K}$ can be reached through revision by one of the inputs in $\mathbb{I}$. This means that for every descriptor Ψ we can identify $\mathcal{K} \circ \Psi$ with a belief state $\mathcal{K} \circledcirc \imath$ for some $\imath \in \mathbb{I}$. In other words, we should have $\mathcal{K} \circ \Psi \in \mathbb{K}_{\mathcal{K}}$. Furthermore, since the operation $\circ$ should be successful, the outcome $\mathcal{K} \circ \Psi$ should satisfy Ψ, i.e. we should have $\mathfrak{s}_{\mathcal{L}}(\mathcal{K} \circledcirc \imath) \Vdash \Psi$ (unless, of course, there is no set $\mathcal{K} \circledcirc \imath$ with this property). Combining the two conditions, we obtain:

$$\mathcal{K} \circ \Psi \in \{\mathcal{K}' \in \mathbb{K}_{\mathcal{K}} \mid \mathfrak{s}_{\mathcal{L}}(\mathcal{K}') \Vdash \Psi\} \text{ if } \Psi \text{ is satisfiable within } \mathbb{K}_{\mathcal{K}}.$$

In order to construct such an operation we need to select an element of $\{\mathcal{K}' \in \mathbb{K}_{\mathcal{K}} \mid \mathfrak{s}_{\mathcal{L}}(\mathcal{K}') \Vdash \Psi\}$. Typically that set will have more than one element, i.e. there will be more than one element of $\mathbb{K}_{\mathcal{K}}$ that satisfies Ψ. For instance, if Ψ represents the belief that the old vase in my family's living-room is broken, then Ψ is satisfied in a large number of potential belief change outcomes, including far-fetched ones with various additional beliefs such as that a wild bird flew in through an open window and knocked down the vase. Revision by Ψ should not result in one of these far-fetched outcomes but rather in a "minimally changed" belief state that is, intuitively speaking, as close or similar to my previous belief state as is compatible with the assimilation of Ψ. We can expect $K \circ \Psi$ to have as few features as possible that are not shared by all the reasonably credible revision outcomes that satisfy Ψ. The crucial assumption that we have to make when modelling deterministic belief change is that one among the various potential outcomes satisfying Ψ is singled out to be *the* outcome of revision *by* Ψ. In the formal language, this singling out is most conveniently represented by a choice function that extracts *only one* element from the set it is applied to. (In indeterministic belief change, we instead have an operation $\check{\circ}$ such that $K \check{\circ} \Psi$ is a non-empty set of belief sets, each of which is equal to $\mathfrak{s}(\mathcal{K}')$ for some $\mathcal{K}' \in \mathbb{K}_{\mathcal{K}}$ with $\mathfrak{s}(\mathcal{K}') \Vdash \Psi$.)

A function that singles out a single element can be constructed as a special case of the definition of a choice function. Then the formal object that we obtain will be a set with the chosen belief state as its only element. Alternatively we can construct a function that directly delivers this belief state (instead of a set in which it is the only element). It does not make much of a difference which of these two formal constructions we employ. They are both introduced in the following definition:

Definition 4.19 ([120]) *Let $\mathbb{Y}$ be a set. A* monoselective choice function *for $\mathbb{Y}$ is a choice function C for $\mathbb{Y}$ such that if $\varnothing \subset \mathbb{Y}' \subseteq \mathbb{Y}$ then $C(\mathbb{Y}')$ has exactly one element. Alternatively it can be represented by a function $\widehat{C}$ such that $\widehat{C}(\mathbb{Y}') \in \mathbb{Y}'$ whenever $\varnothing \subset \mathbb{Y}' \subseteq \mathbb{Y}$, and otherwise $\widehat{C}(\mathbb{Y}')$ is undefined.*

We will apply monoselective choice functions to a predetermined set of potential outcomes, namely the set $\mathbb{K}_{\mathcal{K}}$ of belief states that are directly reachable from $\mathcal{K}$. In this way the select-and-intersect method is replaced by a direct choice among the potential outcomes. We can use this method to construct our first version of descriptor revision:

Definition 4.20 *Let $\mathbb{K}$ be a set of belief states, $\mathbb{I}$ a set of inputs, $\circledcirc$ an input assimilation function on $\mathbb{K} \times \mathbb{I}$, $\mathfrak{s}$ a support function for $\mathbb{K}$ in a language $\mathcal{L}$, and $\widehat{C}$ a monoselective choice function for $\mathbb{K}$. The* (deterministic) descriptor revision[22] *based on $\langle \mathbb{K}, \mathbb{I}, \circledcirc, \mathfrak{s}, \widehat{C} \rangle$ is the operation $\circ$ such that for all $\mathcal{K} \in \mathbb{K}$ and all descriptors Ψ for the language $\mathcal{L}$:*

(i) If Ψ is satisfiable within $\mathbb{K}_{\mathcal{K}}$, then $\mathcal{K} \circ \Psi = \widehat{C}(\{\mathcal{K}' \in \mathbb{K}_{\mathcal{K}} \mid \mathfrak{s}(\mathcal{K}') \Vdash \Psi\})$, and

(ii) otherwise $\mathcal{K} \circ \Psi = \mathcal{K}$.

This definition introduces a uniformity property for descriptor revision. If it holds for a belief state $\mathcal{K}$ and two descriptors Ψ_1 and Ψ_2 that

$$\{\mathcal{K}' \in \mathbb{K}_{\mathcal{K}} \mid \mathfrak{s}(\mathcal{K}') \Vdash \Psi_1\} = \{\mathcal{K}' \in \mathbb{K}_{\mathcal{K}} \mid \mathfrak{s}(\mathcal{K}') \Vdash \Psi_2\},$$

then $\mathcal{K} \circ \Psi_1 = \mathcal{K} \circ \Psi_2$. To see why this is a plausible principle, it may be helpful to consider the special case when there are sentences p and q such that $\Psi_1 = \{\mathfrak{B}p\}$ and $\Psi_2 = \{\mathfrak{B}q\}$. It then follows from

$$\{\mathcal{K}' \in \mathbb{K}_{\mathcal{K}} \mid \mathfrak{s}(\mathcal{K}') \Vdash \mathfrak{B}p\} = \{\mathcal{K}' \in \mathbb{K}_{\mathcal{K}} \mid \mathfrak{s}(\mathcal{K}') \Vdash \mathfrak{B}q\}$$

that exactly those belief changes that make the agent believe in p will also make her believe in q, and vice versa. Therefore, making her believe in p and making her believe in q seems to be essentially the same thing.

Definition 4.20 provides the most general form of (deterministic) descriptor revision. We will now introduce two useful simplifications of the model, both of which were anticipated in the previous sections of this chapter. First, we can assume that the set of reachable belief states is the same irrespective of what belief state we begin with, i.e. that $\mathbb{K}_{\mathcal{K}} = \mathbb{K}_{\mathcal{K}'}$ for all $\mathcal{K}, \mathcal{K}' \in \mathbb{K}$. Since we do not need to consider belief states that are not reachable from anywhere, this is equivalent to adopting the postulate of Direct access, i.e. $\mathbb{K}_{\mathcal{K}} = \mathbb{K}$. This allows us to make a small but important modification of clause (i) in Definition 4.20:

(i_L) If Ψ is satisfiable within $\mathbb{K}$, then $\mathcal{K} \circ \Psi = \widehat{C}(\{\mathcal{K}' \in \mathbb{K} \mid \mathfrak{s}(\mathcal{K}') \Vdash \Psi\})$.

---

[22]The term “descriptor revision” refers to operations that take descriptors as inputs. For clarity, the operations called “revision” in the traditional approach will be called “sentential revision”.

The index of (i_L) stands for "local". Obviously, replacing *(i)* by (i_L) makes no difference in studies of local change, i.e. one-step changes that all start with the same belief state.

The second simplification is somewhat more far-reaching. It consists in adopting the principle of injectivity from Section 4.2. (If $\mathfrak{s}(\mathcal{K}) = \mathfrak{s}(\mathcal{K}')$ then $\mathcal{K} = \mathcal{K}'$.) There will then be a one-to-one correspondence between the set $\mathbb{K}$ of belief states and the set $\{\mathfrak{s}(\mathcal{K}') \mid \mathcal{K}' \in \mathbb{K}\}$ of the support sets of its elements. The following observation shows that the belief states that are reachable with $\circledcirc$ will then coincide with those that are reachable with $\circ$:

Observation 4.21 *Let $\circ$ be the descriptor revision based on $\langle \mathbb{K}, \mathbb{I}, \circledcirc, \mathfrak{s}, \widehat{C} \rangle$. If injectivity holds, then for each $\mathcal{K} \in \mathbb{K}$ and $\imath \in \mathbb{I}$ there is a descriptor Ψ with $\mathcal{K} \circledcirc \imath = \mathcal{K} \circ \Psi$.*

We can use these correspondences to construct a version of descriptor revision that refers directly to belief sets and descriptors, without mentioning the primitive belief states and inputs that we started with. For that purpose, let $\mathbb{X} = \{\mathfrak{s}(\mathcal{K}') \mid \mathcal{K}' \in \mathbb{K}\}$ and $\mathbb{X}_{\mathfrak{s}(\mathcal{K})} = \{\mathfrak{s}(\mathcal{K}') \mid \mathcal{K}' \in \mathbb{K}_{\mathcal{K}}\}$:

Definition 4.22 *Let $\mathcal{L}$ be a language, $\mathbb{X}$ a set of belief sets in $\mathcal{L}$, a an* accessibility function *that assigns to each K in $\mathbb{X}$ a set $\mathbb{X}_K$ with $K \in \mathbb{X}_K \subseteq \mathbb{X}$, and $\widehat{C}$ a monoselective choice function for $\mathbb{X}$. The* descriptor revision $\circ$ *based on $\langle \mathcal{L}, \mathbb{X}, a, \widehat{C} \rangle$ is the operation $\circ$ such that for all $K \in \mathbb{X}$ and all descriptors Ψ for the language $\mathcal{L}$:*

(i_S) If Ψ is satisfiable within $\mathbb{X}_K$, then $K \circ \Psi = \widehat{C}(\{X \in \mathbb{X}_K \mid X \Vdash \Psi\})$, and

(ii_S) otherwise $K \circ \Psi = K$.

The index of (i_S) and (ii_S) stands for "sentential". We can of course combine the two simplifications. This amounts to letting a in Definition 4.22 have the property $a(K) = \mathbb{X}$ for all K. We can then replace (i_S) by the following:

(i_{SL}) If Ψ is satisfiable within $\mathbb{X}$, then $K \circ \Psi = \widehat{C}(\{X \in \mathbb{X} \mid X \Vdash \Psi\})$.

4.6 Conclusion

In this chapter we have done two things in parallel. First, we have removed all references to sentences in the belief state model, and then reintroduced them in a step-by-step fashion, identifying the assumptions required at each stage. At the completion of this process, we have a fully sentential model. However, we have not reintroduced the more problematic assumptions related to possible worlds and remainders that are usually associated with sentential models. In particular, we now have the means to perform belief change through a choice among a finite set of logically finite potential outcomes rather than among an infinite set of logically infinite entities that are not themselves potential outcomes. The expansion property does not hold, and (as will be shown in detail later on) neither does the recovery property.

Secondly, we have introduced the two major formal elements of descriptor revision, namely: (1) the use of belief descriptors as a general representation of the success conditions of belief change, and (2) the direct application of a choice function to the set of potential outcomes of the operation. The rest of this book is devoted to the further development of belief change models employing these two principles.

Chapter 5
Local Descriptor Revision

An operation of belief change is *local* if it can only take one particular belief state (or belief set) as its starting-point. As we saw in Section 3.7, the original AGM operations are local in this sense. Such an operation tells us how to revise or contract a particular belief set K that is interpreted as representing the current belief state. However, it has nothing to say about changes of other belief sets than K. For that purpose we need a *global* operation, i.e. one than can perform changes on all admissible belief states (or belief sets).

We will begin the exposition of descriptor revision with a presentation of its local variant, which will be the topic of the present chapter (to be followed in the next chapter by a treatment of the global version). Section 5.1 investigates and characterizes the basic construction with a monoselective choice function that was introduced in the previous chapter. Section 5.2 is devoted to the important special case in which the choice function is based on a relation on belief sets that can be interpreted as representing distances from the current belief set. Section 5.3 deals with the generalization to indeterministic descriptor revision, in which the operation does not specify a single outcome for each input, but only a set of possible outcomes. In Section 5.4, descriptor revision is constructed from a blockage relation on the set of potential outcomes. A belief set blocks another belief set if the latter is ineligible as a revision outcome whenever the first is available. Finally, in Section 5.5 we introduce a binary relation on descriptors, the relation of epistemic proximity. A descriptor is more epistemically proximate than another descriptor if its satisfaction is closer at hand for the agent, or it can be satisfied with a less far-reaching change. Relations of epistemic proximity (between descriptors) are a generalization of the relations of epistemic entrenchment (between sentences) that have been developed in the AGM framework.

S.O. Hansson, *Descriptor Revision*, Trends in Logic 46,
DOI 10.1007/978-3-319-53061-1_5

5.1 Monoselective Descriptor Revision

In this chapter we will use Definition 4.22, employing the clause i_{SL} (instead of i_S). Since our focus is on local change, we have $a(K) = \mathbb{X}$, and we can therefore replace all references to $a(K)$ by direct references to $\mathbb{X}$. Furthermore, we can simplify the notation for choice functions as follows:

Definition 5.1 *For any set* $\mathbb{X}$ *of belief sets and any descriptor* Ψ*:*
$[\![\Psi]\!]_{\mathbb{X}} = \{X \in \mathbb{X} \mid X \Vdash \Psi\}$
The index can be omitted if no ambiguity follows from doing so, i.e. we can then write $[\![\Psi]\!]$ *instead of* $[\![\Psi]\!]_{\mathbb{X}}$.

We can now write $\widehat{C}([\![\Psi]\!])$ as an abbreviation of $\widehat{C}(\{X \in \mathbb{X} \mid X \Vdash \Psi\})$.[1] With these changes we arrive at the following simplified definition of local descriptor revision:

Definition 5.2 ([124]) *An operation* $\circ$ *on a belief set* K *is a* (deterministic) local monoselective descriptor revision *if and only if there is a set* $\mathbb{X}$ *of belief sets, with* $K \in \mathbb{X}$*, and a monoselective choice function* $\widehat{C}$ *on the descriptor-definable subsets of* $\mathbb{X}$*, such that (i)* $K \circ \Psi = \widehat{C}([\![\Psi]\!])$ *if* Ψ *is satisfiable within* $\mathbb{X}$*, and (ii) otherwise* $K \circ \Psi = K$.

The set $\mathbb{X}$ in Definition 5.2 is a *repertoire*, i.e. a set of potential (or viable) belief sets among which the outcomes of the operation $\circ$ have to be chosen. Intuitively, we can think of the repertoire as consisting of all those belief sets that are coherent, stable, and/or plausible enough to be suitable as outcomes of a change in belief. [105] In a cognitively realistic model all elements of $\mathbb{X}$ should be finite-based. By the *outcome set* is meant the set of actually chosen outcomes, i.e. the set of belief sets X for which there exists a descriptor Ψ with $X = K \circ \Psi$.[2]

The outcome set is a subset of the repertoire. Furthermore, if the operation is a monoselective revision, then the outcome set and the repertoire are one and the same. The reason for this is that for each element X of $\mathbb{X}$ there is a maxispecified descriptor Π_X that is satisfied by X and by no other belief set. (See Definition 4.14.) It follows that $\widehat{C}([\![\Pi_X]\!]) = X$ and $K \circ \Pi_X = X$. Consequently, X is an element of the outcome set.

The following representation theorem characterizes monoselective descriptor revision in terms of four quite plausible postulates:

Theorem 5.3 *Let* $\circ$ *be an operation on a consistent belief set* K*, with descriptors as inputs and belief sets as outputs. Then the following two conditions are equivalent:*

[1] In [124] the abbreviation $\widehat{C}(\Psi)$ was used for $\widehat{C}(\{X \in \mathbb{X} \mid X \Vdash \Psi\})$. The notation used here has the advantage of making it more clear among what objects the choice is made.

[2] On outcome sets, see also [121].

(I) $\circ$ *is a (deterministic) local monoselective descriptor revision.*

(II) $\circ$ *satisfies the postulates:*

$K \circ \Psi = \mathrm{Cn}(K \circ \Psi)$ *(closure)*

$K \circ \Psi \Vdash \Psi$ *or* $K \circ \Psi = K$. *(relative success)*

If $K \circ \Xi \Vdash \Psi$, *then* $K \circ \Psi \Vdash \Psi$. *(regularity)*

If $K \circ \Xi \Vdash \Psi$ *if and only if* $K \circ \Xi \Vdash \Psi'$ *for all* Ξ, *then* $K \circ \Psi = K \circ \Psi'$.

(uniformity)

The closure property comes with our use of belief sets as (idealized) representations of belief states.

Uniformity says essentially that if two success conditions are satisfied by exactly the same elements of the outcome set, then the changes that have these two conditions as inputs yield the same outcome.[3] For an example, suppose that there has been a large explosion in an underground mine. My friend Alfredo works in the mine, but I do not know if he was in the mine when the explosion took place. I believe it to be impossible that he survived if he was in the mine at the time, and I am equally sure that he was not killed in the explosion if he was not there. Let Ψ_1 denote belief that he was in the mine at the time of the explosion and Ψ_2 belief that he was killed in the explosion. Under the background conditions just given, any of my potential belief states satisfies Ψ_1 if and only if it satisfies Ψ_2. Uniformity tells us that in such a case, Ψ_1 and Ψ_2 yield the same revision outcome, i.e. $K \circ \Psi_1 = K \circ \Psi_2$.[4]

Uniformity implies the following, arguably less controversial property:

If $\Psi \dashVdash \Psi'$, then $K \circ \Psi = K \circ \Psi'$. (extensionality)

In analogy with the success condition $K * p \vdash p$ for sentential revision, one might consider the postulate $K \circ \Psi \Vdash \Psi$ for descriptor revision. However, such a condition is quite implausible for descriptor revision. It would require that all logically closed subsets of the language are elements of the outcome set.[5] (This is because, as we just noted, for all belief sets X there is a descriptor Π_X that is satisfied by X and by no other belief set.) Furthermore, such an operation would make the epistemic agent totally credulous in the sense that a single input can make her believe or disbelieve anything.[6] We should expect a rational epistemic agent to have some convictions that she does not give up that easily.[7]

Both relative success and regularity can be seen as weaker and more plausible versions of the implausible success condition $K \circ \Psi \Vdash \Psi$. They are also both generalizations of conditions that have been used for the characterization of sentential

---

[3]Properties similar to uniformity have been employed in studies of operations on belief bases, see for instance [83, 88, 89].

[4]If we wish to avoid uniformity, then we can replace $\widehat{C}$ by a function S on descriptors such that $S(\Psi) \in [\![\Psi]\!]$ whenever $[\![\Psi]\!] \neq \varnothing$.

[5]Similar density properties of the outcome sets of AGM operations are discussed in [121].

[6]Sentential belief revision and contraction have been criticized for credulity in this sense, see Section 3.2.

[7]Cf. Section 7.6.

semi-revision (not always successful sentential revision) and shielded contraction (not always successful contraction).[8]

Regularity may not be very plausible if we interpret $K \circ \Psi$ as the belief set that will result if the agent is told that Ψ is truthful.[9] For instance, let p denote that the Pope has ordered St. Peter's Basilica to be torn down, and q that St. Peter's Basilica has been irreparably damaged by a major earthquake. With the interpretation just referred to we can then have $K \circ \mathfrak{B}(p\&q) \Vdash \mathfrak{B}p$ but $K \circ \mathfrak{B}p \nVdash \mathfrak{B}p$ (and similarly $K * (p\&q) \vdash p$ but $K * p \nvdash p$ for the derived sentential revision $*$) since p alone is so hard to believe that the information will be rejected. Regularity is more plausible if we stick to the interpretation recommended in Section 4.5, namely that $K \circ \Psi$ has been singled out among the potential outcomes satisfying Ψ because it deviates as little as possible from K and/or because it is a kind of least common denominator for comparatively credible revision outcomes that satisfy Ψ.

The successfulness of an operation of change can be described in terms of its *success set*, i.e. the set of inputs that are satisfied in the outputs that they give rise to.[10] A related concept is the *local access set*, the set of inputs that can be satisfied after a single application of the operation. These two notions can be defined more precisely as follows:

Definition 5.4 *Let $\circ$ be a descriptor revision on a belief set K. Then:*

$\{\Psi \mid K \circ \Psi \Vdash \Psi\}$ *is its* success set, *and*

$\{\Psi \mid (\exists \Xi)(K \circ \Xi \Vdash \Psi)\}$ *is its* local access set.[11]

The regularity postulate has important effects on the success set:

Observation 5.5 *Let $\circ$ be a descriptor revision on a belief set K and let S be its success set. If $\circ$ satisfies regularity, then:*

(1) S coincides with the local access set of $\circ$.

(2) If $\Psi \in S$ and $\Psi \Vdash \Xi$, then $\Xi \in S$. (closure under single-premiss inference)

(3) If $\Psi \veebar \Xi \in S$, then either $\Psi \in S$ or $\Xi \in S$.

[8]Relative success for sentential revision, "either $K * p \vdash p$ or $K * p = K$", was introduced in [213, p. 54] and used for instance in [137]. The corresponding postulate for contraction, "$K \div p \nvdash p$ or $K \div p = K$" was introduced in [51]. Regularity for sentential revision, "If $K * q \vdash p$ then $K * p \vdash p$" was introduced in [137]. It is used in Chapter 8. The corresponding principle for contraction, "If $K \div q \nvdash p$ then $K \div p \nvdash p$" was introduced in [51] under the name "persistence". It is used in Chapters 9 and 10. Related properties are discussed in [179].

[9]A descriptor Ψ is truthful if and only if $T \Vdash \Psi$, where T is the set of all true sentences.

[10]Similar constructions have been used in studies of non-prioritized sentential change, e.g. the set of retractable sentences in [51] and the set of credible sentences in [137]. David Makinson employed the converse notion of a core protected from change in [179].

[11]The *global access set* is the set of descriptors that can be satisfied after a finite number of applications of $\circ$, $\{\Psi \mid (\exists \Xi_1)...(\exists \Xi_n)(K \circ \Xi_1 \circ \ldots \circ \Xi_n \Vdash \Psi)\}$. For a modal account of this notion of access, see Section 7.6.

Part (2) of the observation cannot be generalized to closure under multiple-premiss inference. Even if regularity is satisfied we can have $\Psi_1 \in S$, $\Psi_2 \in S$ and $\Psi_1 \cup \Psi_2 \Vdash \Xi$, but $\Xi \notin S$. (For a simple example let $\Psi_1 = \{\mathfrak{B}p\}$, $\Psi_2 = \{\neg\mathfrak{B}p\}$, and $\Xi = \Psi_1 \cup \Psi_2$.)[12]

Monoselective descriptor revision includes most plausible patterns of belief change, but it also includes some utterly implausible ones. It is for instance compatible with the "absolutely stubborn" pattern such that $\mathbb{X} = \{K\}$ and $K \circ \Psi = K$ for all Ψ. This is at least as implausible as the pattern of "total credulity" discussed above. We should expect the behaviour of a rational epistemic agent to be somewhere between these two extremes, but unfortunately that is a desideratum not easily expressed in axiomatic terms.

However, other plausible properties can be expressed axiomatically, such as the following two:

If $K \Vdash \Psi$, then $K \circ \Psi = K$. (confirmation)

If $K \circ \Xi \Vdash \Psi$ and $K \circ \Xi \nvdash \perp$, then $K \circ \Psi \nvdash \perp$. (inconsistency avoidance)

Both these postulates are easily characterizable in terms of properties of the choice function:

Definition 5.6 *Let C be a choice function on $\mathbb{X}$. C is X-*favouring *if and only if it holds for all $\mathbb{Y}$ that if $X \in \mathbb{Y} \subseteq \mathbb{X}$ then $X \in C(\mathbb{Y})$.*

*If $\mathbb{X}$ is a set of belief sets, then C is X-*favouring for descriptor-definable arguments *if and only if it holds for all descriptors Ψ that if $X \in [\![\Psi]\!]$, then $X \in C([\![\Psi]\!])$.*

Observation 5.7 *Let $\circ$ be the descriptor revision on a consistent belief set K that is based on the monoselective choice function $\widehat{C}$. Then:*

(1) $\circ$ satisfies confirmation if and only if $\widehat{C}$ is K-favouring for descriptor-definable arguments.

(2) $\circ$ satisfies inconsistency avoidance if and only if $\widehat{C}$ satisfies:

If $\widehat{C}([\![\Psi]\!]) \vdash \perp$, then $\bigcap [\![\Psi]\!] \vdash \perp$.

The condition characterizing confirmation states that the choice function always selects K if it is available, whereas the condition characterizing inconsistency avoidance says that the choice function never selects the inconsistent belief set if some other belief set is available.

5.2 Linear and Centrolinear Revision

The usual way to make a choice function orderly is to base it on a binary relation that can be interpreted as representing preference, closeness, or degrees of choice-worthiness. When the choice function is applied to a set, then the outcome is interpreted as consisting of those of its elements that are most preferred, closest at hand, or

[12] This is analogous to single sentence logical closure in sentential revision, see [137, p. 1583] and [99, p. 31].

most choiceworthy. [191, 237, 238] We can apply this construction to monoselective choice functions as well. For that purpose we can use a relation $\leqq$ on $\mathbb{X}$, to be called a *belief set ordering*. We will require that $\widehat{C}([\![\Psi]\!])$ is $\leqq$-minimal in $[\![\Psi]\!]$. In order to obtain a deterministic operation, all descriptor-definable subsets of $\mathbb{X}$ must have a unique $\leqq$-minimal element:

Definition 5.8 *([124]) (1) Let $\leqq$ be a relation on $\mathbb{X}$ and let $\mathbb{Y} \subseteq \mathbb{X}$. Then X is* $\leqq$-minimal *in $\mathbb{Y}$ if and only if $X \in \mathbb{Y}$ and $X \leqq Y$ for all $Y \in \mathbb{Y}$.*

(2) A relation $\leqq$ on a set $\mathbb{X}$ of belief sets is descriptor-wellfounded *if and only if each non-empty descriptor-definable subset of $\mathbb{X}$ has a $\leqq$-minimal element.*

Descriptor-wellfoundedness is a weakened form of standard well-foundedness that would require every subset of $\mathbb{X}$ to have a $\leqq$-minimal element. Descriptor-wellfoundedness only requires this to hold for those subsets of $\mathbb{X}$ that correspond exactly to a descriptor. The difference is substantial; we saw above in Observation 4.16 that some subsets of $\mathbb{X}$ may not have a corresponding descriptor.

We can define the relational variant of monoselective descriptor revision as follows:

Definition 5.9 (modified from [124]) *An operation $\circ$ on a belief set K is a* relational descriptor revision, *based on a relation $\leqq$ on its outcome set $\mathbb{X}$, if and only if it holds for all descriptors Ψ that $K \circ \Psi$ is the unique $\leqq$-minimal element of $\mathbb{X}$ that satisfies Ψ, unless Ψ is unsatisfiable within $\mathbb{X}$, in which case $K \circ \Psi = K$.*

Theorem 5.10 (modified from [124]) *Let $\circ$ be an operation on a consistent belief set K, with descriptors as inputs and belief sets as outputs. Then the following three conditions are equivalent:*

(I) There is a set $\mathbb{X}$ of belief sets with $K \in \mathbb{X}$ and a relation $\leqq$ on $\mathbb{X}$, such that $\circ$ is the relational descriptor revision based on $\leqq$.

(II) There is a set $\mathbb{X}$ of belief sets with $K \in \mathbb{X}$ and a complete, transitive, antisymmetric, and descriptor-wellfounded relation $\leqq$ on $\mathbb{X}$, such that $\circ$ is the relational descriptor revision based on $\leqq$.

(III) $\circ$ satisfies the postulates:

$K \circ \Psi = \mathrm{Cn}(K \circ \Psi)$ *(closure)*

$K \circ \Psi \Vdash \Psi$ *or* $K \circ \Psi = K$. *(relative success)*

If $K \circ \Xi \Vdash \Psi$, *then* $K \circ \Psi \Vdash \Psi$. *(regularity)*

If $K \circ \Psi \Vdash \Xi$, *then* $K \circ \Psi = K \circ (\Psi \cup \Xi)$. *(cumulativity)*

The equivalence of conditions (I) and (II) shows that a relational descriptor revision can only be obtained from a relation that satisfies the properties listed in condition (II). It follows from this that one important technical distinction in the AGM framework, namely that between relational and transitively relational operations [1, 217] cannot be transferred to descriptor revision. The underlying reason for this is that descriptors can carry much more information than sentences; in particular a descriptor can tell us both what sentences should and what sentences should not be elements of the outcome. As follows from Theorem 5.10, this forces the relation to be a linear ordering. In what follows, the class of operations characterized in that theorem will therefore be called *linear descriptor revisions*.

The uniformity postulate from Theorem 5.3 does not appear in Theorem 5.10, but that is only because it follows from three of the other postulates.

Observation 5.11 *Let* $\circ$ *be a descriptor revision on a belief set* K*. If* $\circ$ *satisfies relative success, regularity and cumulativity, then it satisfies uniformity.*

Cumulativity, the new postulate in Theorem 5.10, is a generalization of a similar postulate for sentential revision (If $K * p \vdash q$ then $K * p = K * (p\&q)$).[13] Provided that $*$ satisfies closure and success, it satisfies cumulativity if and only if it satisfies the following postulate [73, p. 54]:

If $K * p \vdash q$ and $K * q \vdash p$, then $K * p = K * q$. (reciprocity)[14]

The analogous postulate for descriptor revision,

If $K \circ \Psi \Vdash \Xi$ and $K \circ \Xi \Vdash \Psi$, then $K \circ \Psi = K \circ \Xi$. (reciprocity)

is also exchangeable for cumulativity in the presence of the other postulates of Theorem 5.10:

Observation 5.12 ([124]) *Let* $\circ$ *be a descriptor revision on a consistent belief set* K*. If* $\circ$ *satisfies relative success and regularity, then it satisfies cumulativity if and only if it satisfies reciprocity.*

Linear descriptor revision does not in general satisfy confirmation (If $K \Vdash \Psi$, then $K \circ \Psi = K$), but it satisfies the following weakened version of that postulate:

Observation 5.13 *Let* $\circ$ *be a descriptor revision on a consistent belief set* K*. If* $\circ$ *satisfies cumulativity, then it satisfies:*

There is a belief set $K^+ \in \mathbb{X}$ *such that for all* Ψ*: If* $K^+ \Vdash \Psi$*, then* $K \circ \Psi = K^+$*.* *(pseudo-confirmation)*

According to pseudo-confirmation, there is some belief set that is the outcome of all revisions whose success condition it satisfies. We can think of it as the element of the outcome set that is closest at hand or most easily available. According to confirmation, that belief set is equal to our starting-point K. Confirmation corresponds to a simple property of the underlying relation:

---

[13] The sentential cumulativity postulate seems to have appeared in the belief revision literature for the first time in [184, p. 198]. It has often been divided into two parts,

If $K * p \vdash q$ then $K * p \subseteq K * (p\&q)$ (cautious monotony) and

If $K * p \vdash q$ then $K * (p\&q) \subseteq K * p$ (cut),

whose names derive from their close relationships with patterns of nonmonotonic reasoning with the same names. [213, p. 49] On these postulates, see also [217].

[14] Reciprocity seems to have been introduced independently in [3, p. 32] and [67, p. 97]. It has been further discussed for instance in [174, p. 354] where it was called the Stalnaker property, and in [184, p. 198], [217, p. 110], and [222].

Observation 5.14 *Let $\circ$ be the linear descriptor revision on a belief set K that is based on the relation $\leqq$ on its outcome set $\mathbb{X}$. Then $\circ$ satisfies confirmation if and only if $K \leqq X$ for all $X \in \mathbb{X}$.*

In what follows, linear descriptor revision satisfying confirmation will be called *centrolinear descriptor revision*. The morpheme "centro-" refers to the central position of the original belief set K. We can think of centrolinear revision as an operation whose outcome stays as close to the original belief set as the success condition allows. This can be illustrated in a spatial model by positioning the elements of $\mathbb{X}$, i.e. the potential belief sets, at different distances from the original belief set K, for instance around K on a surface or on a straight line with K at one end. When we revise K by a success condition (descriptor) Ψ, the outcome $K \circ \Psi$ is the belief set closest to K in which Ψ is satisfied.

5.3 Indeterministic Descriptor Revision

In the previous sections we have constructed descriptor revision as a deterministic operation, which means that for each given input, the operation delivers exactly one belief state as its output. This is well in line with the tradition in studies of belief change, but it is nevertheless important to discuss whether the use of deterministic models is adequate, or whether they should be replaced by indeterministic models in which the outputs are sets of equally choiceworthy belief states that the operation does not help us to choose between. The answer to that question depends on the purpose for which the models are constructed.

We can develop belief revision models with the objective to express the requirements of *rational* belief change as accurately as possible. Even if we have very strict requirements of rationality it is reasonable to assume that there will sometimes be more than one maximally rational way to perform an operation of change. This speaks in favour of indeterminism for models intended to represent what rationality requires.

However, belief revision models can also have the purpose of representing *actual* belief change (possibly the actual belief changes performed by a sufficiently rational agent). An indeterministic model will not be sufficient for that purpose. In the cases when it yields more than one outcome we will have to supplement it with some other mechanism that narrows down the choice to a single outcome. It would then seem preferable to combine the two mechanisms into one, and such a complete mechanism will be deterministic.

Since both types of models are justified, we have reasons to complement our investigations of deterministic operations with some consideration of what indeterministic descriptor revision would look like. As already mentioned in Section 4.1, we can generalize the deterministic generic model $\langle \mathbb{K}, \mathbb{I}, \circledcirc \rangle$ to an indeterministic model $\langle \mathbb{K}, \mathbb{I}, \check{\circledcirc} \rangle$ such that $\varnothing \neq \mathcal{K} \check{\circledcirc} \imath \subseteq \mathbb{K}$ for all $\mathcal{K} \in \mathbb{K}$ and $\imath \in \mathbb{I}$. Similarly, the basic definition of descriptor revision in Definition 4.20 can be generalized to indeterministic revision by just relaxing the requirement that the choice function be monoselective. This means that the condition $\mathcal{K} \circ \Psi = \widehat{C}(\{\mathcal{K}' \in \mathbb{K}_{\mathcal{K}} \mid \mathfrak{s}(\mathcal{K}') \Vdash \Psi\})$ in that definition should be replaced by $\mathcal{K} \breve{\circ} \Psi = C(\{\mathcal{K}' \in \mathbb{K}_{\mathcal{K}} \mid \mathfrak{s}(\mathcal{K}') \Vdash \Psi\})$ where C satisfies the usual requirements on a (not necessarily monoselective) choice function. The following examples illustrate how the postulates for a deterministic descriptor revision $\circ$ can be generalized to an indeterministic operation $\breve{\circ}$:

If $X \in K \breve{\circ} \Psi$, then $X = \mathrm{Cn}(X)$. (closure)
Either $X \Vdash \Psi$ for all $X \in K \breve{\circ} \Psi$, or $K \breve{\circ} \Psi = \{K\}$. (relative success)
If $Y \Vdash \Psi$ for some $Y \in \mathbb{X}$, then $X \Vdash \Psi$ for all $X \in K \breve{\circ} \Psi$. (regularity)
If $X \Vdash \Psi$ if and only if $X \Vdash \Psi'$ for all $X \in \mathbb{X}$, then $K \breve{\circ} \Psi = K \breve{\circ} \Psi'$. (uniformity)
If $K \Vdash \Psi$, then $K \breve{\circ} \Psi = \{K\}$. (confirmation)
If $X \Vdash \Xi$ for all $X \in K \breve{\circ} \Psi$, then $K \breve{\circ} \Psi = K \breve{\circ} (\Psi \cup \Xi)$. (cumulativity)

It could be argued against deterministic descriptor operations that they put implausible demands on the selection mechanism. The requirements that choice functions should be monoselective and – in particular – that belief set orderings should be antisymmetric are arguably not very plausible. However, it should be noted that in important cases, deterministic operations can be obtained with selection mechanisms that do not satisfy these requirements. For instance, from each descriptor revision $\breve{\circ}$ we can define an operation $\breve{*}$ of sentential revision according to the simple recipe $K \breve{*} p = K \breve{\circ} \mathfrak{B} p$.[15] Let $\breve{\circ}$ be based on a belief set ordering $\leqq$. One might perhaps believe that $\leqq$ has to be antisymmetric in order for $\breve{*}$ to be deterministic, but it turns out that the following condition is sufficient (and necessary):

If p is satisfiable within $\mathbb{X}$, then there is some $Z \in \mathbb{X}$ such that $p \in Z$ and that $p \notin Z'$ for all Z' such that $Z' \leqq Z$ and $Z' \neq Z$.

To see how this works we can use the following simple example: Let the language be based on the three logical atoms p, q, and r. Let $K = \mathrm{Cn}(\{\neg p, \neg q\})$ and let $\mathbb{X}$ contain only three belief sets in addition to K, namely $\mathrm{Cn}(\{r, p \vee q\})$, $\mathrm{Cn}(\{p\})$, and $\mathrm{Cn}(\{q\})$. Let $\leqq$ be based in the obvious way on the distances to K in Fig. 5.1. There is a tie between $\mathrm{Cn}(\{p\})$ and $\mathrm{Cn}(\{q\})$, i.e. $\mathrm{Cn}(\{p\}) \leqq \mathrm{Cn}(\{q\}) \leqq \mathrm{Cn}(\{p\})$, but in spite of this breach of antisymmetry, the sentential revision that is based on $\leqq$ is deterministic. The reason for this is that all sentences that are present in both $\mathrm{Cn}(\{p\})$ and $\mathrm{Cn}(\{q\})$ are also present in $\mathrm{Cn}(\{r, p \vee q\})$ which is closer to K and will therefore be the outcome of revising by these sentences.[16]

Interestingly, we can obtain exactly the same sentential revision as in Fig. 5.1 with an antisymmetric belief set ordering, i.e. one without ties. Such a construction is illustrated in Fig. 5.2. It can easily be shown that the sentential revisions obtained

[15] The deterministic variant of this construction will be investigated in detail in Chapter 8.

[16] To prove this note that $\mathrm{Cn}(\{p\}) \cap \mathrm{Cn}(\{q\}) = \mathrm{Cn}(\{p \vee q\})$, thus $\mathrm{Cn}(\{p\}) \cap \mathrm{Cn}(\{q\}) \subseteq \mathrm{Cn}(\{r, p \vee q\})$. –Too be more precise: For all sentences $x \in \mathrm{Cn}(\{p\}) \cap \mathrm{Cn}(\{q\})$: (1) If $x \in \mathrm{Cn}(\varnothing)$, then $K * x = K$. (2) If $x \notin \mathrm{Cn}(\varnothing)$, then $K * x = \mathrm{Cn}(\{r, p \vee q\})$.

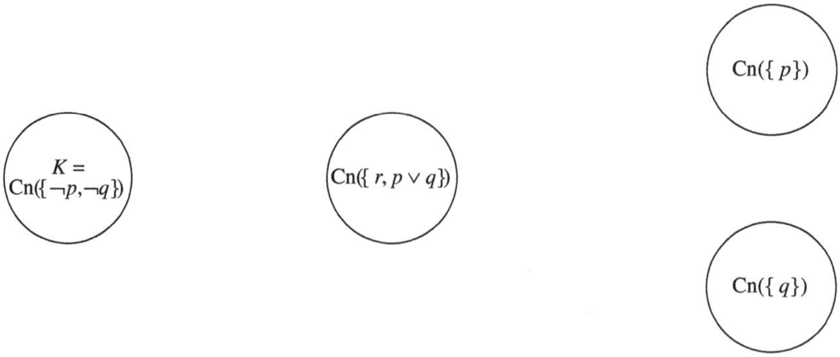

Fig. 5.1 An example of a deterministic sentential revision that is based on a belief set ordering not satisfying antisymmetry.

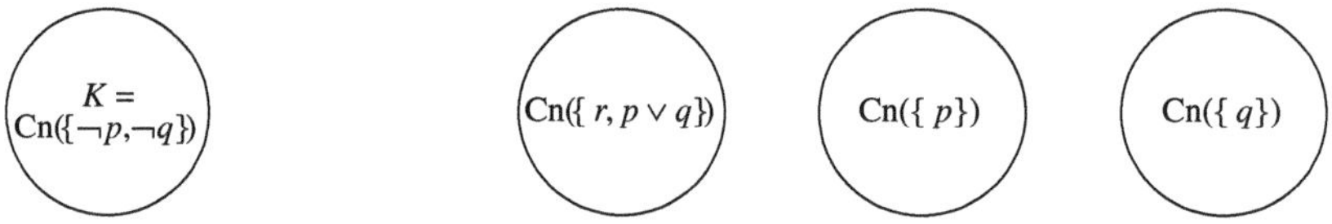

Fig. 5.2 The same sentential revision as in Fig. 5.1, obtained with an antisymmetric belief set ordering.

in the two figures coincide.[17] This is remarkable, since it shows that the use of an antisymmetric belief set ordering (as in Fig. 5.2) does not prevent us from a faithful rendering of an operation that we first constructed with a belief set ordering that is not antisymmetric, i.e. one that has ties. The crucial feature in both constructions is that everything that is common between the stalemated belief sets $\text{Cn}(\{p\})$ and $\text{Cn}(\{q\})$ has already been covered by some belief set that comes before them in the ordering. It might be countered that the example is oversimplified and therefore of limited interest, but the method of *pre-empting stalemates* that it illustrates can in fact be applied to a wide range of sentential revisions. We will see in Section 8.2 that all transitively relational partial meet (AGM) revisions can in fact be reconstructed as linear descriptor revisions with an antisymmetric belief set ordering.[18]

However, although this works for sentential revision ($K * x$ or equivalently $K \circ \mathfrak{B}x$), it does not work for descriptor revision in general ($K \circ \Psi$). This is exemplified by

[17]Suppose there is some sentence x such that $K * x$ will not be the same in the two constructions. (1) x must be an element of at least one of the four elements of $\mathbb{X}$ since otherwise $K * x = K$ in both constructions. (2) x cannot be an element of either K or $\text{Cn}(\{r, p \vee q\})$, since then the outcome would be the same in both constructions. (3) x cannot be an element of both $\text{Cn}(\{p\})$ and $\text{Cn}(\{q\})$ since then either $K * x = K$ in both constructions (if x is a tautology) or (else) $K * x = \text{Cn}(\{r, p \vee q\})$ in both constructions, for the reasons given in footnote 16. (4) If $\text{Cn}(\{p\})$ is the only element of $\mathbb{X}$ that contains x, then $K * x = \text{Cn}(\{p\})$ in both constructions. (5) If $\text{Cn}(\{q\})$ is the only element of $\mathbb{X}$ that contains x, then $K * x = \text{Cn}(\{q\})$ in both constructions.

[18]This follows from Theorem 8.8.

the construction shown in Fig. 5.1. Although it gives rise to a deterministic sentential revision, the general descriptor revision that is based on it is indeterministic. This can be seen from the descriptor $\mathfrak{B}p \vee \mathfrak{B}q$ that is caught in the stalemate between $\text{Cn}(\{p\})$ and $\text{Cn}(\{q\})$.

Descriptor revision has the advantage of providing faithful representations of both deterministic and indeterministic operations. In particular, indeterministic descriptor revision has the realistic feature that the stalemates that it represents are stalemates between the potential outcomes of the operation, not between larger, cognitively inaccessible objects into which these potential outcomes can be imbedded. In this book, the primary focus is on deterministic operations. This has the advantage of maintaining close comparability with previous approaches, including AGM, that have been almost exclusively devoted to deterministic operations. However, indeterministic descriptor revision is a fruitful field for future investigations.

5.4 Blockage Revision

Alternatively, descriptor revision can be based on a formal representation of how potential outcomes of belief change can block each other. A belief set X blocks another belief set Y if and only if: whenever X satisfies a success condition Ψ, then Y cannot be the outcome of revision by Ψ. This can be represented by a *blockage relation* $\rightharpoondown$, such that $X \rightharpoondown Y$ denotes that X blocks Y.[19]

One possible interpretation of $X \rightharpoondown Y$ is that X is at least as plausible as Y. This would mean that X and Y block each other ($X \rightharpoondown Y \rightharpoondown X$) if and only if they are exactly equally plausible. However, examples can be found in which two potential outcomes seem to block each other although one of them is somewhat more plausible than the other.[20] For instance, consider Linda who believed that both her parents were still in good health. One day when visiting them she discovered a pill organizer full with medicines that seemed to have been stowed away to prevent her from seeing it. This made her hesitate between believing that her mother was ill and that her father was ill. The former option was somewhat more plausible, but the latter was plausible enough to put her in a state of doubt between the two options. Thus two potential belief states blocked each other, although one was more plausible than another.

There are two ways to introduce a blockage relation. We can either derive it from a descriptor revision, or introduce it as a primitive notion from which a descriptor revision can be derived. Beginning with the former option, for any descriptor revision $\circ$ we can derive a blockage relation $\rightharpoondown$ as follows:

> For all $X, Y \in \mathbb{X}$: $X \rightharpoondown Y$ if and only if it holds for all descriptors Ψ that if $X \Vdash \Psi$, then $K \circ \Psi \neq Y$.

---

[19] Blockage relations were first introduced in [116] where they were used to construct an operation of contraction.

[20] Cf. [116, pp. 418-419], [130, 222].

However, with this construction we cannot in general regain the operation $\circ$ from the blockage relation $\rightharpoondown$ that was derived from it.[21] This makes formal developments difficult, and it is therefore often preferable to introduce the blockage relation as a primitive notion. This can be done as follows:

Definition 5.15 ([130]) *Let $\rightharpoondown$ be a binary relation on the set $\mathbb{X}$ of belief sets with $K \in \mathbb{X}$. The* (deterministic) blockage revision *on K generated by $\rightharpoondown$ is the operation $\circ$ such that for all descriptors Ψ:*

(i) If X is the unique element of $[\![\Psi]\!]$ that is not blocked by any other element of $[\![\Psi]\!]$, then $K \circ \Psi = X$.

(ii) If there is no such unique unblocked element in $[\![\Psi]\!]$, then $K \circ \Psi = K$.

We can arrive at clause (ii) of the definition either because (1) $[\![\Psi]\!]$ is empty, (2) all elements of $[\![\Psi]\!]$ are blocked by some other element, or (3) at least two elements of $[\![\Psi]\!]$ are unblocked within $[\![\Psi]\!]$.[22]

The set $\mathbb{X}$ in Definition 5.15 is a repertoire for the operation $\circ$. The following observation identifies the cases when the repertoire and the outcome set of a blockage revision coincide.

Observation 5.16 ([130]) *Let $\circ$ be the blockage revision on K that is generated by the relation $\rightharpoondown$ on the set $\mathbb{X}$ of belief sets with $K \in \mathbb{X}$. Then $\mathbb{X}$ is the outcome set of $\circ$ if and only if $\rightharpoondown$ satisfies irreflexivity within $\mathbb{X}\backslash\{K\}$.*

In what follows it will be assumed that $\rightharpoondown$ is irreflexive and that consequently, the outcome set coincides with the repertoire.[23]

The following two observations show that blockage revision satisfies three of the four postulates used above to characterize monoselective descriptive revision, but not in general the fourth. Furthermore, it does not in general satisfy the postulates that we used to characterize linear and centrolinear revision.

Observation 5.17 *Let $\circ$ be the blockage revision on K that is generated by the relation $\rightharpoondown$ on its outcome set $\mathbb{X}$. It satisfies closure, relative success, and uniformity.*

Observation 5.18 ([130]) *Let $\circ$ be the blockage revision on K that is generated by the relation $\rightharpoondown$ on its outcome set $\mathbb{X}$. Then:*

---

[21] To see this, let $\mathbb{X} = \{K, X, Y, Z, W\}$, and let $\circ$ be based on a monoselective choice function $\widehat{C}$ such that $\widehat{C}(\mathbb{Y}) = K$ whenever $K \in \mathbb{Y} \subseteq \mathbb{X}$ and that $\widehat{C}(\{X, Y\}) = X$, $\widehat{C}(\{X, Z\}) = Z$, $\widehat{C}(\{X, W\}) = X$, $\widehat{C}(\{Y, Z\}) = Y$, $\widehat{C}(\{Y, W\}) = W$, $\widehat{C}(\{Z, W\}) = Z$, $\widehat{C}(\{X, Y, Z\}) = X$, $\widehat{C}(\{X, Z, W\}) = W$, $\widehat{C}(\{Y, Z, W\}) = Z$, $\widehat{C}(\{X, Y, W\}) = W$, and $\widehat{C}(\{X, Y, Z, W\}) = Y$. Furthermore, let $\widehat{C}'$ coincide with $\widehat{C}$ with the sole exception that $\widehat{C}'(\{X, Y\}) = Y$, and let $\circ'$ be based on $\widehat{C}'$. Then $\circ$ and $\circ'$ give rise to the same blockage relation namely $\rightharpoondown = \{\langle K, X, \rangle, \langle K, Y\rangle, \langle K, Z\rangle, \langle K, W\rangle\}$. However, $K \circ (\Pi_X \veebar \Pi_Y) = X$ and $K \circ' (\Pi_X \veebar \Pi_Y) = Y$.

[22] This applies to deterministic revision. In indeterministic blockage revision all unblocked elements of $[\![\Psi]\!]$ will be elements of $K \breve{\circ} \Psi$, and case (3) does not apply.

[23] If this assumption is not made, then the outcome set is equal to $\{K\} \cup \{X \in \mathbb{X} \mid X \not\rightharpoondown X\}$, where $\mathbb{X}$ is the repertoire.

(1) It does not hold in general that $\circ$ satisfies regularity.

(2) It does not hold in general that $\circ$ satisfies cumulativity.

(3) It does not hold in general that $\circ$ satisfies reciprocity.

(4) It does not hold in general that $\circ$ satisfies confirmation.

It follows that blockage revision is not a subcase of monoselective descriptor revision. The following observation introduces two postulates that exhibit the opposite pattern: they are satisfied by all blockage revisions but not by all monoselective descriptor revisions.

Observation 5.19 ([130]) *(1) Let $\circ$ be the blockage revision on K that is generated by the relation $\rightharpoondown$ on its outcome set $\mathbb{X}$. Then $\circ$ satisfies:*

If $K \circ \Psi \neq K \neq K \circ (\Psi \cup \Xi)$ and $K \circ \Psi \Vdash \Xi$, then $K \circ \Psi = K \circ (\Psi \cup \Xi)$.
(peripheral cumulativity),

and

If $K \neq K \circ \Psi = K \circ \Xi$, then $K \circ \Psi = K \circ (\Psi \veebar \Xi)$.
(peripheral disjunctive identity).

(2) Let $\circ$ be a monoselective descriptor revision. It does not hold in general that $\circ$ satisfies peripheral cumulativity. Furthermore, it does not hold in general that $\circ$ satisfies peripheral disjunctive identity.

The term "peripheral" excludes the original belief set K from the scope of these postulates. Peripheral cumulativity is a restriction of cumulativity, and peripheral disjunctive identity is a restriction of:

If $K \circ \Psi = K \circ \Xi$, then $K \circ \Psi = K \circ (\Psi \veebar \Xi)$. (disjunctive identity)

that holds for linear revision.[24]

In the remainder of this section we are going to investigate what properties a blockage relation has to satisfy in order for the corresponding blockage revision to also be either a monoselective, linear, or centrolinear descriptor revision. The following two formal results answer that question for monoselective descriptor revision:

Theorem 5.20 ([130]) *Let $\circ$ be the blockage revision on K that is generated by the relation $\rightharpoondown$ on its outcome set $\mathbb{X}$. Then $\circ$ satisfies regularity if and only if $\rightharpoondown$ satisfies the two postulates*

If $K \notin [\![\Psi]\!] \neq \varnothing$, then there is at least one unblocked element within $[\![\Psi]\!]$.
(peripheral non-occlusion),

and

---

[24]Its sentential analogue, "If $K * p = K * q$ then $K * p = K * (p \vee q)$", holds for transitively relational AGM revision. It does not seem to have been referred to in the belief revision literature, but it is a trivial consequence of the postulate of disjunctive factoring (Either $K * (p \vee q) = K * p$ or $K * (p \vee q) = K * q$ or $K * (p \vee q) = K * p \cap K * q$) that holds for transitively relational AGM revision. (Disjunctive factoring was proved by Hans Rott and first reported in [69, pp. 57, 212, and 244].)

If $X \neq Y \neq K \neq X$, *then either* $X \rightharpoondown Y$ *or* $Y \rightharpoondown X$.

(peripheral weak connectedness[25]*)*

Observation 5.21 ([130]) *Let* $\circ$ *be the blockage revision on* K *that is generated by the relation* $\rightharpoondown$ *on its outcome set* $\mathbb{X}$. *Then* $\circ$ *is a monoselective descriptor revision on* K *if and only if* $\rightharpoondown$ *satisfies peripheral non-occlusion and peripheral weak connectedness.*

The following two results show that linear descriptor revision coincides exactly with blockage revision that is generated from a blockage relation $\rightharpoondown$ satisfying three conditions.

Theorem 5.22 ([130]) *Let* $\circ$ *be the blockage revision on* K *that is generated by the relation* $\rightharpoondown$ *on its outcome set* $\mathbb{X}$. *Then* $\circ$ *satisfies cumulativity if and only if* $\rightharpoondown$ *satisfies peripheral non-occlusion, peripheral weak connectedness, and:*

If $X \neq K \neq Y$ *and* $X \rightharpoondown Y \rightharpoondown K$, *then* either $X \rightharpoondown K \not\rightharpoondown X$ or *both* $K \rightharpoondown X$ *and* $K \rightharpoondown Y$. *(top adjacency)*

Observation 5.23 *Let* $\circ$ *be a descriptor revision on* K. *Then* $\circ$ *is a linear revision if and only if it is a blockage revision generated by a relation* $\rightharpoondown$ *that satisfies peripheral non-occlusion, peripheral weak connectedness, and top adjacency.*

It follows from Theorems 5.20 and 5.22 that if a blockage revision satisfies cumulativity, then it satisfies regularity. The following two formal results show that the converse relationship does not hold.

Observation 5.24 ([130]) *A blockage revision* $\circ$ *satisfies cumulativity if and only if it satisfies both regularity and reciprocity.*

Observation 5.25 ([130]) *Let* $\circ$ *be the blockage revision on* K *that is generated by the relation* $\rightharpoondown$ *on its outcome set* $\mathbb{X}$. *Then: (1) It does not hold in general that if* $\circ$ *satisfies regularity then it satisfies reciprocity, and (2) it does not hold in general that if* $\circ$ *satisfies reciprocity then it satisfies regularity.*

Finally, let us turn to centrolinear revision. The following two formal results show that a descriptor revision is centrolinear if and only if it is a blockage revision whose generating relation satisfies the three properties required for linear revision, and in addition a fourth property that ensures the satisfaction of the confirmation postulate.

Theorem 5.26 ([130]) *Let* $\circ$ *be the blockage revision on* K *that is generated by the relation* $\rightharpoondown$ *on its outcome set* $\mathbb{X}$. *Then* $\circ$ *satisfies confirmation if and only if* $\rightharpoondown$ *satisfies the postulate*

If $X \neq K \neq Y$ *and* $X \rightharpoondown K$, *then (1)* $K \rightharpoondown X$ *and (2) either* $K \rightharpoondown Y$ *or* $X \rightharpoondown Y$.

(near-superiority)

---

[25] A relation $\multimap$ satisfies weak connectedness if and only if it holds for all X and Y that if $X \neq Y$, then either $X \multimap Y$ or $Y \multimap X$. See [56, p. 11].

Observation 5.27 *Let $\circ$ be a descriptor revision on K. The following three conditions are equivalent:*

(I) *$\circ$ is a centrolinear revision.*

(II) *$\circ$ is a blockage revision generated by a relation $\rightharpoondown$ that satisfies peripheral non-occlusion, peripheral weak connectedness, top adjacency, and near-superiority.*

(III) *$\circ$ is a blockage revision generated by a relation $\rightharpoondown$ that satisfies transitivity, weak connectedness, irreflexivity, and:*

If $K \neq X$, then $K \rightharpoondown X$. *(superiority)*

In this context, the following well-known connections should be pointed out:

Observation 5.28 *(1) If a relation $\rightharpoondown$ satisfies asymmetry, then it satisfies irreflexivity. (2) If a relation $\rightharpoondown$ satisfies transitivity and irreflexivity, then it satisfies asymmetry.*

As the following two observations show, confirmation is independent of regularity, reciprocity and cumulativity.

Observation 5.29 ([130]) *Let $\circ$ be a blockage revision that satisfies cumulativity. It does not hold in general that $\circ$ satisfies confirmation.*

Observation 5.30 ([130]) *Let $\circ$ be a blockage revision that satisfies confirmation. (1) It does not hold in general that $\circ$ satisfies regularity. (2) It does not hold in general that $\circ$ satisfies reciprocity.*

Top adjacency and near-superiority that are used in Theorems 5.22 and 5.26 represent different ways to ensure that K has a strong position within $\mathbb{X}$ in terms of $\rightharpoondown$. The following observation shows that they are logically independent of each other but both weaker than each of the following two simpler properties:

If $K \neq X$, then $K \rightharpoondown X$. (superiority)
$X \not\rightharpoondown K$ (non-inferiority)

Observation 5.31 ([130]) *Let $\rightharpoondown$ be an irreflexive relation on a set $\mathbb{X}$ with $K \in \mathbb{X}$. Then:*

(1) *If $\rightharpoondown$ satisfies superiority, then it satisfies top adjacency and near-superiority.*

(2) *If $\rightharpoondown$ satisfies non-inferiority, then it satisfies top adjacency and near-superiority.*

(3) *It does not hold in general that if $\rightharpoondown$ satisfies top adjacency, then it satisfies near-superiority.*

(4) *It does not hold in general that if $\rightharpoondown$ satisfies near-superiority, then it satisfies top adjacency.*

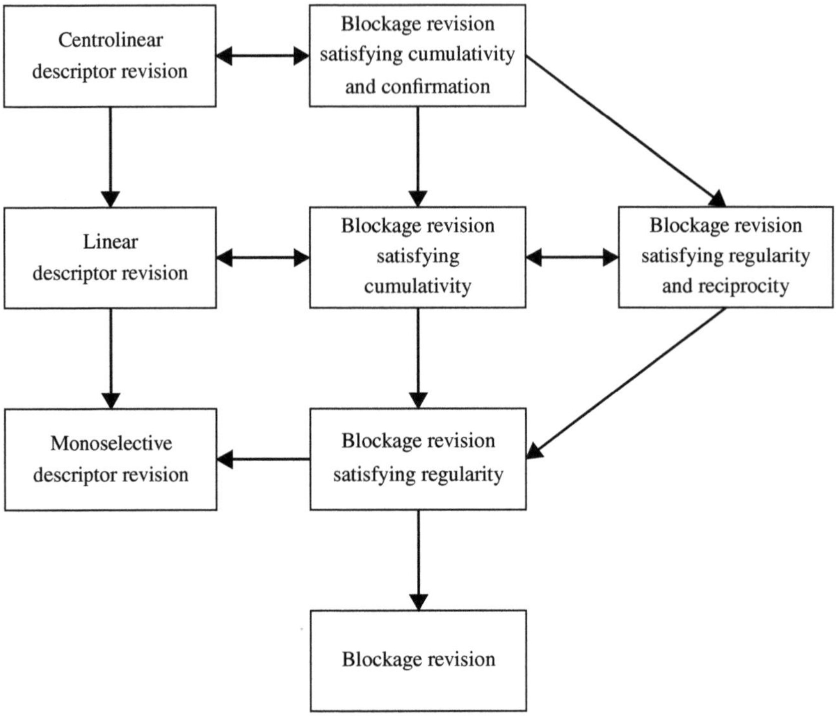

Fig. 5.3 Relationships of inclusion among some major classes of blockage revision and their connections with monoselective, linear, and centrolinear descriptive revision. An arrow from one category to another indicates that the first is a subset of the second.

The results of this section are summarized in Fig. 5.3. They confirm that although blockage revision is neither a subclass nor a superclass of monoselective descriptor revision, the more orderly classes of monoselective revision coincide with interesting classes of blockage revisions.

5.5 Relations of Epistemic Proximity

The construction and investigation of relations of epistemic entrenchment is rightly considered to be one of the major achievements in the AGM tradition. The approach was briefly summarized in Section 1.5. Entrenchment relations are relations between sentences; here they are denoted by the symbol $\leq$. We can read $p \leq q$ ("p is at most as entrenched as q") as saying that p has an at most as strong standing as q in the agent's set of beliefs, or that p is at least as easily given up as q. In Section 1.5 we saw that if $\leq$ satisfies five fairly plausible properties, then an operation of contraction can be defined from it. The operations of contraction that are derivable from entrenchment relations in this way coincide exactly with the standard AGM contractions (transitively relational partial meet contractions). Not surprisingly, only tautologies have the highest degrees of entrenchment; they are the sentences that the agent is assumed to be least willing to give up, to wit, not willing at all. Furthermore, all sentences outside of the belief set (elements of $\mathcal{L}\backslash K$) have the lowest degree of entrenchment; these are the sentences that the agent does not to believe in.

In the context of descriptor revision we can talk of entrenchment not as a relation between two sentences p and q to be removed but as a relation between two descriptors $\neg\mathfrak{B}p$ and $\neg\mathfrak{B}q$. Instead of saying that p is at most as entrenched as q ($p \leq q$) we can then say that the agent is at least as inclined to have a belief state satisfying $\neg\mathfrak{B}p$ as one satisfying $\neg\mathfrak{B}q$, or that having beliefs satisfying $\neg\mathfrak{B}p$ is at least as close at hand as having beliefs satisfying $\neg\mathfrak{B}q$. We can introduce a relation $\succeq$ on descriptors to denote this relationship, and thus write $\neg\mathfrak{B}p \succeq \neg\mathfrak{B}q$ instead of $p \leq q$. It is a small but important step to extend this relation to all descriptors, not only those that are negations of an atomic descriptor.

The relation $\succeq$ on descriptors will be called a relation of *epistemic proximity*. [126] Its strict part will be denoted $\succ$ and its symmetric part $\simeq$. It differs from epistemic entrenchment in having much more expressive power. This can be seen from cases when the agent neither believes in a sentence p nor in its negation ($p \notin K$ and $\neg p \notin K$, or equivalently $K \nVdash \mathfrak{B}p$ and $K \nVdash \mathfrak{B}\neg p$). The agent may then be more easily convinced that p is the case than that it is not the case ($\mathfrak{B}p \succ \mathfrak{B}\neg p$), or the other way around ($\mathfrak{B}\neg p \succ \mathfrak{B}p$). We can also distinguish between open issues that are easily settled (at least one of $\mathfrak{B}p$ and $\mathfrak{B}\neg p$ ranks high in the order expressed by $\succeq$) and issues that are difficult to settle in either direction (both $\mathfrak{B}p$ and $\mathfrak{B}\neg p$ rank low in that order). To exemplify the latter case, I consider it to be extremely difficult to settle issues about mental experiences of non-human creatures. If p denotes a statement such as "dogs experience human speech as a form of barking", then neither $\mathfrak{B}p$ nor $\mathfrak{B}\neg p$ is close at hand for me to believe, i.e. they both rank low in the order expressed by $\succeq$.

The following formal definition introduces five properties of a relation of epistemic proximity that will turn out to make the relation workable for belief revision purposes.

Definition 5.32 ([126]) *A relation* $\succeq$ *on descriptors is a relation of* epistemic proximity *if and only if it satisfies:*

If $\Psi \succeq \Xi$ *and* $\Xi \succeq \Sigma$*, then* $\Psi \succeq \Sigma$. *(transitivity)*
If $\Psi \Vdash \Xi$*, then* $\Xi \succeq \Psi$. *(counter-dominance)*
If $\Psi \simeq \Xi$*, then* $\Psi \simeq \Psi \cup \Xi$. *(coupling)*
Either $\Psi \cup \{\mathfrak{B}p\} \succeq \Psi$ *or* $\Psi \cup \{\neg\mathfrak{B}p\} \succeq \Psi$. *(amplification)*
If $\nvdash p$ *then* $\neg\mathfrak{B}p \succ \bot$. *(absurdity avoidance)*

Transitivity and counter-dominance are comparatively uncontroversial properties. According to counter-dominance, a logically weaker descriptor is always at least as close at hand as a logically stronger one. For example, $\mathfrak{B}p$ is logically weaker than $\mathfrak{B}p \,\&\, \neg\mathfrak{B}q$. It cannot therefore, according to this axiom, be more difficult for me to adopt the belief pattern represented by $\mathfrak{B}p$ than that represented by $\mathfrak{B}p \,\&\, \neg\mathfrak{B}q$. If not believing in q would make it easier for me to believe in p, then I would presumably give up q in order to adopt the belief pattern $\mathfrak{B}p$.

Absurdity avoidance is a fairly weak property. It says that disbelief in a sentence is possible if that sentence is not a tautology. (Note that $\perp$ refers to inconsistencies in the metalanguage, such as $\mathfrak{B}p \,\&\, \neg\mathfrak{B}p$. Contrary to inconsistent beliefs such as $\mathfrak{B}(p \& \neg p)$, such absurd belief patterns are not even representable in the AGM framework.)

Amplification says that for any belief pattern Ψ and any sentence p, Ψ can either be combined with belief in p or with lack of belief in p, without loss in epistemic proximity. This is plausible even if Ψ and p are contentually unrelated. For instance, let Ψ denote some belief pattern concerning the reproductive behaviour of the blue-and-yellow macaw and let p denote that Robespierre was married. Presumably, how close at hand Ψ is (how high it ranks in $\succeq$) depends on how much resistance I have to adopting the belief set satisfying Ψ that I am least unwilling to adopt. Clearly, in that belief set p is either believed or not believed. If it is believed, then Ψ is as close at hand as $\Psi \cup \{\mathfrak{B}p\}$. If it is not believed, then Ψ is as close at hand as $\Psi \cup \{\neg\mathfrak{B}p\}$. What makes amplification plausible is that it refers to the addition of either $\mathfrak{B}p$ or $\neg\mathfrak{B}p$, and every belief set supports one of them. Replacing $\neg\mathfrak{B}p$ by $\mathfrak{B}\neg p$ would turn this very plausible postulate into a highly implausible one.

Coupling is arguably the least plausible of the five postulates. We can see it as a price we pay for making the operation deterministic. Coupling can perhaps best be understood if we assume that the ranking of descriptors in terms of epistemic proximity correlates with an antisymmetric ranking of belief sets satisfying these descriptors. We can then interpret $\Psi \simeq \Xi$ as saying that the most credible (or most proximate) belief set satisfying Ψ is equally credible (equally proximate) as the most credible (most proximate) belief set satisfying Ξ. In deterministic descriptor revision, ties between belief sets in terms of their credibility or their closeness to the present belief set have to be excluded.[26] Therefore, if $\Psi \simeq \Xi$, then the most proximate belief set satisfying Ψ must be identical to the most proximate belief set satisfying Ξ. It is then also the most proximate belief set satisfying $\Psi \cup \Xi$, which explains why $\Psi \simeq \Psi \cup \Xi$ holds.

Contrary to the standard definition of epistemic entrenchment [69, 71], Definition 5.32 does not mention the original belief set (K) to which the relation is associated. However, K can easily be retrieved as the belief set specified by the maxispecified descriptor $\bigcup\{\Psi \mid \Psi \simeq \mathfrak{B}\top\}$.[27]

[26]Cf. Sections 4.5 and 5.3.

[27]To see that $\bigcup\{\Psi \mid \Psi \simeq \mathfrak{B}\top\}$ is maxispecified, note that for all $p \in \mathcal{L}$, due to amplification either $\{\mathfrak{B}\top, \mathfrak{B}p\} \simeq \{\mathfrak{B}\top\}$ or $\{\mathfrak{B}\top, \neg\mathfrak{B}p\} \simeq \{\mathfrak{B}\top\}$. It follows that for all $p \in \mathcal{L}$, either $\mathfrak{B}p \in \bigcup\{\Psi \mid \Psi \simeq \mathfrak{B}\top\}$ or $\neg\mathfrak{B}p \in \bigcup\{\Psi \mid \Psi \simeq \mathfrak{B}\top\}$. It remains to show that $\bigcup\{\Psi \mid \Psi \simeq \mathfrak{B}\top\}$ specifies a belief set. If not, then there is some p with $\mathfrak{B}\top \simeq \mathfrak{B}p$ and $\mathfrak{B}\top \simeq \neg\mathfrak{B}p$. Due to coupling and transitivity, $\mathfrak{B}\top \simeq \perp$. Due to counter-dominance, $\mathfrak{B}\top \succeq \Psi$ for all Ψ, and transitivity yields $\perp \succeq \Psi$ for all Ψ, contrary to absurdity avoidance.

The following observation reports two important properties of $\succeq$ that follow from the ones already given.

Observation 5.33 ([126]) *Let $\succeq$ be a relation on descriptors.*

(1) If $\succeq$ satisfies transitivity and counter-dominance, then it satisfies:

If $\Psi \dashVdash \Psi'$ and $\Xi \dashVdash \Xi'$, then $\Psi \succeq \Xi$ if and only if $\Psi' \succeq \Xi'$. (intersubstitutivity)

(2) If $\succeq$ satisfies transitivity, counter-dominance, coupling, and amplification, then it satisfies:

$\Psi \succeq \Xi$ or $\Xi \succeq \Psi$. (completeness)

A useful class of maxispecified descriptors can be defined with the use of epistemic proximity:

Definition 5.34 ([126]) *Let $\succeq$ be a relation on descriptors and let Ψ be an element of its domain. For each $p \in \mathcal{L}$, let $\beta(p) = \mathfrak{B}p$ if $\Psi \simeq \Psi \cup \{\mathfrak{B}p\}$, and otherwise let $\beta(p) = \neg\mathfrak{B}p$. If $\Psi \nVdash \bot$ then the* maximal amplification *of Ψ is the descriptor $\widehat{\Psi}$ such that $\widehat{\Psi} = \{\beta(p) \mid p \in \mathcal{L}\}$. If $\Psi \Vdash \bot$ then $\widehat{\Psi} = \{\mathfrak{B}p \mid p \in \mathcal{L}\} \cup \{\neg\mathfrak{B}p \mid p \in \mathcal{L}\}$.*

Observation 5.35 ([126]) *Let $\succeq$ be a relation on descriptors that satisfies transitivity, counter-dominance, coupling, and amplification. Then:*

(1) $\widehat{\Psi} \simeq \Psi$, and

(2) $\widehat{\Psi} \simeq \Psi \cup \Xi$ if and only if $\widehat{\Psi} \Vdash \Xi$.

Furthermore:

(3) For any maxispecified descriptors Ψ and Ψ': $\Psi \simeq \Psi'$ if and only if $\Psi \dashVdash \Psi'$.

The following definition and observation confirm that epistemic proximity is indeed a generalization of epistemic entrenchment.

Definition 5.36 *([126]) Let $\succeq$ be a relation of epistemic proximity. The relation $\leqq$ on sentences, such that*

$p \leqq q$ if and only if $\neg\mathfrak{B}p \succeq \neg\mathfrak{B}q$

is the relation of (epistemic) entrenchment *that is based on $\succeq$.*

The strict part of $\leqq$ is denoted $<$ and its symmetrical part $\doteqdot$.

Observation 5.37 (modified from [126]) *Let $\succeq$ be a relation on descriptors and let $\leqq$ be the relation on sentences such that $p \leqq q$ if and only if $\neg\mathfrak{B}p \succeq \neg\mathfrak{B}q$. If $\succeq$ satisfies transitivity, counter-dominance, coupling, and amplification, then $\leqq$ satisfies:*

If $p \leqq q$ and $q \leqq r$, then $p \leqq r$. (transitivity)

If $p \vdash q$, then $p \leqq q$. (dominance)

Either $p \leqq p\&q$ or $q \leqq p\&q$. (conjunctiveness)

$p \notin \{r \mid \bot < r\}$ if and only if $p \leqq q$ for all q. (minimality)

Furthermore, if $\succeq$ *also satisfies absurdity avoidance, then* $\leqslant$ *satisfies:*

If $q \leqslant p$ *for all* q*, then* $\vdash p$*. (maximality)*

The standard definition of minimality for entrenchment relations is:

$p \notin K$ if and only if $p \leqslant q$ for all q.

The definition used in Observation 5.37 differs from this in not mentioning the belief set K. However, K is derivable from $\leqslant$ as $\{r \mid \perp < r\}$, and therefore it is not necessary to mention K in the axiomatization.[28] The relationship between a relation of epistemic proximity ($\succeq$) and its derived relation of epistemic entrenchment ($\leqslant$) will be further investigated in Section 9.3.

The above explication of the axioms of amplification and coupling indicated that relations of epistemic proximity on descriptors can be correlated with orderings of belief sets. More precisely, if we have an ordering $\leqq$ of belief sets giving rise to a centrolinear descriptor revision, then we can define a proximity relation $\succeq$ such that the position of a descriptor in $\succeq$ is determined by the position of the $\leqq$-closest belief set that satisfies the descriptor:

$\Xi \succeq \Psi$ if and only if either (i) there is some X with $X \Vdash \Xi$ such that $X \leqq Y$ for all Y with $Y \Vdash \Psi$, or (ii) Ψ is not satisfiable within the domain of $\leqq$.

Conversely, we can define $\leqq$ from $\succeq$ as follows:

$X \leqq Y$ if and only if there is some descriptor Ψ with $Y \Vdash \Psi$ such that $\Xi \succeq \Psi$ for all Ξ with $X \Vdash \Xi$.

Given $\succeq$ we can define a *proximity-based descriptor revision* as follows:

$p \in K \circ \Psi$ if and only if either (i) $\Psi \cup \{\mathfrak{B}p\} \simeq \Psi \succ \perp$ or (ii) $p \in K$ and $\Psi \simeq \perp$.

The following definition and two theorems show that there is a one-to-one relationship between the two types of binary relation ($\leqq$ and $\succeq$), and also between each of them and operations ($\circ$) of descriptor revision. The connections among $\leqq$, $\succeq$, and $\circ$ are expressed in terms of transformation functions whose names are based on the abbreviations b for relations on belief sets and d for relations on descriptors. Thus t^{db} is a transformation function that takes us from a relation (of epistemic proximity) on *d*escriptors to a relation on *b*elief sets, and $t^{\circ d}$ one that takes us from an operation of change to a relation on *d*escriptors. These interdefinabilities are also summarized in Fig. 5.4.

Definition 5.38 ([126]) *Let* $\leqq$ *be a descriptor-wellfounded linear ordering on belief sets. Let* $\mathbb{X}$ *be its domain and let* K *be the* $\leqq$*-minimal element of* $\mathbb{X}$*. Then:*

$t^{b\circ}(\leqq)$ *is the operation* $\circ$ *on* K *such that (i) if* Ψ *is satisfiable within* $\mathbb{X}$*, then* $K \circ \Psi$ *is the unique* $\leqq$*-minimal element of* $\mathbb{X}$ *that satisfies* Ψ*, and (ii) otherwise* $K \circ \Psi = K$*.*

---

[28] This is explained in Section 9.3.

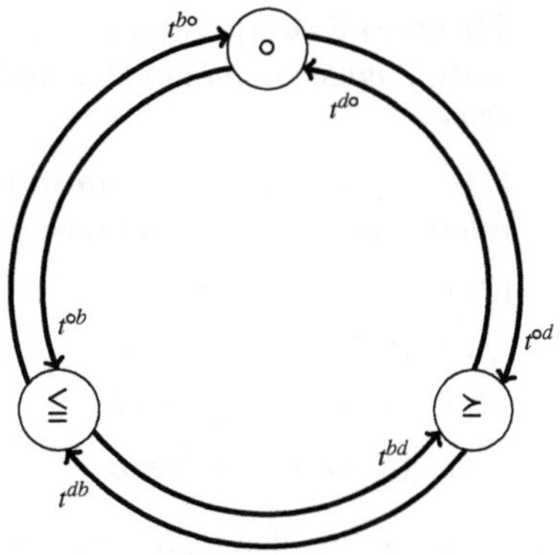

Fig. 5.4 The transformation functions for descriptor revision.

$t^{bd}(\leqq)$ *is the relation* $\succeq$ *on descriptors such that* $\Xi \succeq \Psi$ *if and only if either (i) there is some* X *with* $X \Vdash \Xi$ *such that* $X \leqq Y$ *for all* Y *with* $Y \Vdash \Psi$*, or (ii)* Ψ *is not satisfiable within the domain of* $\leqq$.

Let $\succeq$ *be a relation on descriptors with the strict part* $\succ$ *and the symmetric part* $\simeq$*, and let* $K = \{p \mid \mathfrak{B}p \simeq \mathfrak{B}\top\}$*. Then:*

$t^{db}(\succeq)$ *is the relation* $\leqq$ *on sets constructible as* $\{p \mid \widehat{\Phi} \Vdash \mathfrak{B}p\}$ *for some* Φ *with* $\Phi \succ \bot$*, such that* $X \leqq Y$ *if and only if there is some descriptor* Ψ *with* $Y \Vdash \Psi$ *such that* $\Xi \succeq \Psi$ *for all* Ξ *with* $X \Vdash \Xi$.

$t^{d\circ}(\succeq)$ *is the descriptor operation* $\circ$ *on* K *such that* $p \in K \circ \Psi$ *if and only if either (i)* $\Psi \cup \{\mathfrak{B}p\} \simeq \Psi \succ \bot$ *or (ii)* $p \in K$ *and* $\Psi \simeq \bot$.

Let $\circ$ *be a descriptor operation. Then:*

$t^{\circ b}(\circ)$ *is the relation* $\leqq$ *on belief sets obtainable as* $K \circ \Psi$ *for some* Ψ*, such that* $X \leqq Y$ *if and only if there are* Ξ *and* Ψ *such that* $X = K \circ \Xi$, $K \circ \Xi \Vdash \Xi$, $Y = K \circ \Psi$, $K \circ \Psi \Vdash \Psi$*, and* $K \circ \Xi = K \circ (\Xi \veebar \Psi)$.

$t^{\circ d}(\circ)$ *is the relation* $\succeq$ *on descriptors such that* $\Xi \succeq \Psi$ *if and only if either (i)* $K \circ \Xi \Vdash \Xi$, $K \circ \Psi \Vdash \Psi$*, and* $K \circ \Xi = K \circ (\Xi \veebar \Psi)$*, or (ii)* $K \circ \Psi \nVdash \Psi$.

Theorem 5.39 ([126]) *Let* $\leqq$ *be a descriptor-wellfounded linear ordering on a non-empty set of belief sets. Then:*

(1) $t^{bd}(\leqq)$ *is a relation on descriptors that satisfies transitivity, counter-dominance, coupling, amplification, and absurdity avoidance,*

(2) $t^{db}(t^{bd}(\leqq)) = \leqq$,

(3) $t^{d\circ}(t^{bd}(\leqq)) = t^{b\circ}(\leqq)$,

(4) $t^{\circ b}(t^{b\circ}(\leqq)) = \leqq$*, and*

(5) $t^{\circ d}(t^{b\circ}(\leqq)) = t^{bd}(\leqq)$.

Theorem 5.40 ([126]) *Let $\succeq$ be a relation on descriptors that satisfies transitivity, counter-dominance, coupling, amplification, and absurdity avoidance. Then:*

(1) $t^{db}(\succeq)$ is a complete, transitive, antisymmetric, and descriptor-wellfounded relation on a set of belief sets,

(2) $t^{bd}(t^{db}(\succeq)) = \succeq$,

(3) $t^{b\circ}(t^{db}(\succeq)) = t^{d\circ}(\succeq)$

(4) $t^{\circ d}(t^{d\circ}(\succeq)) = \succeq$, and

(5) $t^{\circ b}(t^{d\circ}(\succeq)) = t^{db}(\succeq)$.

In summary, this chapter has introduced local descriptor revision in several equivalent or closely related forms. Monoselective, linear, and centrolinear descriptor revision are all based on simple and intuitively plausible constructions that are characterizable with equally plausible sets of postulates. Blockage revision is based on quite different construction principles, but it has versions that coincide exactly with linear, respectively centrolinear descriptor revision. Finally, although proximity-based revision employs a relation on descriptors rather than on belief sets, it turns out to coincide exactly with centrolinear revision. Thus we obtain the same operations from different constructions, each of which has a plausible intuitive justification. These constructions, and the axioms that characterize them, all lend support to each other.

Chapter 6
Global Descriptor Revision

Descriptor revision can easily be extended to global (and therefore also iterated) belief change. In local change, the operation $\circ$ is specific for the original belief set K. Formally it is a function that takes us from a descriptor Ψ to an element $K \circ \Psi$ of the set $\mathbb{X}$ of belief sets. Therefore it can only represent changes that have K as their starting-point. In global change, the operation $\circ$ can be applied to any potential belief set. Formally it is a function that takes us from a pair consisting of a belief set X and a descriptor Ψ to a new belief set $X \circ \Psi$:

Definition 6.1 *Let $\mathbb{X}$ be a set of belief sets. A* (deterministic) global descriptor revision *on $\mathbb{X}$ is a function $\circ$ that takes pairs of an element of $\mathbb{X}$ and a belief descriptor as inputs, and has elements of $\mathbb{X}$ as outputs.*

Repeated uses of a global operation $\circ$ allow us to represent sequences of belief change, thereby opening up a host of questions about the relationships between different such sequences. For instance, $X \circ \neg\mathfrak{B}\neg p \circ \mathfrak{B}p$ is the outcome of first removing belief in $\neg p$ and then adding belief in p. We may ask under what conditions it coincides with the operation $X \circ \mathfrak{B}p$ that revises directly with belief in p. $K \circ (\mathfrak{B}p \vee \mathfrak{B}\neg p) \circ (\mathfrak{B}q \vee \mathfrak{B}\neg q)$ is a sequence in which one first makes up one's mind about p and then about q. We may ask how it relates to the sequence $K \circ (\mathfrak{B}q \vee \mathfrak{B}\neg q) \circ (\mathfrak{B}p \vee \mathfrak{B}\neg p)$ in which this is done in the reverse order.

Section 6.1 provides global versions of monoselective and centrolinear revision, and in Section 6.2 a distance model for global revision is investigated. In Section 6.3 blockage relations are generalized for global revision.

S.O. Hansson, *Descriptor Revision*, Trends in Logic 46,
DOI 10.1007/978-3-319-53061-1_6

6.1 Global Monoselective and Centrolinear Revision

The most obvious way to construct a global operation is to combine a set of local operations, one for each element of the outcome set. When performing this generalization, we need to pay close attention to the outcome set. In local revision, there is only need for one outcome set, namely the set containing the belief sets that are accessible from the original belief set K. In global revision it cannot be taken for granted that the accessible sets are the same from all starting-points. The belief set Z may be inaccessible from the viewpoint of X in the sense that there is no Ψ such that $X \circ \Psi = Z$, while at the same time Z is accessible from Y since there is some Ψ such that $Y \circ \Psi = Z$. The simplest way to express such patterns of accessibility is to use an accessibility function a, as introduced in Definition 4.22. To each belief set X in the global repertoire $\mathbb{X}$ it assigns the set $a(X)$ of belief sets that are directly accessible from X, i.e.:

$$a(X) = \{Y \in \mathbb{X} \mid (\exists\Psi)(X \circ \Psi = Y)\}$$

For simplicity we assume that each element of $\mathbb{X}$ is accessible from itself.[1] It follows that $\mathbb{X}$ consists of the sets that are accessible from at least one of its elements, i.e. $\mathbb{X} = \bigcup\{a(X) \mid X \in \mathbb{X}\}$.

In this way, monoselective descriptor revision can be straight-forwardly generalized to a global setting:

Definition 6.2 *A global descriptor revision* $\circ$ *on a set* $\mathbb{X}$ *of belief sets is a* (deterministic) global monoselective revision *if and only if (1) there is a function a from* $\mathbb{X}$ *to* $\wp(\mathbb{X})$ *(the* accessibility function*) such that* $X \in a(X) \subseteq \mathbb{X}$ *for all* $X \in \mathbb{X}$*, and (2) for each* $X \in \mathbb{X}$ *there is a monoselective choice function* $\widehat{C}_X$ *on* $a(X)$*, such that for all descriptors* Ψ*:*

(i) If Ψ *is satisfiable within* $a(X)$*, then* $X \circ \Psi = \widehat{C}_X([\![\Psi]\!]_{a(X)})$*, and*

(ii) otherwise $X \circ \Psi = X$*.*

Furthermore, if $a(X) = \mathbb{X}$ *for all* $X \in \mathbb{X}$*, then* $\circ$ *is a* coextensive *(global) monoselective revision.*

We will focus here on the most orderly version of monoselective revision that was introduced in Chapter 5, namely centrolinear revision. It can be globalized as follows:

Definition 6.3 *A global descriptor revision* $\circ$ *on a set* $\mathbb{X}$ *of belief sets is a* (deterministic) global centrolinear revision *if and only if (1) there is a function a from* $\mathbb{X}$ *to* $\wp(\mathbb{X})$ *(the accessibility function) such that* $X \in a(X) \subseteq \mathbb{X}$ *for all* $X \in \mathbb{X}$*, and (2) for each* $X \in \mathbb{X}$ *there is a relation* $\leqq_X$ *on* $a(X)$ *with* $X \leqq_X Y$ *for all* $Y \in a(X)$*, such that for all descriptors* Ψ*:*

(i) If Ψ *is satisfiable within* $a(X)$*, then* $X \circ \Psi$ *is the unique* $\leqq_X$*-minimal element of* $a(X)$ *that satisfies* Ψ*, and*

[1] This will be the case if $X \circ \Pi_X = X$ holds, cf. Section 4.4.

(ii) otherwise, $X \circ \Psi = X$.

Furthermore, if $a(X) = \mathbb{X}$ *for all* $X \in \mathbb{X}$, *then* $\circ$ *is a* coextensive *(global) centrolinear revision.*

The two classes of operations introduced in Definition 6.3 can both be characterized axiomatically in a way that is very similar to that of the corresponding local operation that was introduced in Section 5.2:

Theorem 6.4 *Let* $\circ$ *be a global descriptor revision on the set* $\mathbb{X}$ *of belief sets. Then the following two conditions are equivalent:*

(A) $\circ$ *is a global centrolinear revision.*

(B) $\circ$ *satisfies the postulates:*

$X \circ \Psi = \mathrm{Cn}(X \circ \Psi)$ *(closure)*

$X \circ \Psi \Vdash \Psi$ *or* $X \circ \Psi = X$. *(relative success)*

If $X \circ \Xi \Vdash \Psi$, *then* $X \circ \Psi \Vdash \Psi$. *(local regularity)*

If $X \circ \Psi \Vdash \Xi$, *then* $X \circ \Psi = X \circ (\Psi \cup \Xi)$. *(cumulativity)*

If $X \Vdash \Psi$, *then* $X \circ \Psi = X$. *(confirmation)*

Theorem 6.5 *Let* $\circ$ *be a global descriptor revision on the set* $\mathbb{X}$ *of belief sets. Then the following two conditions are equivalent:*

(A^e) $\circ$ *is a coextensive (global) centrolinear revision.*

(B^e) $\circ$ *satisfies the postulates*

$X \circ \Psi = \mathrm{Cn}(X \circ \Psi)$ *(closure)*

$X \circ \Psi \Vdash \Psi$ *or* $X \circ \Psi = X$. *(relative success)*

If $X \circ \Xi \Vdash \Psi$, *then* $Y \circ \Psi \Vdash \Psi$. *(global regularity)*

If $X \circ \Psi \Vdash \Xi$, *then* $X \circ \Psi = X \circ (\Psi \cup \Xi)$. *(cumulativity)*

If $X \Vdash \Psi$, *then* $X \circ \Psi = X$. *(confirmation)*

There is only one difference between the postulates used in Theorems 6.4 and 6.5: To characterize global centrolinear revision in general we use the same regularity postulate as in Theorem 5.10 (now calling it "local regularity"). To characterize the coextensive variant of global centrolinear revision we use the stronger postulate of global regularity. As the above two theorems show, the latter postulate has the effect of ensuring that the whole of $\mathbb{X}$ is accessible from each of its elements.

6.2 Distance-Based Global Revision

In centrolinear revision we can think of the relation $\leq_X$ as representing distance from X, so that $Y \leq_X Z$ holds if and only if Y is at least as close to X as Z is. Each

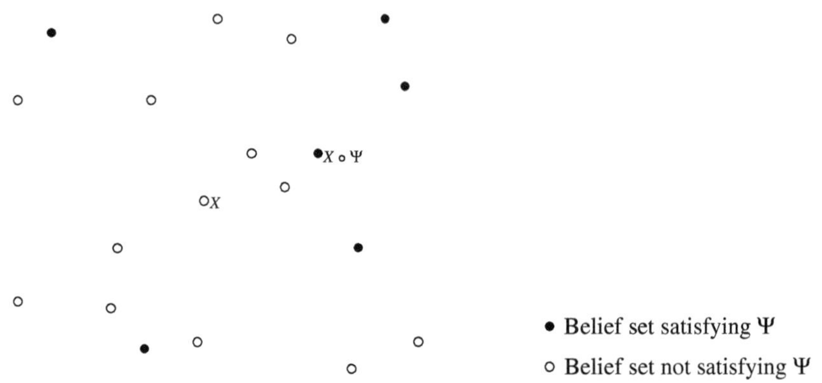

Fig. 6.1 Global distance-based descriptor revision. $X \circ \Psi$ is the belief set closest to X among those that satisfy Ψ. Note that the circles denote belief sets, not possible worlds.

application of $\circ$ to a belief set X and a descriptor Ψ will then take us to the belief set $(X \circ \Psi)$ that is closest to X among those belief sets that are accessible from X and satisfy Ψ, unless there is no such belief set in which case we stay in X. This is illustrated in Fig. 6.1 where the belief sets that are accessible from X are placed at different distances from it. (Belief sets in $\mathbb{X}$ that are inaccessible from X can be represented as having a distance from X that is larger than a certain limit.) If the revision that took us from X to $X \circ \Psi$ is followed by another revision, then that revision will of course be based on distances from $X \circ \Psi$.

By a distance we usually mean a non-negative number that is assigned to a pair of objects and represents how far they are from each other. The mathematical representation is a function that takes us from a pair of two objects X and Y to a non-negative number. It is commonly denoted $\delta(X, Y)$. The distance from X to Y is usually assumed to be the same as that from Y to X; in other words the distance measure δ has the following property:

$$\delta(X, Y) = \delta(Y, X) \qquad \text{(symmetry)}$$

However, in the belief revision context this property is not at all plausible. It can be "a small step" to go from a belief set X to a belief set Y, but a much larger step to go back from Y to X. For instance, I was once in a belief state in which I did not have any opinion on whether 361 is a prime number or not. It was very easy to bring me to a belief state in which I believe it not to be a prime number. (It was sufficient to convince me that $361 = 19 \times 19$.) However, it would be far from easy to bring me back to a state with no belief in the matter. Let X be the belief set representing the first of these two belief states and Y that representing the second. In a distance-based representation, the "distance" from X to Y should be smaller than that from Y to X, i.e. $\delta(X, Y) < \delta(Y, X)$.

There is nothing strange with such asymmetric distances. We also see them in spatial applications. In a city with many one-way streets, the distance by motor vehicle from the railway station to the City Hall may differ from the corresponding distance in the other direction. And as pedestrians we often measure distances in walking time, "the railway station is five minutes from here". Measured in that way, the distance from the bottom to the top of a hill is usually longer than that from the top to the bottom. "Asymmetric" distances, i.e. distances not satisfying the axiom $\delta(X, Y) = \delta(Y, X)$, are often called "pseudodistances".[2] This is also the term that should preferably be used for measures representing how easily a belief set is reached from another belief set.

The following are plausible properties of a pseudo-distance measure that gives rise to a centrolinear revision:

$\delta(X, X) = 0$ (self-closeness)
$\delta(X, Y) \geq 0$ (non-negativity)
If $\delta(X, Y) = \delta(X, Y')$, then $Y = Y'$. (righthand uniqueness)

Self-closeness is a convenient property if we wish to make the operation *centro*linear, i.e. ensure that the confirmation postulate is satisfied.[3] Just like non-negativity it contributes to make the pseudodistance measure intuitively plausible.

Righthand uniqueness does not share the intuitive plausibility of the other two postulates, but it is needed to ensure that all revisions have a unique outcome. In (deterministic) descriptor revision, the selection mechanism (in this case pseudodistance minimization) has to avoid ties.[4] In an indeterministic variant of global revision, righthand uniqueness would not hold. It should be noted that the following property:

If $\delta(X, Y) = \delta(X', Y)$, then $X = X'$. (lefthand uniqueness)

is *not* required to hold even in the deterministic variant of the operation.

The following theorem shows that the three properties of δ listed above are what we need to ensure that the pseudodistance measure gives rise to a coextensive (global) centrolinear revision:

Theorem 6.6 ([132]) *Let $\circ$ be a global descriptor revision on the set $\mathbb{X}$ of belief sets. If it satisfies:*

(C^e) *There is a real-valued pseudodistance measure δ on $\mathbb{X}$ that satisfies self-closeness, non-negativity, and right-hand uniqueness, and such that for all $X \in \mathbb{X}$:*

(i) *if Ψ is satisfiable within $\mathbb{X}$, then $X \circ \Psi \Vdash \Psi$ and $\delta(X, X \circ \Psi) \leq \delta(X, Y)$ whenever $Y \Vdash \Psi$, and*

(ii) *if Ψ is unsatisfiable within $\mathbb{X}$, then $X \circ \Psi = X$,*

---

[2]The term "pseudodistance" has been used since the 19th century for various weakenings of, and alternatives to, standard Euclidean distance. See [155, p. 300] for a useful general definition of pseudodistances in this sense.

[3]We can achieve the same effect with the postulate
$\delta(X, X) < \delta(X, Y)$ if $X \neq Y$,
but the self-closeness property is easier to work with.

[4]Cf. Section 5.3.

then it satisfies:

(A^e) $\circ$ is a coextensive (global) centrolinear revision.

Furthermore, if $\mathbb{X}$ is countable, then (A^e) and (C^e) are equivalent.

For the general (not necessarily coextensive) case of global revision we need to derive an accessibility function from the distance measure. This can be done by assigning to each belief set X in $\mathbb{X}$ a limit in the form of a positive real number $l(X)$ such that Y is accessible from X if and only if its distance from X is smaller than $l(X)$, i.e.:

$Y \in a(X)$ if and only of $\delta(X, Y) < l(X)$.

Since the distances from different elements of $\mathbb{X}$ are not compared, nothing is lost by letting $l(X)$ be the same for all X. The most convenient choice is to let the limit be equal to 1, i.e. to let $Y \in a(X)$ if and only of $\delta(X, Y) < 1$.[5] This means that distances above 1 signalize inaccessibility. This convention provides us with a convenient construction of global centrolinear revision:

Theorem 6.7 *Let $\circ$ be a global descriptor revision on the set $\mathbb{X}$ of belief sets. If it satisfies:*

(C) There is a real-valued pseudodistance measure δ on $\mathbb{X}$ that satisfies self-closeness, non-negativity, and right-hand uniqueness, and such that for all $X \in \mathbb{X}$:

(i) if there is some $Y \in \mathbb{X}$ with $Y \Vdash \Psi$ and $\delta(X, Y) < 1$, then $X \circ \Psi \Vdash \Psi$ and $\delta(X, X \circ \Psi) \leq \delta(X, Y)$ whenever $Y \Vdash \Psi$, and

(ii) otherwise $X \circ \Psi = X$,

then it satisfies:

(A) $\circ$ is a global centrolinear revision.

Furthermore, if $\mathbb{X}$ is countable, then (A) and (C) are equivalent.

As argued in Section 2.5, in a cognitively realistic model all elements of $\mathbb{X}$ have to be finite-based. If they are, then $\mathbb{X}$ is countable, and consequently (C^e) and (A^e) in Theorem 6.6 are equivalent, and so are (C) and (A) in Theorem 6.7. In both cases, pseudodistance measures provide us with alternative characterizations of global centrolinear revision.

[5]Let δ be a measure that satisfies self-closeness and non-negativity. Let l be a limit function with $l(X) > 0$ for all $X \in \mathbb{X}$. Furthermore, let a be an accessibility function such that $Y \in a(X)$ if and only if $\delta(X, Y) < l(X)$. We can then define δ' as the measure such that $\delta'(X, Y) = \delta(X, Y)/l(X)$ for all $X, Y \in \mathbb{X}$. Then $Y \in a(X)$ if and only of $\delta'(X, Y) < 1$.

6.3 Global Blockage Revision

The blockage relations introduced in Section 5.4 can be straightforwardly generalized to a global framework. We can introduce a ternary (three-place) relation $\rightharpoondown$ such that $Y \rightharpoondown_X Z$ denotes that when we revise X, the potential outcome Y blocks the potential outcome Z. Let $X \in \mathbb{Y}$ and $\mathbb{Y} \subseteq \mathbb{X}$. Then a belief set $Z \in \mathbb{Y}$ is $\rightharpoondown_X$-*blocked* within $\mathbb{Y}$ if and only if there is some $Y \in \mathbb{Y}$ such that $Y \rightharpoondown_X Z$; otherwise Z is $\rightharpoondown_X$-*unblocked* within $\mathbb{Y}$. Global blockage revision is defined as follows:

Definition 6.8 *A global descriptor revision* $\circ$ *on a set* $\mathbb{X}$ *of belief sets is a* (deterministic) global blockage revision *if and only if there is a ternary relation* $\rightharpoondown$ *(the blockage relation) such that:*

(i) If Y *is the unique element of* $[\![\Psi]\!]_{a(X)}$ *that is not* $\rightharpoondown_X$*-blocked by any other element of* $[\![\Psi]\!]_{a(X)}$*, then* $X \circ \Psi = Y$*, and*
(ii) otherwise $X \circ \Psi = X$.

The following are some interesting properties of a ternary blockage relation[6]:

If $Y \rightharpoondown_X Z$, then $Z \not\rightharpoondown_X Y$. (asymmetry)
If $Y \rightharpoondown_X Z$ and $Z \rightharpoondown_X V$, then $Y \rightharpoondown_X V$. (transitivity)
If $Y \neq Z$, then $Y \rightharpoondown_X Z$ or $Z \rightharpoondown_X Y$. (weak connectedness)
$Y \not\rightharpoondown_X Y$ (irreflexivity)
$X \not\rightharpoondown_X X$ (self access)
If $X \neq Y$, then $X \rightharpoondown_X Y$. (superiority)
If $X_1 \not\rightharpoondown_{X_2} X_3$, $X_2 \not\rightharpoondown_{X_3} X_4, \ldots, X_{n-2} \not\rightharpoondown_{X_{n-1}} X_n$ and $X_2 = X_{n-1}$, then $X_1 \not\rightharpoondown_{X_2} X_n$. (negative transmission)

Obviously, irreflexivity implies self access. Negative transmission is a somewhat complicated condition that prohibits a type of cycles of non-blocking. It is related with the symmetry requirement for δ ($\delta(X, Y) = \delta(Y, X)$), as can be seen from the fact that negative transmission holds if there is some real-valued measure δ such that $\delta(X, Y) = \delta(Y, X)$ for all X and Y and that $X \not\rightharpoondown_Y Z$ if and only if $\delta(X, Y) \geq \delta(Y, Z)$.[7] In combination with two of the other postulates it implies transitivity:

---

[6] Properties of binary relations are transferred to ternary relations by keeping the middle term constant. Hence, a ternary relation $\rightharpoondown$ satisfies asymmetry if and only if $\rightharpoondown_X$ satisfies asymmetry for all X.

[7] Negative transmission is also closely related with an axiom introduced under the name "loop" in [155, p. 306]. To see the connection with "loop", note that Lehmann et al. refer to distances between sets of objects. In their notation, $X|Y$ is the set of elements y of Y such that $min_{x \in X}\delta(x, y)$ is at least as small as is $min_{x \in X}\delta(x, y')$ for any other element y' of Y. Therefore their formula $(X_1 \mid (X_0 \cup X_2)) \cap X_0 \neq \varnothing$ can be interpreted in the singleton case ($X_0 = \{x_0\}$, $X_1 = \{x_1\}$, and $X_2 = \{x_2\}$) as saying that x_1 is at least as close to x_0 as it is to x_2.

Observation 6.9 ([132]) *Let $\rightharpoondown$ be a ternary relation on $\mathbb{X}$ that satisfies asymmetry, weak connectedness, and negative transmission. Then it satisfies transitivity.*

The following theorems show that global centrolinear revision can be reconstructed as global blockage revision.

Theorem 6.10 *Let $\circ$ be a global descriptor revision on the set $\mathbb{X}$ of belief sets. Then the following two conditions are equivalent:*

(A^e) *$\circ$ is a coextensive (global) centrolinear revision.*

(D^e) *$\circ$ is a coextensive (global) blockage revision based on a blockage relation that satisfies transitivity, weak connectedness, irreflexivity, and superiority.*

Furthermore, the connection indicated above between negative transmission and symmetrical distances is confirmed in the following theorem:

Theorem 6.11 ([132]) *Let $\circ$ be a global descriptor revision on the set $\mathbb{X}$ of belief sets. If it satisfies:*

(C^s) *There is a real-valued distance measure δ over $\mathbb{X}$ that satisfies self-closeness, symmetry, non-negativity, and right-hand uniqueness, and such that for all $X \in \mathbb{X}$:*

(i) *If Ψ is satisfiable within $\mathbb{X}$, then $X \circ \Psi \Vdash \Psi$ and $\delta(X, X \circ \Psi) \leq \delta(X, Y)$ whenever $Y \Vdash \Psi$, and*

(ii) *if Ψ is unsatisfiable within $\mathbb{X}$, then $X \circ \Psi = X$.*

Then it satisfies:

(D^s) *$\circ$ is a coextensive (global) blockage revision based on a blockage relation that satisfies negative transmission, weak connectedness, asymmetry, and superiority.*

Furthermore, if $\mathbb{X}$ is countable, then (C^s) and (D^s) are equivalent.

The equivalent characterizations of (pseudo)distance-based global revision reported in this chapter are summarized in Fig. 6.2. As in other contexts, when several different characterizations of an operation turn out to be equivalent, they can be seen as mutually supportive.

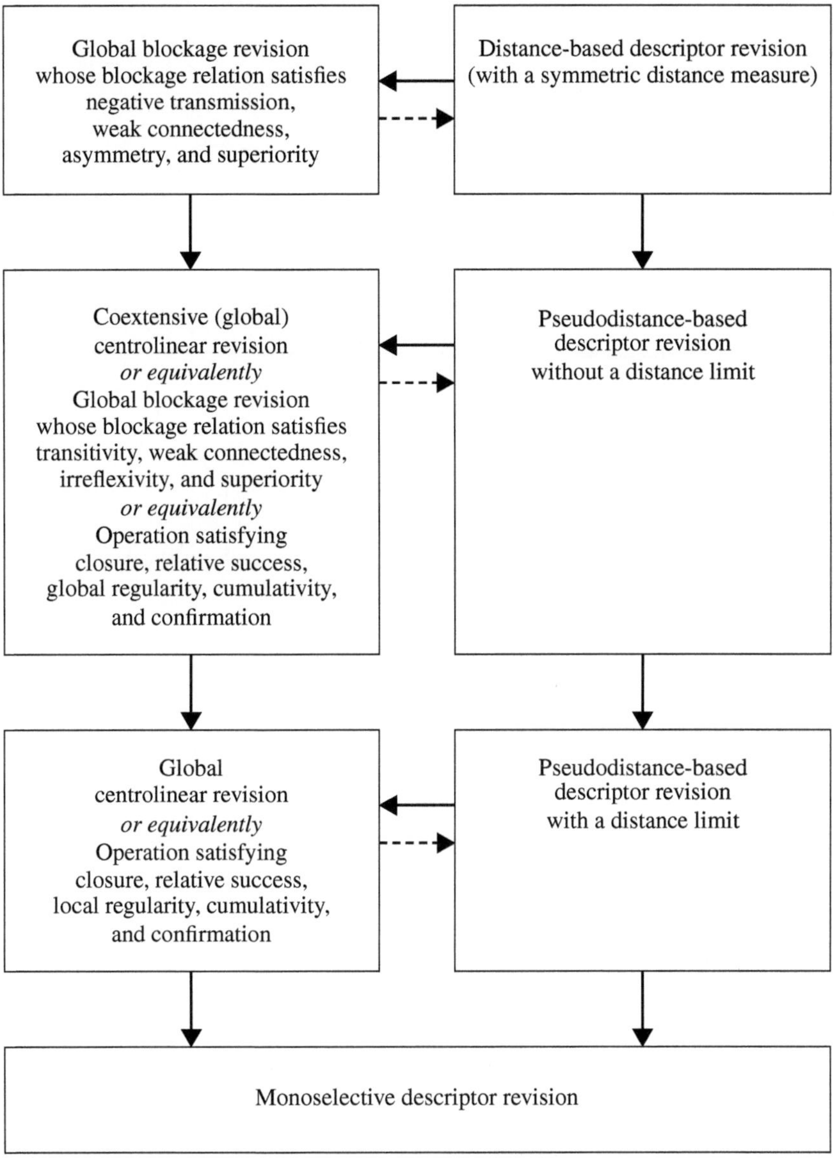

Fig. 6.2 Relations among some major categories of global descriptor revisions. A solid arrow from one category to another indicates that the first is a subset of the second. A dashed line from one category to another indicates that for countable outcome sets, the first category is a subset of the second.

Chapter 7
Dynamic Descriptors

It is common in belief revision theory to distinguish between *static* and *dynamic* information about a belief state. Static information refers to what the agent believes in that belief state. Dynamic information refers to what changes the agent's beliefs are disposed to undergo, in particular in response to various external inputs. For example, I know that my friend Sarah believes herself to be in excellent health. This is a piece of static information about her belief state. I am also convinced that she will give up that belief if her doctor tells her that she has the early signs of a serious autoimmune disease. This is a piece of dynamic information about her belief state. The following are other examples of such dynamic information:

"If the agent receives the input ..., then she will believe that ...."
"If the agent comes to believe that ..., then she will also believe that ...."
"The agent might in the future come to believe that ...."
"Nothing can bring the agent to believe that ...."

We can introduce expressions into the formal language that represent these and other patterns of change. Such expressions will be called *dynamic descriptors* in contrast to the *static descriptors*, formed with $\mathfrak{B}$, that we have been concerned with up until now.

In Section 7.1 we will introduce dynamic and autoepistemic descriptors and discuss whether they should preferably be parts of the object language and included in the belief sets, or only be parts of the metalanguage (which is how we have treated the predicate $\mathfrak{B}$). In Section 7.2 an important class of sentences carrying dynamic information, namely Ramsey test conditionals, are generalized to a class of dynamic descriptors called Ramsey descriptors. In Section 7.3 the logical properties of Ramsey descriptors are determined. Section 7.4 puts focus on standard (sentential) conditionals and introduces two alternatives to the Ramsey test. In Section 7.5 we turn to the logic of non-monotonic inference. It is usually considered to be a fragment of the logic of conditional sentences, but that turns out not to be quite true. Finally, in Section 7.6 we introduce modal expressions into our belief change framework.

S.O. Hansson, *Descriptor Revision*, Trends in Logic 46,
DOI 10.1007/978-3-319-53061-1_7

7.1 Representing Autoepistemic Beliefs

By an *autoepistemic belief* is meant a belief that an agent has about her or his own beliefs. Autoepistemic beliefs can be either static or dynamic. In the example above, Sarah is aware that she considers herself to be in excellent health. This is a belief that she has about her own beliefs at the same point in time, in other words a *static autoepistemic belief*. If I were inconsiderate enough to ask her whether she would retain that belief if her doctor told her she has the early signs of a serious autoimmune disease, then her answer would be in the negative. That answer would report a *dynamic autoepistemic belief*, a belief about how she would change her own beliefs in response to new information.

Should (static and dynamic) autoepistemic beliefs be included in belief sets, or does their special nature require that they be kept out? As indicated in Section 3.6, this has been a difficult and sometimes controversial issue in the belief change literature. In descriptor revision, autoepistemic beliefs already have a representation in the form of descriptors, so all we have to do is to move these descriptors from the metalanguage to the object language and extend the belief sets to contain some of them. Let $\mathbb{X}$ be an outcome set. We can form an augmented version $\overline{\mathbb{X}}$ of it, such that $\overline{\mathbb{X}} = \{\overline{X} \mid X \in \mathbb{X}\}$ where each $\overline{X}$ contains exactly the sentences that are introduced with (sequential) use of the following rules:

1. If $\alpha \in \text{Cn}(X)$, then $\alpha \in \overline{X}$
2. If $\alpha \in \overline{X}$ then $\mathfrak{B}\alpha \in \overline{X}$
3. If $\alpha \notin \overline{X}$ then $\neg\mathfrak{B}\alpha \in \overline{X}$

This means that if $p \in X$ and X is consistent, then $\mathfrak{B}p$, $\mathfrak{B}\mathfrak{B}p$, $\mathfrak{B}\neg\mathfrak{B}\neg\mathfrak{B}p$ and an infinite number of other such composite autoepistemic sentences are all elements of X.

Next, let us turn to dynamic descriptors. There is a well-known recipe for including conditional sentences into (extended) belief sets, namely the Ramsey test that was introduced in Section 3.6. It prescribes that $p \rightarrowtail q$ holds at K if and only if $q \in K * p$. Just like the above recipe for $\mathfrak{B}$, this one can be used repeatedly:

1. If $\alpha \in \text{Cn}(X)$ then $\alpha \in \overline{X}$
2. If $\beta \in \overline{X} * \alpha$ then $\alpha \rightarrowtail \beta \in \overline{X}$

In this way, nested conditionals of unlimited length can be included in extended belief sets, for instance[1]:

$p_1 \rightarrowtail (p_2 \rightarrowtail (p_3 \rightarrowtail p_4)) \in \overline{X}$
if and only if $p_2 \rightarrowtail (p_3 \rightarrowtail p_4) \in \overline{X} * p_1$
if and only if $p_3 \rightarrowtail p_4 \in \overline{X} * p_1 * p_2$
if and only if $p_4 \in \overline{X} * p_1 * p_2 * p_3$

[1] At this point we can set aside the problems with the Ramsey test referred to in Section 3.6. It will be shown in Section 7.2 how these problems can be overcome.

However, from the logical feasibility of these constructions it does not follow that they are philosophically plausible.[2] Some authors have claimed that it would be a conceptual mistake to include autoepistemic beliefs in belief sets. For instance, Alvaro del Val has argued that belief sets should be kept free from dynamic autoepistemic information since there is a "need to separate the specification of the agent's beliefs from the specification of the agent's revision policy, which are fully orthogonal, independent issues" [39, p. 223].[3] This view can be called the *thesis of autoepistemic ignorance*; it claims that the formal representation of actual (static) beliefs should contain no information pertaining to how the agent's beliefs will change in response to new inputs.

From a philosophical point of view this is not a plausible standpoint, for the simple reason that an agent's current beliefs and the ways in which she tends to change these beliefs are far from independent or "orthogonal" issues. For instance, there are strong connections between the justificatory structure of an agent's beliefs and how they will change in response to various inputs. If p represents my only justification for believing that q, and I am aware of this justificatory relationship, then we should expect q to be lost if p is given up.[4] Furthermore, and perhaps more importantly, some forms of modal and conditional beliefs seem to be strongly connected with how one would change one's (non-modal and non-conditional) beliefs upon receipt of certain inputs. For a simple example, suppose that you believe the bog bilberry to be acutely poisonous. If you revise your belief set to accommodate the information that I have just picked and eaten a considerable amount of bog bilberries you will, presumably, also believe that I will soon be sick. Other such examples are easily found; it may in fact be more difficult to find examples of "purely static" beliefs that have no impact on how our beliefs will be modified in response to any type of new information. Our static and dynamic beliefs are closely interwoven and in practice often inseparable. A belief set that adequately represents the beliefs held at a particular moment will of necessity contain an abundance of information pertaining to how it will change in response to different inputs. Therefore the thesis of autoepistemic ignorance is untenable.

According to the diametrically opposite standpoint, the agent is fully aware of her own belief state. This can be called the *thesis of autoepistemic omniscience*.[5] For static autoepistemic beliefs it means that the agent is assumed to have perfectly accurate beliefs about what she believes and does not believe at present (even with respect to complex sentences such as $\mathfrak{B}\neg\mathfrak{B}\neg\mathfrak{B}p$). This is not plausible if we take the belief set to represent her actual beliefs, but it may be plausible if we follow Isaac Levi in taking it to represent the beliefs that she is committed to hold. With the former

---

[2]On what it means to know one's own beliefs, see [231]. On logics employing an autoepistemic belief operation, see [158, 190, 241, 259].

[3]Isaac Levi has expressed a similar view with respect to the dynamic information contained in conditional sentences. See [161] and [163, pp. 49–50]. See also [61, 65, 102].

[4]The effects of justificatory relationships on patterns of belief change has been investigated with models employing belief bases, see for instance [91, 104].

[5]This term was used with essentially the same meaning by Hans Rott [212].

interpretation it would be desirable for static autoepistemic beliefs to be includible in belief sets, but they should not be automatically included whenever they are true. Then an agent who believes in p may or may not believe in $\mathfrak{B}\neg\mathfrak{B}\neg\mathfrak{B}p$.

For dynamic autoepistemic beliefs, the thesis of autoepistemic omniscience is much more demanding. It requires that the agent has perfectly accurate beliefs not only about her present belief state but also about how her beliefs will change in response to any (arbitrarily long) series of inputs that she may be exposed to. Needless to say, this is an utterly unrealistic feature of a formal model.[6] Therefore, a realistic treatment of dynamic autoepistemic beliefs requires that we find a middle way between autoepistemic ignorance and autoepistemic omniscience. We can call this a *thesis of autoepistemic incompleteness*: agents should be modelled as having belief sets that answer some but not all questions about what they will believe after various (series of) operations of belief change.

We can expect the coverage of dynamic autoepistemic beliefs to be highest for single-step changes that do not require the retraction of highly entrenched beliefs. I have fairly well-developed (and probably accurate) beliefs about how my belief set will change if I receive some unsurprising piece of information such as that Real Madrid won their latest match against Granada CF with $5-1$. I am much less certain about how my belief set will change if I learn that Granada CF won over Real Madrid with $13-0$. And if you provide me with a longish list of statements, each of which contradicts a strong belief of mine, then I will be at a loss for what my belief set would look like after I had successively revised by all of them. In a distance-based model such as that developed in Section 6.2 this can be approximated by the assumption that the agent has true autoepistemic beliefs about the belief changes that only take her to belief sets within a certain (small) distance from the present belief set, but not in general for belief changes that take her further away from the present belief state.

Most treatments of autoepistemic beliefs have assumed that these beliefs are all truthful. However, there is no reason to take that for granted. We are no more infallible in these issues than in any others [119]. Therefore, false autoepistemic beliefs should be includible in belief sets, and their inclusion should not (or at least not always) make the belief set inconsistent.

7.2 Ramsey Descriptors

There are many varieties of conditional sentences ("if... then..."-sentences), and several ways to classify them in terms of their meanings have been put forward [6, 92]. For our purposes, a simple typology proposed by Lindström and Rabinowicz is particularly useful. They divided conditionals into two groups: ontic and epistemic (doxastic) conditionals. The crucial difference is that "ontic conditionals concern

[6]The same type of cognitive unrealism is inherent in standard probability theory. If an agent with a probability function $\mathfrak{p}$ learns that q, then (provided that $\mathfrak{p}(q) \neq 0$) her new probability function $\mathfrak{p}'$ is derivable from $\mathfrak{p}$ through the simple formula $\mathfrak{p}'(x) = \mathfrak{p}(x \mid q) = \mathfrak{p}(x\&q)/\mathfrak{p}(q)$.

hypothetical modifications of the *world*, but epistemic conditionals have to do with hypothetical modifications of our *beliefs* about the world" [170, p. 225]. To exemplify the difference, suppose that one late night you have just arrived in a small town that has only two snackbars.[7] You meet a man eating a hamburger. This makes you believe in the following conditional sentence:

(1) If snackbar A is closed, then snackbar B is open.

Soon afterwards, you see that bar A is in fact open. You would then probably not assent to the following conditional sentence:

(2) If snackbar A were closed, then snackbar B would be open.

(1) is most naturally interpreted as an epistemic conditional, i.e. it signalizes that belief in the antecedent would make you believe in the consequent. In contrast, (2) is an ontic conditional, expressing patterns in the world rather than in your beliefs about it. Grammatically, the antecedent of (1) is expressed with a verb in the indicative mood ("is") and that of (2) with a verb in the subjunctive mood ("were"). The grammatical difference reflects common usage in the English language: the antecedents of epistemic conditionals are typically expressed with an indicative verb form and those of ontic conditionals with a subjunctive verb form. However, this connection between meaning and mood only holds in some languages. The difference between ontic and epistemic conditionals is also present in languages that do not express it by shifting between indicative and subjunctive verb forms. Furthermore, as was pointed out by Michael R. Ayers and more recently by Hans Rott, the correlation between meaning and mood is far from perfect in English [9, 216]. Consider the following examples:

(3) If everyone in this room is legally married to someone else in the room, then there is an even number of persons in the room.
(4) If everyone in this room were legally married to someone else in the room, then there would have been an even number of persons in the room.

(3) and (4) differ in grammatical form but they do not differ in terms of the ontic–epistemic distinction. Instead, the indicative form in (3) ("is") imparts the impression that the antecedent is reasonably plausible, whereas the subjunctive form in (4) signals an assumption that it does not hold.

Due to its higher philosophical relevance, the distinction between epistemic and ontic conditionals should preferably replace that between conditionals expressed in the subjunctive respectively indicative mood in the English language.

The Ramsey test has its origin in a famous footnote by Frank Ramsey:

> If two people are arguing 'If p will q?' and are both in doubt as to p, they are adding p hypothetically to their stock of knowledge and arguing on that basis about q. [210, p. 247]

This rather sketchy proposal was developed by Robert Stalnaker into a general principle that is now commonly called the Ramsey test [240, pp. 101–105].[8] The test

[7] This example is an improvement by Hans Rott [216] of an example first published in [81].
[8] On the Ramsey test, see also Section 3.6 and [6, 85].

is intended for epistemic conditionals, and it can be expressed as an equivalence between on the one hand the epistemic agent's acceptance of the conditional "If p, then q" and on the other hand her propensity to believe in q after revising by p. In formal notation:

$p \rightarrowtail q$ holds at K *if and only if* $q \in K * p$.

In the framework of descriptor revision the right-hand part of this formula can be written $K \circ \mathfrak{B}p \Vdash \mathfrak{B}q$. This opens up for an obvious generalization: we can replace $\mathfrak{B}p$ and $\mathfrak{B}q$ by more general (i.e. not necessarily atomic) descriptors to serve as antecedent respectively consequent. This results in *Ramsey descriptors*, a generalization of (sentential) Ramsey test conditionals that will be denoted $\Psi \Rightarrow \Xi$ [132]. This formula means that if the belief set is revised by Ψ, then the outcome will satisfy Ξ. The Ramsey test can be straightforwardly generalized as follows:

$\Psi \Rightarrow \Xi$ holds at K *if and only if* $K \circ \Psi \Vdash \Xi$.

Standard (sentential) Ramsey test conditionals are of course a special case of Ramsey descriptors, obtainable by defining $p \rightarrowtail q$ as $\mathfrak{B}p \Rightarrow \mathfrak{B}q$. But more interestingly, other forms of conditional belief patterns can also be expressed, such as the following:

> "If he gives up his belief that his wife is faithful to him, then he will also lose his belief that she loves him." ($\neg\mathfrak{B}p \Rightarrow \neg\mathfrak{B}q$)
>
> "If she gives up her belief that the first chapter of Genesis is literally true, then she will still believe that God exists." ($\neg\mathfrak{B}p \Rightarrow \mathfrak{B}q$)
>
> "If she makes up her mind on whether this painting is a genuine Picasso or not, then she will come to believe that it is genuine." ($\mathfrak{B}p \vee \mathfrak{B}\neg p \Rightarrow \mathfrak{B}p$)[9]

Dorothy Edgington has pointed out that the conventional form of conditional sentences (represented here as $p \rightarrowtail q$) is insufficient to cover the wide variety of conditionalities that are expressible in ordinary language:

> Any kind of propositional attitude can occur within the scope of a supposition... and hence... a theory of conditionals should be applicable to more than conditional statements. [45, p. 177]

Ramsey descriptors can hopefully facilitate investigations of the wide range of conditional expressions, in addition to standard "if p then q" conditionals, that are available in ordinary language.

7.3 The Logic of Ramsey Descriptors

In the above explication of $\Rightarrow$, $\Psi \Rightarrow \Xi$ was said to hold at K if and only if $K \circ \Psi \Vdash \Xi$. Importantly, whether $\Psi \Rightarrow \Xi$ holds at a belief set K is not a property of that belief

---

[9]It is an interesting issue whether a rational agent can have the autoepistemic belief $\mathfrak{B}p \vee \mathfrak{B}\neg p \Rightarrow \mathfrak{B}p$ without also having the (static) belief p. This relates to the discussion in Section 7.1 on the connection between static and dynamics beliefs.

set alone. There may be two different operations $\circ$ and $\circ'$ such that $K \circ \Psi \Vdash \Xi$ but $K \circ' \Psi \nVdash \Xi$. Then $\Psi \Rightarrow \Xi$ holds at K according to $\circ$ but not according to $\circ'$. Consequently, sentences formed with $\Rightarrow$ have to be evaluated both *at a specific belief set* and *in relation to a specific operation*. In this they differ from sentences with $\mathfrak{B}$ that represent static beliefs. For any static belief p, whether $\mathfrak{B}p$ holds at K is not influenced by the operation of revision we use. It is a property of K alone. This difference between $\Rightarrow$ and $\mathfrak{B}$ is, of course, the defining difference between static and dynamic descriptors.

It follows that whereas $K \Vdash \Psi$ is an adequate representation of what it means for a static descriptor Ψ to be "held" at K, it would be misleading to substitute $\Psi \Rightarrow \Xi$ for Ψ in that formula to express that $\Psi \Rightarrow \Xi$ holds at K. For dynamic descriptors we need to mention the operation of revision. The symbol $\twoheadleftarrow$ will be used to denote that a dynamic descriptor holds. The truth condition associated with $\twoheadleftarrow$ will have to refer both to the belief set and to the operation of revision:

Definition 7.1 *Let $\circ$ be a descriptor revision on K and let $\mathbb{X}$ be its outcome set. The* Ramsey descriptor *associated with $\circ$ is the relation $\Rightarrow$ on descriptors such that for all descriptors Ψ and Ξ:*

$\langle K, \circ \rangle \twoheadleftarrow \Psi \Rightarrow \Xi$ *if and only if* $K \circ \Psi \Vdash \Xi$.

Obviously, it does not follow from $\langle K, \circ \rangle \twoheadleftarrow \Psi \Rightarrow \Xi$ that the agent is aware that $\Psi \Rightarrow \Xi$ holds at K (or, more precisely, at $\langle K, \circ \rangle$). Using Definition 7.1 we will develop a logic of Ramsey descriptors that does *not* assume that sentences containing $\Rightarrow$ are believed by the agent or included in belief sets. Whether they should be so is a separate issue that we will return to at the end of this section.

If $\circ$ satisfies confirmation then Definition 7.1 has the following special case:

Observation 7.2 *Let $\circ$ be a descriptor revision on K that satisfies confirmation ($K \circ \Psi = K$ whenever $K \Vdash \Psi$), and let $\Rightarrow$ be the Ramsey descriptor associated with $\circ$. Then:*

$\langle K, \circ \rangle \twoheadleftarrow \mathfrak{B}\top \Rightarrow \Xi$ *if and only if* $K \Vdash \Xi$.

In studies of Ramsey descriptors it is useful to assume that the underlying operation of revision satisfies confirmation. This makes it possible to regain the belief set from the set of satisfied Ramsey descriptors. It follows from Observation 7.2 that $p \in K$ holds if and only if $\langle K, \circ \rangle \twoheadleftarrow \mathfrak{B}\top \Rightarrow \mathfrak{B}p$.

Definition 7.1 also has the following implications:

Observation 7.3 *Let $\circ$ be a centrolinear revision on K and let $\mathbb{X}$ be its outcome set. Let $\Rightarrow$ be the Ramsey descriptor and $\succeq$ the relation of epistemic proximity that are associated with $\circ$.*

If $\Psi \cup \Xi$ is satisfiable within $\mathbb{X}$, then:

(1) $\langle K, \circ \rangle \twoheadleftarrow \Psi \Rightarrow \Xi$ *if and only if* $K \circ \Psi = K \circ (\Psi \cup \Xi)$.

(2) $\langle K, \circ \rangle \twoheadleftarrow \Psi \Rightarrow \Xi$ *if and only if* $\Psi \cup \Xi \succeq \Psi$.

If Ψ *is satisfiable within* $\mathbb{X}$, *then:*

(3) $\Psi \succeq \Xi$ *if and only if* $\langle K, \circ\rangle \twoheadleftarrow (\Psi \veebar \Xi \Rightarrow \Psi)$.

Based on Definition 7.1 one would expect $\circ$ and $\Rightarrow$ to be interdefinable so that if we have one of them, then the other can be derived from it. Such a one-to-one relationship between two areas of logic, belief revision and conditional logic, is of course highly interesting.[10] However, there is a limiting case that creates problems for the interdefinability, namely the case of unsatisfiable inputs respectively antecedents. If Ψ is unsatisfiable, then $\Psi \Rightarrow \Xi$ cannot be evaluated with reference to belief sets in which Ψ is satisfied. Belief revision and the logic of conditionals tend to treat this case in different ways. In belief revision, when the input cannot be satisfied, the standard solution is to let the outcome be equal to the original belief set.[11] In conditional logic, the tradition is instead to follow the *ex falso quodlibet* principle according to which a false sentence implies all other sentences.[12] To exemplify this practice, let p be a sentence that is not included in any element of $\mathbb{X}$. Then $\mathfrak{B}p$ is not satisfied at any element of $\mathbb{X}$. It follows from Definition 7.1 that $\mathfrak{B}p \Rightarrow \mathfrak{B}p$ does not hold at K, and the same applies to its translation into sentential conditional logic, $p \rightarrowtail p$. This is contrary to a well-established tradition in conditional logic, where $p \rightarrowtail p$ is almost universally assumed to hold. (See for instance [217, pp. 33, 112–114] and [257, p. 294].)

In order to avoid the convention-bound translation problems in this rather uninteresting limiting case, it is preferable to relate the two frameworks to each other only in the main case. As the following theorem shows, a one-to-one correspondence can then be obtained with plausible postulates for the Ramsey descriptors:

Theorem 7.4 ([132], modified) *Let* $\Rightarrow$ *be a Ramsey descriptor. Then the following three conditions are equivalent:*

(I) There is a coextensive centrolinear *revision* $\circ$ *such that the restriction of* $\circ$ *to inputs that are satisfiable within its outcome set has an associated Ramsey descriptor that coincides with* $\Rightarrow$.

(II) There is a coextensive linear *revision* $\circ$ *such that the restriction of* $\circ$ *to inputs that are satisfiable within its outcome set has an associated Ramsey descriptor that coincides with* $\Rightarrow$.

(III) $\Rightarrow$ *satisfies:*

[10] In the AGM framework such a connection was introduced in [184]. See also [13, 176, 214, 217, 228].

[11] This is the solution commonly chosen for contraction by a tautology [1], for shielded contraction in which some non-tautologous sentences are not contractible [51], and for non-prioritized revision in which some sentences cannot be incorporated into the belief set [137, 179]. In our presentation of descriptor revision, we have followed this tradition. (See for instance Definitions 5.2 and 5.9.)

[12] The *ex falso quodlibet* principle is seldom mentioned in presentations of conditional logic, but it follows from the common principle that if p logically implies q, then $p \rightarrowtail q$ holds in all belief states. See e.g. [26].

If $\Psi \dashv\Vdash \Psi'$, *then* $\langle K, \circ\rangle \twoheadleftarrow\!\!\!-\Psi \Rightarrow \Xi$ *if and only if* $\langle K, \circ\rangle \twoheadleftarrow\!\!\!-\Psi' \Rightarrow \Xi$. *(left logical equivalence)*

For all Ψ *there is some belief set* $Y \subseteq \mathcal{L}$ *such that for all* Ξ: $\langle K, \circ\rangle \twoheadleftarrow\!\!\!-\Psi \Rightarrow \Xi$ *if and only if* $Y \Vdash \Xi$. *(unitarity*[13]*)*

$\langle K, \circ\rangle \twoheadleftarrow\!\!\!-\Psi \Rightarrow \Psi$ *(reflexivity), and*

If $\langle K, \circ\rangle \twoheadleftarrow\!\!\!-\Psi \Rightarrow \Xi$, *then* $\langle K, \circ\rangle \twoheadleftarrow\!\!\!-\Psi \Rightarrow \Phi$ *if and only if* $\langle K, \circ\rangle \twoheadleftarrow\!\!\!-\Psi \cup \Xi \Rightarrow \Phi$. *(cumulativity)*

The postulates used in the theorem are all generalizations of properties commonly referred to in the logic of (sentential) conditionals.[14]

Reflexivity and left logical equivalence have been given the same names as properties of sentential conditionals that they generalize, namely:

$p \rightarrowtail p$ (reflexivity)
If $\vdash p \leftrightarrow p'$, then $p \rightarrowtail q$ if and only if $p' \rightarrowtail q$. (left logical equivalence)

Cumulativity also has a direct analogue in sentential conditional logic:

If $p \rightarrowtail q$, then $p \rightarrowtail r$ if and only if $p\&q \rightarrowtail r$. (cumulativity)

In sentential conditional logic, cumulativity is usually split into the following two conditions ([177, p. 43], [12]):

If $p \rightarrowtail q$ and $p\&q \rightarrowtail r$, then $p \rightarrowtail r$. (cut)
If $p \rightarrowtail q$ and $p \rightarrowtail r$, then $p\&q \rightarrowtail r$. (cumulative monotony)[15]

The connection between unitarity and well-known properties of sentential connectives is somewhat less obvious but can easily be brought to light. The following is a sentential variant of unitarity:

For all p there is some $Y \subseteq \mathcal{L}$ such that for all q: $p \rightarrowtail q$ if and only if $Y \vdash q$.

It is equivalent with the following property:

$\{q \mid p \rightarrowtail q\} = \mathrm{Cn}(\{q \mid p \rightarrowtail q\})$ (left absorption, left logical absorption)[16]

In a compact logic, left absorption it is equivalent with the combination of the following two, well-known properties of conditionals:

If $p \rightarrowtail q$ and $q \vdash r$, then $p \rightarrowtail r$. (right weakening)

---

[13]Technically, in the logic of descriptors a belief set X is interchangeable with a descriptor Π_X that is satisfied by X but not by any other belief set. (See Definition 4.14.) Therefore unitarity can equivalently be expressed by a requirement that the descriptor $\bigcup\{\Xi \mid \Psi \Rightarrow \Xi\}$ is satisfied by exactly one belief set; this is also why the name "unitarity" was chosen for this postulate.

[14]See [177] or [217, pp. 111–119] for useful overviews of properties of sentential conditionals.

[15]This postulate is called "cautious monotonicity" in [153, p. 178].

[16]See [177, p. 45] and [57, pp. 164–165].

If $p \rightarrowtail q_1$ and $p \rightarrowtail q_2$, then $p \rightarrowtail q_1 \& q_2$. *(And)*

Thus, in summary, the logical properties of Ramsey descriptors ($\Rrightarrow$) used in Theorem 7.4 generalize the following properties of sentential conditionals ($\rightarrowtail$): reflexivity, left logical equivalence, cut, cumulative monotony, right weakening, and *And*. (*And* can be omitted since it follows from the other postulates.[17]) These are exactly the logical principles for sentential conditionals that characterize the system C (cumulative reasoning) proposed by Kraus, Lehmann, and Magidor [153, p. 176].[18] However, the parallel between the two analogous systems of postulates is not complete. In particular, the following result has no counterpart for sentential conditionals:

Observation 7.5 ([132]) *If a Ramsey descriptor* $\Rrightarrow$ *satisfies left logical equivalence, unitarity, reflexivity, and cumulativity, then it satisfies:*

If $\langle K, \circ \rangle \twoheadleftarrow \Psi_1 \Rrightarrow \Xi$ *and* $\langle K, \circ \rangle \twoheadleftarrow \Psi_2 \Rrightarrow \Xi$, *then* $\langle K, \circ \rangle \twoheadleftarrow \Psi_1 \veebar \Psi_2 \Rrightarrow \Xi$.
(Or)

This is analogous to a well-known postulate for sentential conditionals:

If $p \rightarrowtail r$ and $q \rightarrowtail r$, then $p \vee q \rightarrowtail r$. *(Or)*

However, the sentential *Or* does not hold in system C. To the contrary, its addition to C gives rise to the stronger system P (preferential reasoning) [153, p. 190]. This confirms again that the logic of descriptors is distinctly different from that of sentences.

We have constructed Ramsey descriptors as metalinguistic objects. They are not included in the object language from which belief sets are formed, and therefore they are not elements of the belief sets at which they are supported. The reason for this was given in Section 7.1: Including all Ramsey descriptors as elements of the belief sets at which they hold (relative to $\circ$) is tantamount to assuming that the agent is completely and correctly informed about how her beliefs will develop in response to any chain of inputs that she may receive in the future. However, it should be mentioned that in spite of the philosophical counterarguments to such an assumption, it can easily be implemented in the formal system. The same construction can be used that was mentioned in Section 7.1 for representations of autoepistemic beliefs. We can extend each belief set $X \in \mathbb{X}$ into a set $\overset{\Rrightarrow}{X}$ that contains, in addition to X, all Ramsey descriptors $\Psi \Rrightarrow \Xi$ such that $X \circ \Psi \Vdash \Xi$. We can also generalize $\circ$ to take such Ramsey descriptors as inputs. For instance, if $\overset{\Rrightarrow}{Y}$ is the $\overset{\Rrightarrow}{X}$-closest extended belief set containing $\Psi \Rrightarrow \Xi$, then $\overset{\Rrightarrow}{X} \circ (\Psi \Rrightarrow \Xi) = \overset{\Rrightarrow}{Y}$. This construction

[17] Let $p \rightarrowtail q_1$ and $p \rightarrowtail q_2$. Cumulative monotony yields $p\&q_1 \rightarrowtail q_2$. Reflexivity yields $p\&q_1\&q_2 \rightarrowtail p\&q_1\&q_2$, and with right weakening we obtain $p\&q_1\&q_2 \rightarrowtail q_1\&q_2$. Applying cut to $p\&q_1 \rightarrowtail q_2$ and $p\&q_1\&q_2 \rightarrowtail q_1\&q_2$ we obtain $p\&q_1 \rightarrowtail q_1\&q_2$. Finally,we apply cut to $p \rightarrowtail q_1$ and $p\&q_1 \rightarrowtail q_1\&q_2$, and obtain $p \rightarrowtail q_1\&q_2$ [153, p. 179].

[18] This was pointed out to me by John Cantwell.

has the formal advantage of allowing conditional sentences satisfying the Ramsey test into belief sets, without being affected by the Gärdenfors impossibility theorem that prevents the inclusion of sentential Ramsey test conditionals into the belief sets of AGM [68].[19] But as already indicated, it would be philosophically much more interesting to investigate constructions in which the extended belief set contains a smaller collection of conditional sentences that has at least some likeness to the set of autoepistemic beliefs that an agent can actually hold.

7.4 Alternative Approaches to Conditionals

Although the Ramsey test provides a highly useful account of conditional sentences, at least for some purposes it should only be seen as a first approximation. Even if we restrict our attention to epistemic conditionals, natural language contains several types of such sentences, and we should not expect a single formal account to cover them all. One important source of this complexity is that we are often reluctant to either approve or disapprove of a conditional sentence. Even if we have no difficulty in understanding the two sentences p and q, we may have great difficulties in taking a stand on the conditional sentence "If p then q". In this section I will sketch out two formal approaches to such reluctance and to the mechanisms by which it is sometimes overcome. In the first of these approaches it is overcome with additional deliberation and in the second with additional information.

For the first approach, consider the following example:

> THE NEW COACH: If we replace Susan by Dorothy as a central defender, will the team as a whole play better?
> THE RECENTLY RETIRED COACH: That is very difficult to say, I do not really know.
> THE NEW COACH: Yes, I know this is difficult, but I really need your opinion. Can you think it over?
> THE RECENTLY RETIRED COACH (after thinking for a while): Well, yes. The team as a whole will play better if you replace Susan by Dorothy. [129]

Let p denote that Susan is replaced by Dorothy and q that the team as a whole improves its play. One way to interpret this dialogue is that for her first answer, the retired coach hypothetically revised her beliefs by $\mathfrak{B}p$. She arrived at a belief set $K \circ \mathfrak{B}p$ that satisfied neither $\mathfrak{B}q$ nor $\mathfrak{B}\neg q$. Then she reconsidered the issue, but now aiming to arrive at a belief set satisfying either $\mathfrak{B}q$ or $\mathfrak{B}\neg q$. We can express this

[19] The Gärdenfors theorem is based on the combination of two properties of a belief revision framework: (1) If a sentence p is logically compatible with a belief set K, i.e. $\neg p \notin K$, then the revision $K * p$ does not remove anything from K, i.e. $K \subseteq K * p$. (2) All Ramsey test conditionals are included in the belief sets at which they hold, i.e. $p \rightarrowtail q \in K$ if and only if $q \in K * p$. The combination of (1) and (2) implies that if $q \in K * r$ and $\neg p \notin K$, then $r \rightarrowtail q \in K \subseteq K * p$, thus $q \in K * p * r$. Counterexamples to this pattern are easily found; see for instance the taxi driver example in Section 3.5. Gärdenfors showed that the combination of (1) and (2) is incompatible with a set of plausible formal properties of a belief revision framework [68]. Descriptor revision avoids these problems since it does not satisfy (1). For arguments against (1), see Section 3.5 and [85, 212].

as an extended success condition. She was no longer searching for the most credible belief set satisfying $\mathfrak{B}p$ but for the most credible belief set satisfying both $\mathfrak{B}p$ and $\mathfrak{B}q \vee \mathfrak{B}\neg q$. We can generalize this pattern by defining the following conditional:

$p \twoheadrightarrow q$ is an abbreviation of $\{\mathfrak{B}p, \mathfrak{B}q \vee \mathfrak{B}\neg q\} \Rightarrow \mathfrak{B}q$.

We can call this an *elicited conditional* [129]. If $\Rightarrow$ is based on centrolinear revision, then $\twoheadrightarrow$ will be weaker than the standard sentential conditional $\rightarrowtail$, i.e. it holds that

If $p \rightarrowtail q$, then $p \twoheadrightarrow q$,

but the reverse implication does not hold.

There are interesting logical differences between the standard sentential Ramsey conditional $\rightarrowtail$ and the elicited conditional $\twoheadrightarrow$. Within the framework of centrolinear revision, the former satisfies the following postulate:

If $p \rightarrowtail q_1$ and $p \rightarrowtail q_2$, then $p \rightarrowtail (q_1 \& q_2)$. *(And)*

However, the corresponding principle for the elicited conditional,

If $p \twoheadrightarrow q_1$ and $p \twoheadrightarrow q_2$, then $p \twoheadrightarrow (q_1 \& q_2)$.

does not hold in general.

To introduce the second approach we can use the following example that was put forward by Lewis [167, p. 1] to illustrate the well-known observation that many conditionals are context dependent[20]:

If kangaroos had no tails, they would topple over.

In a discussion on the principles of mechanics we would have good reasons to assent to this statement. However, in a discussion on evolutionary biology we would probably say to the contrary that if kangaroos had no tails, then their bodies would have had a different weight distribution, so that they would not topple over.

The underlying reason for this variability in interpretation seems to be that the antecedent:

Kangaroos have no tails. (p)

is so indeterminate that epistemic agents cannot be expected to know how to revise by it.[21] In consequence, (sentential) revision by p will be unsuccessful. Provided

[20] The context dependence of conditionals has been referred to as the shiftability problem [79]. Other early discussions can be found in [166, p. 465] and [202, pp.134–135]. Several other examples have been given in the literature: "If frogs were mammals, they would have mammae." – "If frogs were mammals, they would be the only ones not to have mammae." [256]. "If I had been John Keats, I should not have been able to write the *Ode to a Nightingale*." – "If I had been John Keats, then I should have been the man who wrote the *Ode to a Nightingale*." [79, pp. 5–6].

[21] This is also an illustration of the difficulties involved in representing an actual or hypothetical input (element of $\mathbb{I}$) by a single sentence. (Cf. Section 4.1.) Serious considerations of what would happen if kangaroos had no tails do not come out of the blue, but would typically take place in some context that makes it clear whether physical or biological principles are under scrutiny.

that relative success is satisfied, we will then have $K * p = K$ and consequently $p \notin K * p$. Such epistemic behaviour is in conflict with the AGM theory due to its exceptionless success postulate ($p \in K * p$ for all p), but as we saw in Section 3.2 that postulate does not express a realistic general feature of belief revision. There are input sentences that a rational agent may well reject, either because they are too unrealistic or because they are too vague. Our sentence p belongs to the latter category.[22]

The Ramsey test for conditionals requires revision to be successful. It does not make sense to evaluate $p \rightarrowtail q$ based on whether q holds in $K * p$ unless the latter set actually contains p. But now consider the two statements:

Kangaroos have suddenly lost their tails. (s)
Kangaroos have lost their tails in an evolutionary process. (e)

If a stranger at a party suddenly asks me: "Would kangaroos topple over if they had no tails?", then I will not be able to answer the question since I do not know how to revise by the sentence p. However, if I am asked the same question in a physics class, then I will assume that revision by $p\&s$ is intended. In a biology classroom I would instead interpret it as referring to revision by $p\&e$. Arguably, both $p\&s$ and $p\&e$ are specified enough to allow for successful (hypothetical) revision, i.e. $p\&s \in K * (p\&s)$ and $p\&e \in K * (p\&e)$.[23] Both these revisions can be expected to provide us with a belief set that has a clear answer to the question whether kangaroos will topple over (q). Consequently, in both these contexts the kangaroo conditional can be unambiguously evaluated [129].

This solution requires an adjustment of the underlying operation of revision. Most of the operations introduced in Chapters 4–6 satisfy the postulate of regularity, according to which it follows from $p \in K * (p\&s)$ that $p \in K * p$. This would of course block the solution just described. A precise formal development of this approach to the context-dependence of conditionals will have to be based on an operation of descriptor revision that does not satisfy regularity, such as blockage revision. (Cf. Observation 5.18.)

7.5 Non-Monotonic Inference

Inference and conditionality are both expressed with "If… then…". Not surprisingly, it has often been assumed that the logic of non-monotonic inference can be based on that of conditional sentences [19, 40, 76]. There is, however, an important difference that was well expressed by Kraus, Lehmann, and Magidor in their seminal 1990 paper on non-monotonic inference:

---

[22] On inputs that cannot be processed due to vagueness, see also [117, pp. 1021–1025].

[23] In their respective contexts, $p\&s$ and $p\&e$ are more adequate representations than p of the hypothetical input whose effect on the belief state, specifically with respect to q, is under consideration.

> [C]onditional logic considers a binary intensional connective that can be embedded inside other connectives and even itself, whereas we [in non-monotonic reasoning] consider a binary relation symbol that is part of the metalanguage. [153, p. 170]

Therefore, the logic of non-monotonic inference cannot be exactly the same as that of conditionals, but it can be constructed as the "flat (i.e. nonnested) fragment of a conditional logic" [153, p. 171]. In the same vein, Makinson and Gärdenfors suggested that "q follows non-monotonically from p" holds if and only if q is an element of the outcome of revising an "arbitrary but fixed background theory" by p [184, p. 189]. Denoting that theory by K and non-monotonic consequence by $\mid\!\sim$ we obtain:

The Ramsey test for non-monotonic inference
$p \mid\!\sim q$ holds if and only if $q \in K * p$.

This connection between non-monotonic inference and belief revision has been subject to much further study and refinement [39, 72, 149, 217, 258]. It can now be described as the standard view that the logic of non-monotonic inference coincides with a logic of non-nested conditional sentences and that it is connected to belief revision via the Ramsey test.

However, although conditionality and inferribility are related concepts, it is far from obvious that inferribility is nothing else than (non-nested) conditionality. For instance, let p denote that it rains in London today and q that Flamengo wins the match they are playing tonight in the Maracanã Stadium. If I become convinced that both p and q are true, then I may arguably conclude that "if p then q". However, it would be absurd to also conclude that "from p it can be inferred that q". More generally speaking, inferribility seems to imply conditionality, but not the other way around.

To account for the difference I propose that we retain the Ramsey test for conditionals (with the reservations made in the previous section) but apply another test to non-monotonic inference. The Ramsey test is based on the following criterion:

Ramsey's criterion
If the agent revises her beliefs by p, then she will believe that q.

For non-monotonic inference, the following criterion is proposed:

The co-occurrence criterion
If the agent comes to believe that p, then she will believe that q.

The two criteria differ since an agent can come to believe in p not only as the result of revising her beliefs by p but also as the result of revising them by some other input. According to the co-occurrence criterion, q has to be an element not only of $K * p$ but also of other belief sets containing p. The criterion concerns whether we will *in general* (given our present epistemic commitments) believe in q if we come to believe in p, not only whether we will do so in one single case. This seems to make the criterion better aligned with the notion of inferribility than the Ramsey criterion.

The co-occurrence test needs to be specified with respect to which of the belief sets containing p we should include in the analysis. A simple answer would be to include all potential belief change outcomes that contain p, i.e. all belief sets $K \circ \Psi$ such that $p \in K \circ \Psi$, or at least all belief sets $K * r$ such that $p \in K * r$. However, such an approach would be inadequate since it fails to reflect an essential feature of non-monotonic reasoning, namely that comparatively remote possibilities are left out of consideration. When you conclude from "Tweety is a bird" that "Tweety can fly", then that is precisely because you do not take remote possibilities into account. Importantly, the degree of remoteness referred to here is relative to the antecedent. Some of the possibilities that are too far-fetched to be taken into account when considering "Tweety is a bird" would be quite close at hand when considering "Tweety is a bird who was born in Antarctica".

Based on this, we arrive at the following test of inferribility:

The co-occurrence test for non-monotonic inference[24]
$p \mathrel{|\!\sim} q$ holds if and only if q holds in all the p-satisfying belief change outcomes that are reasonably plausible as compared to other p-satisfying belief change outcomes.

We will develop this approach in a centrolinear model that is exhaustive in the sense that $\bigcup \mathbb{X} = \mathcal{L}$. This is what is required for the success postulate for sentential revision ($p \in K * p$) to be satisfied. Although it is not a realistic feature, this condition is adopted here as a simple way to get rid of the rather uninteresting limiting cases of conditionality and inference with non-satisfiable antecedents ($p \rightarrowtail q$ and $p \mathrel{|\!\sim} q$ when the epistemic agent cannot be brought to believe that p is true).

In a centrolinear model, when evaluating non-monotonic inferences with p as the antecedent, we have to consider not only $K * p$ that is the most plausible ($\leqq$-minimal) p-containing belief set, but also a band of other p-containing belief sets that are less plausible than $K * p$ but still reasonably plausible.[25] That band has $K * p$ as its inner limit, and since it does not extend indefinitely we must assign an outer limit to it. In formal terms, for each potential outcome X there will be another potential outcome $\ell(X)$ that is the outer limit of the plausibility band that has X as its inner limit.[26] Intuitively, the plausibility band consists, in addition to X, of all the belief sets that are less plausible than X but only moderately so. In formal terms:

Definition 7.6 *The triple* $\langle \mathbb{X}, \leqq, \ell \rangle$ *is a* dilated centrolinear model *if and only if* $\langle \mathbb{X}, \leqq \rangle$ *is a centrolinear model and* ℓ *(the* delimiter*) is a function from and to* $\mathbb{X}$ *such that* $X \leqq \ell(X)$ *for all* $X \in \mathbb{X}$.

[24]This criterion does not preclude the existence of belief change outcomes in which $p\&\neg q$ holds. There can be some sentence r, less plausible than p, such that $p\&\neg q$ holds in some or all of the r-satisfying belief set outcomes. For an example, let p denote that Bitsy is a female mammal, q that Bitsy can give birth to live young, and r that Bitsy is a platypus.

[25]From a formal point of view, this proposal is related to the proposals by Nute [201] and Schlossberger [230, p. 80] that in possible world semantics, the assessment of a conditional sentence should refer not only to the antecedent-satisfying possible worlds that are most similar to the actual world but to all those that are sufficiently similar.

[26]This construction has the property that if $K * p_1 = K * p_2$ then p_1 and p_2 are evaluated with the same set of belief sets. Another plausible property of ℓ is: If $X \leqq Y$ then $\ell(X) \leqq \ell(Y)$. It will not be needed here.

A belief set $X \in \mathbb{X}$ is self-limited *according to ℓ if and only if $X = \ell(X)$.*

We can now express the co-occurrence test in more precise terms:

The co-occurrence test in an exhaustive and dilated centrolinear model $p \mathrel{|\!\sim} q$ holds if and only if it holds for all $Y \in \mathbb{X}$ that if $K * p \leqq Y \leqq \ell(K * p)$ and $p \in Y$, then $q \in Y$ [131].

In the limiting case when all elements of $\mathbb{X}$ are self-limited according to ℓ, the co-occurrence test coincides with the Ramsey test (and thus $\mathrel{|\!\sim}$ coincides with $\rightarrowtail$).

This recipe can be straightforwardly extended to take sets as antecedents:

$A \mathrel{|\!\sim} q$ if and only if it holds for all $Y \in \mathbb{X}$ that if
$K \circ \{\mathfrak{B}p \mid p \in A\} \leqq Y \leqq \ell(K \circ \{\mathfrak{B}p \mid p \in A\})$
and $A \subseteq Y$, then $q \in Y$.

With this reformulation we can define a non-monotonic inference operation C such that that $q \in \mathrm{C}(A)$ if and only if $A \mathrel{|\!\sim} q$. Such an operation is an important tool for studying non-monotonic inference and its relationship to classical consequence (as expressed by the consequence operation Cn) [177, 181]. The need for this extension to sets of sentences is, by the way, another reason why non-monotonic inference should not be assumed to coincide with the non-nested fragment of conditional logic.

The following theorem provides us with a close connection between the Ramsey test and the co-occurrence test for single-sentence antecedents:

Theorem 7.7 ([131]) *Let $\langle \mathbb{X}, \leqq, \ell \rangle$ be an exhaustive and dilated centrolinear model such that the strict part of $\leqq$ is a well-ordering with an order type that is either finite or ω.*[27] *Furthermore, let $\mathrel{|\!\sim}$ be the non-monotonic inference relation that is based on $\langle \mathbb{X}, \leqq, \ell \rangle$ via the co-occurrence test. Then there is a centrolinear model $\langle \mathbb{X}', \leqq' \rangle$ such that $\mathrel{|\!\sim}$ coincides with the conditional $\rightarrowtail$ that is based on $\langle \mathbb{X}', \leqq' \rangle$ via the Ramsey test.*

Furthermore, if K is the $\leqq$-minimal element of $\mathbb{X}$ and K' is the $\leqq'$-minimal element of $\mathbb{X}'$, then $K' \subseteq K$.

At first glance one might be tempted to see this theorem as an argument against the distinction made above between Ramsey test conditionals and non-monotonic inference based on the co-occurence test. However, that would be too rash a conclusion. Two distinct concepts may have the same logical properties.[28] Furthermore,

[27] A well-ordering is a linear ordering such that every non-empty subset of its domain has at least one minimal element. That the strict part $<$ of $\leqq$ has an order type that is either finite or ω means that $<$ is either isomorphic with a finite string $\langle 0, 1, \dots n \rangle$ of natural numbers or with the full infinite series $\langle 0, 1, 2 \dots \rangle$ of natural numbers. This is a stronger requirement than wellfoundedness. For instance, let $\mathbb{X}$ consist of all sets X_k where k is a natural number, and let $X_k < X_m$ hold if and only if either (a) X_k is even and X_m is odd, or (b) X_k and X_m are either both even or both odd, and $k < m$. (This is the sequence $X_0, X_2, X_4 \dots X_1, X_3, X_5 \dots$.) This relation is well-founded since every subset of $\mathbb{X}$ has a $<$-minimal element. However, it does not satisfy the criterion of Theorem 7.7.

[28] Logical necessity and physical necessity may both have the same (S5) logic, but that is no reason to conflate them. ([74, pp. 104–105], cf. [27, 59].) In social choice theory, we usually assume that

although the theorem provides us with a reconstruction of any co-occurrence test as a Ramsey test, this derived Ramsey test is based on another initial belief set and another operation of belief revision than those employed in the co-occurrence test that we started with. Therefore, although the logical properties of $\mid\!\sim$ alone coincide with those of $\rightarrowtail$ alone, the same cannot be said of the logical properties that connect $\mid\!\sim$ respectively $\rightarrowtail$ to the original beliefs or to the operation of belief revision. That is exemplified by the following property of conditonals:

Property CS
If p and q both hold, then so does $p \rightarrowtail q$.

In our centrolinear model, $\rightarrowtail$ satisfies CS, but $\mid\!\sim$ does not do so in general. This is a highly plausible difference, as shown in the above Flamengo example.[29]

7.6 Modalities of Belief

A common way to construct a modal epistemic logic is to let a belief operation $\mathfrak{B}$ take the role that the necessity operation has in alethic logic (the logic of necessity and possibility). The corresponding possibility operation $\diamond p$ will then be defined as follows:

$\diamond p$ if and only if $\neg\mathfrak{B}\neg p$.

Isaac Levi introduced the term "serious possibility" for this operation. ([159], cf. [61, 163, 164].) It can be translated "p is compatible with what the agent believes". It has also often been interpreted as "the agent considers p possible" [248, p. 23]. However, the latter interpretation is less convincing. If the agent's beliefs are consistent then $\diamond$ satisfies the property $\diamond p \vee \diamond \neg p$ [61, p.120]. This contradicts the common experience of having no opinion on what is possible in some particular matter. For instance, until recently I had no opinion on how high pitches it might be possible for a baboon to hear. Therefore, I did not hold it to be possible that a baboon hears a tone of the highest pitch a human can hear, and neither did I hold the opposite to be possible.[30]

But this is not the only way to introduce modal notions into the logic of belief. Much more expressive power can be obtained by introducing separate representations of necessity and possibility. We can then represent statements about what the agent believes to be possible or necessary, and also about what is possible or necessary for

(Footnote 28 continued)
the preferences of different persons satisfy the same logical rules, but in all non-trivial cases they differ in substance.

29 $\rightarrowtail$ satisfies CS in any model such that $*$ satisfies confirmation. CS holds in many systems of conditional logic, see for instance [167, pp. 26–31], [207, p. 249], and [203]. However, it has also been criticized, for instance by Bennett [11, pp. 386–388] and Nozick [200, p. 176].

30 Actually, baboons can even hear tones that are an octave above the upper limit of what a human can hear [243].

the agent to believe. In what follows I will sketch out how the latter development can be achieved in descriptor revision. We can unproblematically introduce necessity and possibility operations that refer to what the agent can come to believe as a result of belief change[31]:

Direct possibility:
$\Diamond\Psi$ holds at K if and only if there is some Ξ such that $K \circ \Xi \Vdash \Psi$.
Direct necessity:
$\Box\Psi$ holds at K if and only if $K \circ \Xi \Vdash \Psi$ for all Ξ.

To exemplify this, $\Diamond(\neg\mathfrak{B}p \;\&\; \neg\mathfrak{B}\neg p)$ means that there is some (single-step) belief change that will make the agent open-minded about p, and $\Box\neg\mathfrak{B}q$ that no (single) change can make the agent believe in q.

Sometimes it takes a whole series of changes to arrive at a new pattern of belief. To express what is necessary or possible through iterated belief change we need another pair of modal operations:

Iterative possibility:
$\diamondsuit\!\!\!\cdot\,\Psi$ holds at K if and only if there is some series $\Xi_1, \ldots, \Xi_n$ of descriptors such that $K \circ \Xi_1 \circ \ldots \circ \Xi_n \Vdash \Psi$.
Iterative necessity:
$\boxdot\Psi$ holds at K if and only if $K \circ \Xi_1 \circ \ldots \circ \Xi_n \Vdash \Psi$ for all series $\Xi_1, \ldots, \Xi_n$ of descriptors.

A semantic model for these modal operations can be based on the same type of accessibility relation that is used in possible world models. However, that relation will have to operate on the entities that can be accessed through belief change, and these are belief sets rather than possible worlds. We will therefore have use for the following construction:

Definition 7.8 *A* possible theories model[32] *is a pair* $\langle\mathbb{X}, a\rangle$ *where* $\mathbb{X}$ *is a set of logically closed sets and* a *is a binary relation on* $\mathbb{X}$. *A* pointed possible theories model *is a triple* $\langle\mathbb{X}, a, K\rangle$, *where* $K \in \mathbb{X}$.

The evaluation of modal sentences follows standardly:

$\Box\Psi$ holds at X if and only if it holds for all Y that if XaY then $Y \Vdash \Psi$.
$\Diamond\Psi$ holds at X if and only if there is some Y such that XaY and $Y \Vdash \Psi$.

Obviously, possible worlds models are special cases of possible theories models.

The intended epistemic interpretation is that XaY holds if and only if there is some Ξ such that $X \circ \Xi = Y$. Not surprisingly there is a close connection between possible theories models and global monoselective revision as introduced in Definition 6.2.

[31]The introduction of modal notions with similar definitions into the AGM framework is less promising. Due to the success property ($p \in K * p$), $\Diamond\mathfrak{B}p$ would hold in AGM for all sentences p.
[32]Due to the general nature of this definition, the term "theory" for a logically closed set of sentences is used rather than "belief set" that is limited to epistemological interpretations.

Observation 7.9 *Let $\langle \mathbb{X}, a\rangle$ be a possible theories model. Then the following two conditions are equivalent:*

(1) a is reflexive.

(2) There is a global monoselective revision $\circ$ on $\mathbb{X}$ such that a is the accessibility relation on which $\circ$ is based.

The operations $\Box$, $\Diamond$, $\boxdot$, and $\diamondtimes$ all refer to what belief patterns can at all be reached with our operation of change.[33] Alternatively, we can restrict our deliberations to changes that satisfy some particular pattern. For instance, we may ask whether there is, among the belief changes that retain a person's belief in p, some change that would make her give up q. Generalizing this pattern we can write $\langle\Phi\rangle\Psi$ to denote that there is some Ξ such that $X \circ \Xi \Vdash \Phi$ and $X \circ \Xi \Vdash \Psi$, and correspondingly $[\Phi]\Psi$ to denote that $X \circ \Xi \Vdash \Psi$ for all Ξ such that $X \circ \Xi \Vdash \Phi$. However these operations are only of limited interest since $\langle\Phi\rangle\Psi$ is satisfied in X if and only if $\Diamond(\Phi \cup \Psi)$ is satisfied in X, and if Ψ is negatable then $[\Phi]\Psi$ is satisfied in X if and only if $\neg\Diamond(\Phi \cup \overline{\neg}\, \Psi)$ is satisfied in X.

But we can go further in this type of restriction on modalities. We can focus on what is necessary or possible after revision by a specific input:

$[\Xi]\Psi$ holds if and only if Ψ is satisfied in all outcomes of revision by Ξ. (1)

In deterministic descriptor revision there is exactly one outcome that can result from revision by Ξ, and therefore (1) is equivalent to:

$[\Xi]\Psi$ holds if and only if $K \circ \Xi \Vdash \Psi$. (2)

Furthermore, in deterministic descriptor revision the corresponding possibility operation $\langle\ \rangle$ will coincide with the necessity operation, i.e.:

For all Ξ and Ψ: $[\Xi]\Psi$ is satisfied if and only if $\langle\Xi\rangle\Psi$ is satisfied. (3)

It follows that in deterministic descriptor revision, $[\Xi]\Psi$ and $\langle\Xi\rangle\Psi$ are nothing else than alternative notations for $K \circ \Xi \Vdash \Psi$. However, in *in*deterministic descriptor revision this trivialization of the modal operations does not take place, since neither (2) nor (3) will hold. In that case, the following definitions will be adequate:

$[\Xi]\Psi$ holds if and only if $X \Vdash \Psi$ for all $X \in K\check{\circ}\Xi$. (4)
$\langle\Xi\rangle\Psi$ holds if and only if $X \Vdash \Psi$ for some $X \in K\check{\circ}\Xi$. (5)

and clearly $\langle\Xi\rangle\Psi$ can be true while $[\Xi]\Psi$ is false. This notation is useful, not least since it brings us into direct contact with other approaches that use similar notations for belief change, such as the update logics developed by van Benthem, Fuhrmann, and de Rijke [37, 62, 246, 247], and Krister Segerberg's Dynamic Doxastic Logic (DDL) [28, 157, 172, 234, 236]. In Segerberg's notation, the standard operations of belief revision come out as follows:

[33] See [91] for a study of corresponding modal notions in a belief base framework.

$[*p]\mathfrak{B}q$ (q is believed after revision by p)
$[\div p]\mathfrak{B}q$ (q is believed after contraction by p)
$[+p]\mathfrak{B}q$ (q is believed after expansion by p)

These can all be obtained as special cases of the modal descriptor notation $[\Xi]\Psi$. Segerberg was right in pointing out that the modal notation provides us with an account of belief change that is "a generalization of ordinary Hintikka type doxastic logic", whereas strictly speaking, "AGM is not really logic; it is a theory about theories" [235, p. 136].

The descriptor-based modal operations $[\Xi]$ and $\langle\Xi\rangle$ have the advantage of being easily combinable with other modal operations such as $\square$, $\Diamond$, $\boxdot$, and $\Diamonddot$, as introduced above. We can for instance write $[\Xi]\boxdot\neg\mathfrak{B}p$ to express that after revision by Ξ, any series of belief changes will result in a belief set in which p is not believed. The logic of all these modal operations can be explored in the framework of possible theories semantics. This type of semantics may also be useful for studies of other concepts, such as intentions, goals, and various types of inference. However, these are topics that will have to be left for later investigations.

Part III
Sentential Change Revisited

Chapter 8
Sentential Revision

As we saw in Chapter 1, belief change theory has traditionally been concerned with sentential changes, i.e. operations with a single sentence as input. The focus has been on two major types of sentential change, namely contraction in which the input sentence is removed and (sentential) revision in which it is added to the belief set. In the previous four chapters we have explored a more general approach to belief change that has belief descriptors as inputs. We have already noted that sentential changes are special cases of this general approach. Sentential revision by p corresponds to revision by the descriptor $\mathfrak{B}p$, and contraction by p can be reconstructed as revision by the descriptor $\neg\mathfrak{B}p$. In this and the following two chapters we will have a closer look at these special cases, beginning in this chapter with sentential revision.

Any descriptor revision $\circ$ gives rise to a sentential revision:

Definition 8.1 *Let $\circ$ be a descriptor revision on the belief set K. The operation $*$ such that*

$$K * p = K \circ \mathfrak{B}p$$

for all $p \in \mathcal{L}$ is the (deterministic) sentential revision *that is derivable from $\circ$.*

In Section 8.1 we will have a close look at the sentential revisions that are derivable from monoselective revision. We will characterize them axiomatically and also show that they have partial meet revision, the operation characterized by the six basic AGM postulates, as a special case. In Section 8.2 we will investigate another important class of sentential revisions, namely those that are derivable from centrolinear revision. In addition to characterizing them axiomatically we will prove that they have transitively relational partial meet revision, the class of operations characterized by the full set of AGM revision postulates, as a special case. This is an important result since it shows that descriptor revision is a generalization of AGM revision, and also because of the insights on the AGM construction that we can derive from it. In Section 8.3 we will study the properties of global sentential revision. In Section 8.4 we derive relations on sentences from relations of epistemic proximity on descriptors according to the simple principle that p is more believable than q if and only if $\mathfrak{B}p$ is epistemically more proximate than $\mathfrak{B}q$. Finally, in Section 8.5 we will have a brief look at multiple

S.O. Hansson, *Descriptor Revision*, Trends in Logic 46,
DOI 10.1007/978-3-319-53061-1_8

sentential revision, i.e. revision by several sentences, and in particular at the operation of making up one's mind.

8.1 Sentential Monoselective Revision

The sentential revisions that are derivable from monoselective descriptor revision can be axiomatically characterized as follows:

Theorem 8.2 *Let $*$ be a sentential operation on the consistent belief set K. Then the following two conditions are equivalent:*

(I) $*$ is the sentential revision that is derivable from some K-favouring monoselective descriptor revision on K.

(II) $*$ satisfies:

$K * p = \mathrm{Cn}(K * p)$ *(closure)*

If it holds for all q that $K * q \vdash p$ if and only if $K * q \vdash p'$, then $K * p = K * p'$. (uniformity)

Either $K * p \vdash p$ or $K * p = K$. (relative success)

If $K * q \vdash p$, then $K * p \vdash p$. (regularity)

If $p \in K$, then $K * p = K$. (confirmation)

Strictly speaking, the operation axiomatized in the above theorem is a not a revision but a *semirevision*, i.e. a revision-like operation that does not accept all inputs and therefore does not satisfy the success postulate for sentential revision ($p \in K * p$). This may not be a disadvantage. As we noted in Section 3.2, it has repeatedly been pointed out that the success postulate is unrealistic since in real life, cognitive agents sometimes do not accept the new information they receive. From a formal point of view, the operations that satisfy success are easily characterized:

Observation 8.3 *Let $*$ be the sentential revision that is derivable from some monoselective descriptor revision $\circ$ on the consistent belief set K, and let $\mathbb{X}$ be the outcome set of $\circ$. Then $*$ satisfies success ($p \in K * p$) if and only if* $\mathrm{Cn}(\{\perp\}) \in \mathbb{X}$.

But even with this addition, only three of the basic AGM postulates (closure, success, and extensionality) are satisfied. To satisfy them all, more extensive conditions on $\mathbb{X}$ and $\widehat{C}$ have to be added:

Theorem 8.4 *Let $*$ be a sentential operation on the consistent belief set K. Then the following three conditions are equivalent:*

(I) $*$ *is the sentential revision that is derivable from some monoselective descriptor revision* $\circ$ *on* K *that is based on the choice function* $\widehat{C}$, *has* $\mathbb{X}$ *as its outcome set, and satisfies:*

($\mathbb{X}$*1) If* $p \nvdash \bot$, *then* p *is satisfiable within* $\mathbb{X} \setminus \{\mathrm{Cn}(\{\bot\})\}$.

($\mathbb{X}$*2)* $\mathrm{Cn}(K \cup \{p\}) \in \mathbb{X}$

($\widehat{C}$*1) If* $p \in K$, *then* $\widehat{C}([\![\mathfrak{B}p]\!]) = K$.[1]

($\widehat{C}$*2) If* $p \nvdash \bot$, *then* $\widehat{C}([\![\mathfrak{B}p]\!]) \not\Vdash \mathfrak{B}\bot$.

($\widehat{C}$*3) If* $K \nvdash p \rightarrow q$, *then* $\widehat{C}([\![\mathfrak{B}p]\!]) \not\Vdash \mathfrak{B}q$.

($\widehat{C}$*4) If* $K \nvdash \neg p$, *then* $K \subseteq \widehat{C}([\![\mathfrak{B}p]\!])$.

(II) $*$ *is a partial meet revision.*

(III) $*$ *satisfies the basic AGM postulates, i.e.:*

$K * p = \mathrm{Cn}(K * p)$ *(closure)*

$K * p \subseteq K + p$ *(inclusion)*

If $\neg p \notin K$, *then* $K + p \subseteq K * p$. *(vacuity)*

$p \in K * p$ *(success)*

$K * p = K * p'$ *whenever* $\vdash p \leftrightarrow p'$. *(extensionality)*

$K * p$ *is consistent if* p *is consistent. (consistency)*

The equivalence between items (II) and (III) is a well-known result from the AGM paper [1], and (III) is included here only to facilitate comparisons. Two important observations should be made about this theorem. First, it shows that partial meet revision is a special case of monoselective revision. Secondly, we can see from (I) that it is a rather demanding special case with far-reaching requirements on the choice function and, perhaps in particular, on the outcome set. According to ($\mathbb{X}$2), for every sentence p the expansion $\mathrm{Cn}(K \cup \{p\})$ must be an element of the outcome set. We saw in Section 3.5 that this is a highly problematic condition. ($\mathbb{X}$1) is no less questionable. It requires that every consistent sentence be contained in some consistent element of the outcome set. For instance, let p denote the statement that the Eiffel Tower was built in 1887–1889 out of papier mâché and painted with a thin layer of watercolour, and has never been repaired since. According to ($\mathbb{X}$1), there is some input that will take me directly, in a single step, from my present belief set to a consistent belief set that contains p. However, to the extent that I can at all be made to believe in p, such a change would have to take place in several steps, so that some of the background

[1] In other words, $\widehat{C}$ is K-favouring for atomic descriptors.

beliefs that contradict it are removed before p itself is accepted. Indeed, when people come to believe in weird things this usually seems to be the result of such multi-stage processes.

8.2 Sentential Centrolinear Revision

Let us return to the most orderly construction of descriptor revision that we introduced in Chapter 5, namely centrolinear revision. Just like monoselective revision it gives rise to a derived sentential revision in the manner shown in Definition 8.1. The axiomatic characterization of that derived operation is an open question in the general case, but a characterization is available for the case when all elements of the outcome set are finite-based and the belief set ordering $\leqq$ orders them in the same way as the natural numbers are ordered.

Theorem 8.5 *Let $*$ be a sentential operation on the finite-based and consistent belief set K. Then the following two conditions are equivalent:*

(I) $K * p = K \circ \mathfrak{B}p$ for all $p \in \mathcal{L}$, where $\circ$ is the centrolinear revision based on a relation $\leqq$ on the set $\mathbb{X}$ of belief sets, such that:

(a) $K \in \mathbb{X}$ and all elements of $\mathbb{X}$ are finite-based.[2]

(b) $K \leqq X$ for all $X \in \mathbb{X}$, and

(c) $\leqq$ is complete, transitive, and antisymmetric, and its strict part is a well-ordering with an order type that is either finite or ω.[3]

(II) $*$ satisfies the following conditions:

$K * p = \mathrm{Cn}(K * p)$ *(closure)*

$K * p = K * p'$ *whenever* $\vdash p \leftrightarrow p'$. *(extensionality)*

Either $K * p \vdash p$ *or* $K * p = K$. *(relative success)*

If $K * q \vdash p$, *then* $K * p \vdash p$. *(regularity)*

If $p \in K$, *then* $K * p = K$. *(confirmation)*

If $q \in K * p$, *then* $K * p = K * (p \& q)$. *(cumulativity)*

---

[2]It follows that there is a one-to-one correspondence between $\mathbb{X}$ and the set $\{\&X \mid X \in \mathbb{X}\}$. Since the latter set is countable, so is $\mathbb{X}$.

[3]For an explanation, see footnote 27, p. 110.

If K is finite-based, then so is $K * p$. (finite-based outcome)

$\{X \mid (\exists t)(X = K * (p \vee t))\}$ *is finite. (finite gradation)*

If each belief set in the series $K_1, \ldots, K_n$ is a specification of its predecessor, then K_1 is not a specification of K_n. (non-circularity)[4]

(K'' is a specification of K', relative to K and $*$, if and only there are sentences p and q such that $K' = K * p \neq K * (p\&q) = K''$, $p \in K * p$, and $p\&q \in K * (p\&q)$.)

Four of the five first-mentioned postulates in this theorem (closure, relative success, regularity, and confirmation) coincide with four of the five postulates of Theorem 8.2. The fifth postulate in Theorem 8.2, namely uniformity, can be shown to follow from some of the postulates in Theorem 8.5:

Observation 8.6 *Let $*$ be a sentential revision on a belief set K. If it satisfies extensionality, relative success, regularity, and cumulativity, then it satisfies uniformity.*

Thus all the postulates of Theorem 8.2 are satisfied by the operations characterized in Theorem 8.5. This confirms that we are dealing with a special case of the sentential revisions derivable from monoselective descriptor revision.

The most significant "new" postulates in Theorem 8.5 are (sentential) cumulativity and non-circularity. (Sentential) cumulativity follows from (descriptor) cumulativity that holds according to Theorem 5.10, and it was mentioned already in Section 5.2. It connects nicely with other plausible properties of a sentential revision:

---

[4]Zhang Li has proposed a highly illustrative example that shows the need for this postulate: Let the language be based on the three atoms $\{a_0, a_1, a_2\}$ and let $K = \text{Cn}(\varnothing)$. Let $*$ be defined as follows: (1) If $a_0\&a_1 \vdash p$ and $p \vdash a_0$, then $K * p = \text{Cn}(a_0\&a_1)$, (2) If $a_1\&a_2 \vdash p$ and $p \vdash a_1$, then $K * p = \text{Cn}(a_1\&a_2)$, (3) If $a_0\&a_2 \vdash p$ and $p \vdash a_2$, then $K * p = \text{Cn}(a_0\&a_2)$, and (4) otherwise $K * p = \text{Cn}(\{p\})$.

With the exception of cumulativity, it is easy to check that the first eight postulates of the theorem are satisfied by this construction. For cumulativity, let $q \in K * p$. There are two cases: (a) $K * p$ was decided according to one of the first three clauses: We consider clause (1). In this case, $a_0\&a_1 \vdash p$ and $a_0\&a_1 \vdash q$, thus $a_0\&a_1 \vdash p\&q$. Furthermore, $p \vdash a_0$, thus $p\&q \vdash a_0$. Consequently, $K * (p\&q) = \text{Cn}(a_0\&a_1) = K * p$. (b) $K * p$ was decided according to the last clause: Then $q \in K * p = \text{Cn}(\{p\})$, thus $p\&q \leftrightarrow p$ and $K * (p\&q) = \text{Cn}(\{p\}) = K * p$. Non-circularity is refuted by the following cycle: $\text{Cn}(\{a_0\&a_2\})$ is a specification of $\text{Cn}(\{a_0\&a_1\})$ since $K * a_0 = \text{Cn}(\{a_0\&a_1\})$ and $K * (a_0\&a_2) = \text{Cn}(\{a_0\&a_2\})$. $\text{Cn}(\{a_1\&a_2\})$ is a specification of $\text{Cn}(\{a_0\&a_2\})$ since $K * a_2 = \text{Cn}(\{a_0\&a_2\})$ and $K * (a_1\&a_2) = \text{Cn}(\{a_1\&a_2\})$. $\text{Cn}(\{a_0\&a_1\})$ is a specification of $\text{Cn}(\{a_1\&a_2\})$ since $K * a_1 = \text{Cn}(\{a_1\&a_2\})$ and $K * (a_0\&a_1) = \text{Cn}(\{a_0\&a_1\})$.

Observation 8.7 *Let $*$ be an operation on a consistent belief set K.*

(1)[5] *If it satisfies closure, extensionality, relative success, and regularity, then it satisfies cumulativity if and only if it satisfies:*

If $q \in K * p$ and $p \in K * q$ then $K * p = K * q$. (reciprocity)

(2) If it satisfies closure, extensionality and cumulativity, then it satisfies:

If $p \in K * p$ and $K * p = K * (p \vee q \vee r)$, then $K * p = K * (p \vee q)$.
(disjunctive interpolation)

Non-circularity is a new postulate that refers to the relationship between successful sentential revisions whose inputs carry different amounts of information. A revision by the conjunction $p\&q$ conveys more information than one by p; we can describe the former as a specification of the latter. The postulate expresses intuitions about the additivity of internally coherent information. The success conditions referred to in the definition of a specification restricts the application of the postulate to cases in which the new information is accepted.

Success ($p \in K * p$) does not hold for the operations characterized in Theorem 8.5, but as can be seen from Observation 8.3 it holds if and only if $\mathrm{Cn}(\{\perp\}) \in \mathbb{X}$.

The following theorem shows that transitively relational partial meet revisions, the operations characterized by the six basic and two supplementary AGM postulates, are derivable from (a subclass of) centrolinear descriptor revisions:

Theorem 8.8 ([128], modified) *Let $*$ be a sentential operation on a consistent belief set K. Then the following three conditions are equivalent:*

(I) $*$ is the sentential revision that is derivable from some centrolinear descriptor revision $\circ$ on K that is based on the relation $\leqq$, has the outcome set $\mathbb{X}$ and satisfies:

($\mathbb{X}$1) *If $p \nvdash \perp$, then p is satisfiable within $\mathbb{X} \setminus \{\mathrm{Cn}(\{\perp\})\}$.*

($\mathbb{X}$3) *For all $X, Y \in \mathbb{X}$: if $X \cup Y \nvdash \perp$, then $X \cap Y \in \mathbb{X}$.*

($\mathbb{X}$4) *If $X \in \mathbb{X}$, then $\mathrm{Cn}(X \cup \{p\}) \in \mathbb{X}$.*

($\mathbb{X}$5) *If X_1, X_2, X_3, $X_1 \cap X_2$, and $X_2 \cap X_3$ are all elements of $\mathbb{X}$, then so is $X_1 \cap X_3$.*

($\leqq$1) *If $X \subseteq Y$, then $X \leqq Y$.*

[5] This is a slight generalization of a result reported in [73, p. 54].

($\leqq$2) If $X \cap Z \in \mathbb{X}$ and $X \leqq Y \leqq Z$, then $X \cap Y \in \mathbb{X}$ and $Y \cap Z \in \mathbb{X}$.

(II) $*$ is a transitively relational partial meet revision.

(III) $*$ satisfies the basic and supplementary AGM postulates, i.e.:

$K * p = \mathrm{Cn}(K * p)$ *(closure)*

$K * p \subseteq K + p$ *(inclusion)*

If $\neg p \notin K$, then $K + p \subseteq K * p$. (vacuity)

$p \in K * p$ *(success)*

$K * p = K * p'$ *whenever* $\vdash p \leftrightarrow p'$. *(extensionality)*

$K * p$ is consistent if p is consistent. (consistency)

$K * (p\&q) \subseteq (K * p) + q$ *(superexpansion)*

If $\neg q \notin K * p$, then $(K * p) + q \subseteq K * (p\&q)$. (subexpansion)

The equivalence between (II) and (III) is well-known from the AGM paper [1]. Theorem 8.8 is important since it shows that full-blown AGM revision, often described as the gold standard of (sentential) belief revision, is indeed a special case of descriptor revision. But just like Theorem 8.4 it also shows that the conditions needed to obtain AGM operations are far from indisputable. A couple of the conditions are somewhat opaque, but they can be explained in relation to the AGM postulates for revision. For that purpose, we will have use for the following five postulates that all hold for transitively relational AGM revision[6]:

$p \in K * p$ (success)

$K * p = K * q$ if and only if $q \in K * p$ and $p \in K * q$. (reciprocity)

$K * (p \vee q) = K * p$ or $K * (p \vee q) = K * q$ or $K * (p \vee q) = (K * p) \cap (K * q)$. (disjunctive factoring)

If $\neg p \notin K * (p \vee q)$, then $K * (p \vee q) \subseteq K * p$. (disjunctive inclusion)

$(K * p) \cap (K * q) \subseteq K * (p \vee q)$ (disjunctive overlap)

[6]On these postulates, see also [73, p. 54], [99, pp. 270–274], and [217, pp. 107–111].

For ($\mathbb{X}3$), note first that since X and Y are assumed to be elements of the outcome set $\mathbb{X}$ we can assume that there are p and q such that $X = K * p$ and $Y = K * q$. With this substitution, we are going to derive the contrapositive form of ($\mathbb{X}3$), i.e.:

If $(K * p) \cap (K * q) \notin \mathbb{X}$, then $(K * p) \cup (K * q) \vdash \perp$.

Let $(K * p) \cap (K * q) \notin \mathbb{X}$. Then $K * (p \vee q) \neq (K * p) \cap (K * q)$. It follows from disjunctive factoring that either $K * (p \vee q) = K * p$ or $K * (p \vee q) = K * q$. Without loss of generality we can assume that $K * (p \vee q) = K * p$. Now suppose for contradiction that $\neg q \notin K * p$. Then equivalently $\neg q \notin K * (p \vee q)$, and it follows from disjunctive inclusion that $K*(p\vee q) \subseteq K*q$, equivalently $K*p \subseteq K*q$. Then it follows from $K*(p\vee q) = K*p$ and $K*p \subseteq K*q$ that $K*(p\vee q) = (K*p)\cap(K*q)$, contrary to our initial assumption. We can conclude that $\neg q \in K * p$. Due to success, $q \in K * q$, thus $(K * p) \cup (K * q) \vdash \perp$ as desired.

For ($\mathbb{X}4$), we distinguish between two cases. If $X \vdash \neg p$, then $\text{Cn}(X \cup \{p\}) = \text{Cn}(\{\perp\})$. It follows from the AGM postulates success and closure that $K * \perp = \text{Cn}(\{\perp\})$. Thus $\text{Cn}(X \cup \{p\}) = K * \perp$, thus $\text{Cn}(X \cup \{p\}) \in \mathbb{X}$. If $X \nvdash \neg p$, then we note that since $X \in \mathbb{X}$, there is some r such that $X = K * r$. The AGM postulates subexpansion and superexpansion yield $(K * r) + p = K * (r \& p)$, thus $\text{Cn}(X \cup \{p\}) = K * (r \& p)$ and $\text{Cn}(X \cup \{p\}) \in \mathbb{X}$.

($\mathbb{X}5$) can probably be best explicated in terms of the sphere model presented in Fig. 1.3. It follows from $X_1 \in \mathbb{X}$, $X_2 \in \mathbb{X}$, and $X_1 \cap X_2 \in \mathbb{X}$ that X_1 and X_2 belong to the same sphere.[7] It follows similarly that X_2 and X_3 belong to the same sphere. Thus X_1 and X_3 belong to the same sphere. Call it $\mathcal{S}$. Since $X_1, X_3 \in \mathbb{X}$ there are sentences p_1 and p_3 such that $X_1 = K * p_1$ and $X_3 = K * p_3$. We have $[K * p_1] = \mathcal{S} \cap [p_1]$ and $[K * p_3] = \mathcal{S} \cap [p_3]$. Furthermore, $[p_1 \vee p_3] = [p_1] \cup [p_3]$, thus[8]:

$$
\begin{aligned}
&[K * (p_1 \vee p_3)] = \mathcal{S} \cap [p_1 \vee p_3] \\
&= \mathcal{S} \cap ([p_1] \cup [p_3]) \\
&= (\mathcal{S} \cap [p_1]) \cup (\mathcal{S} \cap [p_3]) \\
&= [K * p_1] \cup [K * p_3] \\
&= [X_1] \cup [X_3] \\
&= [X_1 \cap X_3],
\end{aligned}
$$

consequently

$$K * (p_1 \vee p_3) = \bigcap[K * (p_1 \vee p_3)] = \bigcap[X_1 \cap X_3] = X_1 \cap X_3,$$

and we can conclude from $K * (p_1 \vee p_3) \in \mathbb{X}$ that $X_1 \cap X_3 \in \mathbb{X}$.

Next, let us turn to ($\leqq 1$). Since X and Y are assumed to be elements of the outcome set we can replace them by $K * p$ and $K * q$. It would be sufficient to show that:

If $K * p \subset K * q$, then $K * p < K * q$. (maximizingness [86])

---

[7]More precisely: The sphere in which there are worlds containing X_1 coincides with the sphere in which there are worlds containing X_2.

[8]Lemma 8.5 (p. 200) is used here.

($<$ is the strict part of $\leqq$.) However, maximizingness cannot be derived from the AGM postulates since the language of those postulates does not contain $\leqq$ or $<$. Instead we can show that $*$ has a property that is necessary for maximizingness to hold. According to the construction of $*$, $K * q$ is the $\leqq$-minimal element of $\mathbb{X}$ that contains q, and therefore $q \in K * p$ and $K * p < K * q$ are incompatible. Consequently, in order for maximizingness (and ($\leqq 1$)) to hold, the following condition must be satisfied:

If $K * p \subset K * q$, then $q \notin K * p$.

To show that this holds, let $K * p \subset K * q$. Success yields $p \in K * q$. Suppose for contradiction that $q \in K * p$. Then reciprocity yields $K * p = K * q$, contrary to our assumption that $K * p \subset K * q$. We can conclude that $q \notin K * p$, as desired.

Finally, ($\leqq 2$) is somewhat more complex but it can be understood with the help of the following property:

If $K * z = (K * p) \cap (K * q)$, then $K * (p \vee q) = (K * p) \cap (K * q)$.
(cut primacy [124])

To show that cut primacy follows from the AGM postulates, let $K * z = (K * p) \cap (K * q)$. Success yields $p \vee q \in K * z$ and $z \in (K * p) \cap (K * q)$. Due to disjunctive overlap, $z \in K * (p \vee q)$. Finally, we apply reciprocity to $p \vee q \in K * z$ and $z \in K * (p \vee q)$, and obtain $K * z = K * (p \vee q)$.

The following equivalent form of cut primacy will be useful:

$(K * p) \cap (K * q) \in \mathbb{X}$ if and only if $K * (p \vee q) = (K * p) \cap (K * q)$.

We can interpret $K * (p \vee q) = (K * p) \cap (K * q)$ as saying that if we accept the information that either p or q, then we enter a state of hesitation between revising by p and revising by q, presumably because these two alternatives are equally plausible. We can therefore interpret $(K * p) \cap (K * q) \in \mathbb{X}$ as saying that seen from the viewpoint of K, $K * p$ and $K * q$ are equally plausible. In this perspective, ($\leqq 2$) can be read as saying that if the belief sets $K * p$ and $K * r$ are equally plausible, and $K * p \leqq K * q \leqq K * r$, then $K * q$ is equally plausible as $K * p$, and also equally plausible as $K * r$.

Theorems 8.4 and 8.8 show that both the general and the transitively relational variant of AGM revision can be reconstructed as subcases of descriptor revision. In a sense, this mitigates the potential conflict between the dominant AGM tradition and our descriptor-based approach. But at the same time, these theorems indicate that the properties characteristic of the AGM special case are not plausible enough to warrant exclusive attention to that case.

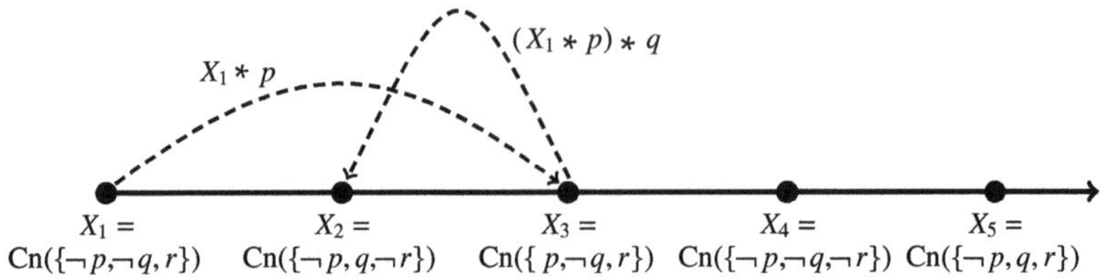

Fig. 8.1 An example of a one-dimensional distance-based sentential revision. Each step in the revision takes us to the closest belief set on the line that satisfies the success condition.

8.3 Global Sentential Revision

In Definition 8.1 we showed how a local sentential revision can be derived from a local descriptor revision. In the same way, a global sentential revision can be derived from a global descriptor revision:

Definition 8.9 *Let $\circ$ be a (deterministic) global descriptor revision on a set $\mathbb{X}$ of belief sets. The operation $*$ such that*

$$X * p = X \circ \mathfrak{B}p$$

for all $X \in \mathbb{X}$ and $p \in \mathcal{L}$ is the (deterministic) global sentential revision *derivable from $\circ$.*

We saw in Section 3.7 that there are reasons why global operations should not be expected to have non-trivial properties in addition to those that follow directly from the properties of their local counterparts. In particular, counterexamples were offered to the four so-called Darwiche–Pearl postulates for iterated revision [36]:

If $q \vdash p$, then $(X * p) * q = X * q$. (DP1)

If $q \vdash \neg p$, then $(X * p) * q = X * q$. (DP2)

If $X * q \vdash p$, then $(X * p) * q \vdash p$. (DP3)

If $X * q \nvdash \neg p$, then $(X * p) * q \nvdash \neg p$. (DP4)

Therefore it should not be seen as a drawback that derived (global) sentential revision does not satisfy any of these postulates. We can indeed prove a much stronger result. None of the Darwiche–Pearl postulates holds even in an oversimplified model of distance-based centrolinear revision in which distances are symmetric and one-dimensional. In such a model, as illustrated in Fig. 8.1, all elements of the outcome set are situated at different points on a line. The revision $X * p$ takes us to the p-containing belief set on the line that is closest to X. In formal terms:

Definition 8.10 *A binary measure δ on a set $\mathbb{X}$ is a* one-dimensional distance measure *if and only if there is a real-valued function $ł$ on $\mathbb{X}$ (the location function) such that for all $X, Y \in \mathbb{X}$:*

$$\delta(X, Y) = |ł(X) - ł(Y)|.$$

A distance-based descriptor revision is one-dimensional *if and only if it is based on a one-dimensional distance measure.*

A sentential revision is a one-dimensional distance-based sentential revision *if and only if it is based in the manner of Definition 8.1 on a one-dimensional distance-based descriptor revision.*

We now have the means for a precise formal statement of the negative result already referred to:

Observation 8.11 ([132]) *None of the four postulates DP1, DP2, DP3, or DP4 holds in general for one-dimensional distance-based sentential revision.*

8.4 Believability Relations

Just as we can derive a sentential revision $*$ from a descriptor revision $\circ$ via the identity $K * p = K \circ \mathfrak{B}p$, we can derive a relation on sentences from a relation $\succeq$ of epistemic proximity:

Definition 8.12 *([126]) Let $\succeq$ be a relation of epistemic proximity. The relation $\stackrel{\star}{\succeq}$ on sentences, such that*

$p \stackrel{\star}{\succeq} q$ *if and only if* $\mathfrak{B}p \succeq \mathfrak{B}q$

is the believability relation *that is based on $\succeq$.*

The strict part of $\stackrel{\star}{\succeq}$ is denoted $\stackrel{\star}{\succ}$ and its symmetrical part $\stackrel{\star}{\simeq}$.

The name "believability" was chosen since the intended interpretation of $p \stackrel{\star}{\succeq} q$ is that belief in p is at least as easily acquired by the epistemic agent as belief in q.

In Section 5.5 close relationships were shown to hold between operations of descriptor revision ($\circ$), relations of epistemic proximity ($\succeq$), and belief set orderings ($\leqq$). From a given belief set ordering $\leqq$ we can derive a descriptor revision $\circ$, and from $\circ$ we can regain $\leqq$. Similarly, given $\leqq$ we can derive a proximity relation $\succeq$, and from $\succeq$ we can regain $\leqq$. Obviously, $\circ$ and $\succeq$ are then also interderivable, so that from any one of these three formal constructs we can derive the other two. Is there a similar triangle of interderivability for the corresponding sentential operations and relations?

It is not difficult to show that $\leqq$ cannot be regained from $*$ or from $\stackrel{\star}{\succeq}$. Consider for simplicity a very small outcome set, consisting of the original belief set $K = \text{Cn}(\{p\})$ and the two additional elements $\text{Cn}(\{r\})$ and $\text{Cn}(\{p \vee r\})$. Let the two (transitive) belief set orderings $\leqq_1$ and $\leqq_2$ (with strict parts $<_1$ and $<_2$) be such that:

$\text{Cn}(\{p\}) <_1 \text{Cn}(\{r\}) <_1 \text{Cn}(\{p \vee r\})$
$\text{Cn}(\{p\}) <_2 \text{Cn}(\{p \vee r\}) <_2 \text{Cn}(\{r\})$

Let $*_1$ and $*_2$ be the sentential revisions that are derivable from the centrolinear descriptor revisions based on $\leqq_1$ respectively $\leqq_2$. Since $\text{Cn}(\{p \vee r\}) \subseteq \text{Cn}(\{p\})$

there is no x with $K *_1 x = \text{Cn}(\{p \vee r\})$ or $K *_2 x = \text{Cn}(\{p \vee r\})$.[9] It follows from this that $*_1 = *_2$ although $\leqq_1 \neq \leqq_2$. This is sufficient to show that the underlying belief set ordering ($\leqq$) cannot in general be regained from the sentential revision it gives rise to. The same example can be used to show that the underlying belief set ordering cannot either be regained from the believability relation it gives rise to.

However, in this example $\text{Cn}(\{p \vee r\})$ is a redundant element of the outcome set since it cannot be the outcome of a sentential revision. It would seem reasonable to remove such belief sets from the ordering before looking for interderivabilities. The following observation and definition show how to do this:

Observation 8.13 *Let $\leqq$ be a belief set ordering with the domain $\mathbb{X}$ and let K be its minimal element. Let $*$ be the sentential revision on K that is based on $\leqq$. A belief set $Z \in \mathbb{X}$ is an element of the outcome set of $*$ if and only if: $Z \nsubseteq \bigcup\{Y \in \mathbb{X} \mid Y < Z\}$.*

Definition 8.14 *([126]) Let $\leqq$ be an ordering on the set $\mathbb{X}$ of belief sets. Then the ordering $\overset{\star}{\leqq}$ such that*

$Z \overset{\star}{\leqq} W$ if and only if $Z \leqq W$, $Z \nsubseteq \bigcup\{Y \in \mathbb{X} \mid Y < Z\}$ and $W \nsubseteq \bigcup\{Y \in \mathbb{X} \mid Y < W\}$

is the additive restriction *of $\leqq$.*

The strict part of $\overset{\star}{\leqq}$ is denoted $\overset{\star}{<}$.

A belief set ordering $\leqq$ is additively restricted *if and only if it is its own additive restriction.*

But even after this adjustment, interderivability among $*$, $\overset{\star}{\leqq}$, and $\overset{\star}{\succeq}$ cannot be obtained. The relations of derivability and underivability among $\circ$, $*$, $\leqq$, $\overset{\star}{\leqq}$, $\succeq$, and $\overset{\star}{\succeq}$ are summarized in the following four observations and in Fig. 8.2.

Observation 8.15 ([126]) *(1) Let $\circ$ be a descriptor revision on the belief set K and let $*$ be the sentential revision on K that is derivable from $\circ$. It does not hold in general that $\circ$ is obtainable from $*$.*

(2) Let $\leqq$ be a belief set ordering and $\overset{\star}{\leqq}$ its additive restriction. It does not hold in general that $\leqq$ is obtainable from $\overset{\star}{\leqq}$.

(3) Let $\succeq$ be a relation of epistemic proximity and $\overset{\star}{\succeq}$ the believability relation that is based on it. It does not hold in general that $\succeq$ is obtainable from $\overset{\star}{\succeq}$.

Observation 8.16 ([126]) *Let K be a belief set, and let $\leqq$ be a descriptor-wellfounded belief set ordering with K as its minimal element. Furthermore, let $\circ = t^{b\circ}(\leqq)$, let $\overset{\star}{\leqq}$ be the additive restriction of $\leqq$, and let $*$ be the sentential revision that is derivable from $\circ$. Then:*

---

[9] $\leqq_1$ and $\leqq_2$ give rise to different descriptor revisions. Let $\circ_1$ and $\circ_2$ be the centrolinear descriptor revisions based on $\leqq_1$ respectively $\leqq_2$. We then have $K \circ_1 \neg\mathfrak{B}p = \text{Cn}(\{r\})$ and $K \circ_2 \neg\mathfrak{B}p = \text{Cn}(\{p \vee r\})$.

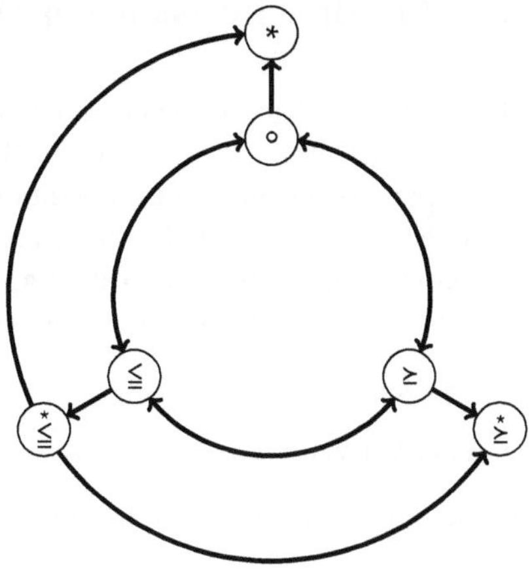

Fig. 8.2 Derivability diagram for sentential revision.

(1) $*$ is obtainable from $\overset{\star}{\leqq}$ as follows: For all p, (i) if p is satisfiable within the domain of $\overset{\star}{\leqq}$, then $K * p$ is the unique $\overset{\star}{\leqq}$-minimal p-containing element of the domain of $\overset{\star}{\leqq}$, and (ii) otherwise, $K * p = K$.

(2) It does not hold in general that $\overset{\star}{\leqq}$ is obtainable from $*$.

Observation 8.17 ([126]) *Let K be a belief set, $\leqq$ a descriptor-wellfounded belief set ordering with K as its minimal element, $\circ = t^{b\circ}(\leqq)$ and $\succeq = t^{bd}(\leqq)$. Furthermore, let $*$ be the sentential revision that is derivable from $\circ$ and $\overset{\star}{\succeq}$ the believability relation that is based on $\succeq$. Then:*

(1) It does not hold in general that $*$ is obtainable from $\overset{\star}{\succeq}$.

(2) It does not hold in general that $\overset{\star}{\succeq}$ is obtainable from $*$.

Observation 8.18 ([126]) *Let $\leqq$ be a descriptor-wellfounded belief set ordering and let $\succeq = t^{bd}(\leqq)$. Furthermore, let $\overset{\star}{\leqq}$ be the additive restriction of $\leqq$ and $\overset{\star}{\succeq}$ the believability relation that is based on $\succeq$. Then:*

(1) $\overset{\star}{\succeq}$ is obtainable from $\overset{\star}{\leqq}$ as follows: $p \overset{\star}{\succeq} q$ if and only if either (i) p and q are both satisfiable within the domain of $\overset{\star}{\leqq}$, and the first p-containing belief set does not come after the first q-containing one, or (ii) q is unsatisfiable within the domain of $\overset{\star}{\leqq}$.

(2) It does not hold in general that $\overset{\star}{\leqq}$ is obtainable from $\overset{\star}{\succeq}$.

In summary, we can obtain $*$ and $\overset{\star}{\succeq}$ from $\overset{\star}{\leqq}$, but in neither case is a derivation in the other direction possible. Furthermore, neither $*$ nor $\overset{\star}{\succeq}$ can be derived from the other. The latter underivability, in particular, marks an important limitation of believability relations.[10]

[10] Some additional results on believability relations can be found in [263].

8.5 Multiple Revision and Making up One's Mind

By multiple revision is meant revision that takes sets of sentences rather than single sentences as inputs. As we saw in Section 4.3 it comes in two major variants. In package revision, the success condition is that all sentences of the input set should be included in the outcome. In choice revision, the success condition is that at least one of them should be included. Denoting package revision by $\ast_{\forall}$ and choice revision by $\ast_{\exists}$, the success conditions are as follows:

$$A \subseteq K \ast_{\forall} A \qquad \text{(package success)}$$

$$A \cap (K \ast_{\exists} A) \neq \varnothing \qquad \text{(choice success)}$$

In the framework of descriptor revision, package revision can easily be obtained as follows:

$$K \ast_{\forall} A = K \circ \{\mathfrak{B}x \mid x \in A\}$$

Choice revision with a finite input set is equally easily obtainable[11]:

$$K \ast_{\exists} \{p_1, \ldots, p_n\} = K \circ \{\mathfrak{B}p_1 \vee \ldots \vee \mathfrak{B}p_n\}$$

One particularly interesting variant of multiple revision has already been mentioned: the operation of making up one's mind. It represents the process of choosing between belief in a sentence and in its negation. It cannot be constructed with the select-and-intersect method of traditional belief revision, but it is representable as the choice revision

$$K \overset{\ast}{\neg} p = K \ast_{\exists} \{p, \neg p\},$$

or equivalently:

Definition 8.19 *Let* $\circ$ *be a descriptor revision on the belief set* K. *The sentential operation* $\overset{\ast}{\neg}$ *on* K *such that*

$$K \overset{\ast}{\neg} p = K \circ \{\mathfrak{B}p \vee \mathfrak{B}\neg p\} \textit{ for all } p \in \mathcal{L}$$

is the operation of resolution (making up one's mind) *that is derivable from* $\circ$.

The following axiomatic characterization has been obtained for a class of operations of resolution that are derivable from monoselective descriptor revision:

Theorem 8.20 ([264], modified) *Let* $\overset{\ast}{\neg}$ *be a sentential operation on the consistent belief set* K. *Then the following two conditions are equivalent:*

(I) $\overset{\ast}{\neg}$ *is the operation of resolution that is derivable from some* K*-favouring monoselective descriptor revision on* K.

[11] Choice revision with an infinite input set would require an extension of the formal language to include infinite disjunctions. Due to their limited epistemological relevance such constructions will not be discussed here.

(II) $\overset{*}{\neg}$ *satisfies*

$K\overset{*}{\neg}p = \mathrm{Cn}(K\overset{*}{\neg}p)$ *(closure)*

If it holds for all q *that* $K\overset{*}{\neg}q \cap \{p, \neg p\} \neq \varnothing$ *if and only if* $K\overset{*}{\neg}q \cap \{p', \neg p'\} \neq \varnothing$, *then* $K\overset{*}{\neg}p = K\overset{*}{\neg}p'$. *(uniformity)*

If $K\overset{*}{\neg}q \cap \{p, \neg p\} \neq \varnothing$, *then* $K\overset{*}{\neg}p \cap \{p, \neg p\} \neq \varnothing$. *(negatable regularity)*

Either $p \in K\overset{*}{\neg}p$, $\neg p \in K\overset{*}{\neg}p$, *or* $K\overset{*}{\neg}p = K$. *(negatable relative success)*

If $p \in K$, *then* $K\overset{*}{\neg}p = K$. *(confirmation)*

A prominent property of operations of resolution, namely:

$$K\overset{*}{\neg}p = K\overset{*}{\neg}\neg p \qquad \text{(negation equivalence)}$$

is not mentioned in the theorem since it follows directly from one of the properties used in the theorem, namely uniformity.

The reader is referred to [264] for a more thorough investigation of operations representing the process of making up one's mind.

Chapter 9
Revocation

Contraction is one of the principal operations of the AGM framework. It satisfies two essential characteristics: (i) an input sentence is removed from the belief set (unless, of course, it is a tautology and so cannot be removed), and (ii) nothing is added to the belief set [99, p. 65]. This means that the following two postulates have to be satisfied:

$$p \notin (K \div p) \setminus \mathrm{Cn}(\varnothing) \qquad \text{(success)}$$
$$K \div p \subseteq K \qquad \text{(inclusion)}$$

These are two of the basic AGM postulates for contraction, and they are also satisfied by most of the weakened versions of AGM contraction that have been proposed in the literature.[1] But as we saw in Section 3.4, inclusion is not a particularly credible postulate. Although many belief changes have the purpose to remove a certain belief, such changes tend to be generated by some new information that is then also added to the belief set. This gives us a good reason to investigate a wider category of operations, namely all those whose success condition is the removal of a specified sentence. Such operations may or may not satisfy inclusion. They will be called operations of *revocation* and denoted by the operation sign $\div$.[2] (The sign $\div$ will be used for contraction, as is customary.)

As we saw in Section 3.2, the success postulate has also been criticized for being unrealistic. However, there is an important difference. The criticism of the success postulate is that it does not hold for all p. It cannot sensibly be denied that in *some* operations with the purpose of giving up a particular belief, that belief is indeed given up. What is denied is that this happens with all non-tautologous input sentences. In other words, $p \notin K \div p$ holds for some but not all non-tautologous sentences p. The

[1] Following Makinson [175] such weakened versions are often called "withdrawals". See also [53, 188, 225].

[2] Booth et al. used the term "retraction" for another type of operations satisfying success but not inclusion [16].

S.O. Hansson, *Descriptor Revision*, Trends in Logic 46,
DOI 10.1007/978-3-319-53061-1_9

criticism against inclusion is much stronger: There does not seem to be any plausible sentential operation $\div$ such that $K \div p \subseteq K$ holds for any sentence p that is removed by $\div$.

Fortunately there is a very simple way to construct revocation by a sentence p in the framework of descriptor revision, namely as revision by the descriptor $\neg\mathfrak{B}p$:

Definition 9.1 *Let* $\circ$ *be a descriptor revision on the belief set* K. *The operation* $\div$ *such that*

$$K \div p = K \circ \neg\mathfrak{B}p$$

for all $p \in \mathcal{L}$ *is the* (deterministic) revocation *that is derivable from* $\circ$.

A comparison of Definitions 8.1 and 9.1 will confirm that revocation and sentential revision are closely related in the present framework. However, their close relationship is different in nature from the close relationship between contraction and (sentential) revision in the AGM framework. In the latter, contraction and revision are interdefinable, which means that the revision operation contains all the information we need to determine how to contract, and vice versa. That is not the case in descriptor revision. Here, the two operations of revocation and sentential revision are both special cases of the more general descriptor revision ($\circ$). As we will see, their properties are more similar to each other than those of the two AGM operations, but they are not interdefinable.

In Section 9.1 we will characterize the operations of revocation that are derivable from monoselective revision and in Section 9.2 those obtainable from centrolinear revision. In Section 9.3 entrenchment relations will be derived from the relations of epistemic proximity that were introduced in Section 5.5. Entrenchment is usually associated with AGM contraction but it turns out to be applicable to revocation as well. In Section 9.4 operations of multiple revocation, i.e. simultaneous revocation of more than one sentence, are introduced, and so are multiple entrenchment relations, i.e. entrenchment relations over sets of sentences, rather than single sentences. The interconnections among these constructs differ in interesting ways from those of the corresponding single-sentence constructions. Finally, in Section 9.5 the alternative approach of revisionary revocation is introduced. By this is meant that operations of sentential revision are used to give up beliefs.

9.1 Monoselective Revocation

The operations of revocation that are derivable from monoselective descriptor revision can be axiomatically characterized as follows:

Theorem 9.2 *Let* $\div$ *be a sentential operation on the consistent belief set* K. *Then the following two conditions are equivalent:*

(I) $\div$ *is the revocation that is derivable from some monoselective descriptor revision on* K.

(II) $\div$ *satisfies:*

$K \div p = \mathrm{Cn}(K \div p)$ *(closure)*

If it holds for all q that $K \div q \vdash p$ if and only if $K \div q \vdash p'$, then $K \div p = K \div p'$. (uniformity)

Either $K \div p \nvdash p$ or $K \div p = K$. (relative success)

If $K \div p \vdash p$, then $K \div q \vdash p$. (persistence)

The operations that satisfy the conditions of this theorem will be called *monoselective revocations*. The four postulates should all be familiar to the reader. Closure is one of the AGM postulates for contraction. Uniformity is a simple adjustment of the corresponding postulate for monoselective descriptor revision. It implies extensionality (If $\vdash p \leftrightarrow q$ then $K \div p = K \div q$). Closure and extensionality are the only two of the AGM postulates for contraction that hold for monoselective revocation.

The remaining two postulates of the theorem, relative success and persistence, both serve as weakenings of the common success postulate for contraction ($p \notin (K \div p) \setminus \mathrm{Cn}(\varnothing)$). Relative success is a version of the postulate with the same name that we have used for descriptor revision, and persistence is a version of the regularity postulate for descriptor revision. A comparison between the postulates in Theorems 8.2 and 9.2 will confirm the close connection in this framework between sentential revision and revocation. (In comparison, the similarities between the basic postulates for AGM revision and contraction are much less prominent.)

The following observation tells us what is required for the AGM postulates success and vacuity to be satisfied.

Observation 9.3 *Let $\div$ be the monoselective revocation on the consistent belief set K that is based on the selection function $\widehat{C}$. Let $\mathbb{X}$ be its outcome set. Then:*

(1) $\div$ satisfies success ($p \notin (K \div p) \setminus \mathrm{Cn}(\varnothing)$) if and only if $\bigcap \mathbb{X} = \mathrm{Cn}(\varnothing)$.
(2) $\div$ satisfies vacuity (If $p \notin K$ then $K \div p = K$) if and only if $\widehat{C}(\llbracket \neg\mathfrak{B}p \rrbracket) = K$ whenever $K \in \llbracket \neg\mathfrak{B}p \rrbracket$.[3]

The two remaining basic AGM postulates, inclusion and recovery, are closely associated with operations of contraction rather than with revocation in general. They will be discussed in Chapter 10.

9.2 Centrolinear Revocation

An operation of revocation can be straightforwardly derived from a centrolinear revision; we just have to focus on descriptors of the form $\neg\mathfrak{B}p$. An alternative way to obtain the same result is to use a partial descriptor revision $\circ$ that is defined for all descriptors of that form but not necessarily for other types of descriptors. The alternative method allows us to relax the wellfoundedness condition on the underlying belief set ordering $\leqq$. We do not have to require that every descriptor-definable subset

[3]We can describe this condition as $\widehat{C}$ being K-favouring for negated atomic descriptors.

of the outcome set $\mathbb{X}$ has a $\leqq$-minimal element. It is sufficient that this holds for those subsets of $\mathbb{X}$ that are definable in terms a descriptor of the form $\neg\mathfrak{B}p$. An axiomatic characterization of this construction is available. The following two definitions serve to introduce it:

Definition 9.4 *A relation $\leqq$ on a set $\mathbb{X}$ of belief sets is* wellfounded for $\neg\mathfrak{B}$ descriptors *(negated atomic descriptors) if and only if it holds for every $p \in \mathcal{L}$ that if $[\![\neg\mathfrak{B}p]\!]$ is non-empty, then it has a $\leqq$-minimal element.*

Definition 9.5 *An operation $\circ$ on a belief set K is a* partial linear descriptor revision *if and only if there is a set $\mathbb{X}$ of belief sets with $K \in \mathbb{X}$ and a relation $\leqq$ on $\mathbb{X}$, such that for all descriptors Ψ:*

(i) If $[\![\Psi]\!]$ has a unique $\leqq$-minimal element, then $K \circ \Psi$ is equal to that element,

(ii) If $[\![\Psi]\!] = \varnothing$, then $K \circ \Psi = K$, and

(iii) If $[\![\Psi]\!]$ is non-empty but has no unique $\leqq$-minimal element, then $K \circ \Psi$ is undefined.

Furthermore, if K is the unique $\leqq$-minimal element of $\mathbb{X}$, then $\circ$ is a partial centrolinear descriptor revision.

With this we are ready for the axiomatic characterization:

Theorem 9.6 *Let $\div$ be a sentential operation on the consistent belief set K. Then the following two conditions are equivalent:*

(I) $\div$ is the revocation that is derivable from some partial centrolinear descriptor revision $\circ$ on K that is based on a belief set ordering $\leqq$ that is wellfounded for $\neg\mathfrak{B}$ descriptors.

(II) $\div$ satisfies the following conditions:

$K \div p = \mathrm{Cn}(K \div p)$ *(closure)*

Either $K \div p \nvdash p$ or $K \div p = K$. (relative success)

If $K \div p \vdash p$, then $K \div q \vdash p$. (persistence)

If $p \notin K$, then $K \div p = K$. (vacuity)

Either $K \div (p\&q) = K \div p$ or $K \div (p\&q) = K \div q$. (decomposition)

If $p \notin K \div q$, then $K \div (p\&q) = K \div p$. (conjunctive adjunction)

If $K \div p = K \div p'$, then $K \div (p\&q) = K \div (p'\&q)$. (conjunctive composition)

Three new postulates appear in this theorem. Decomposition was discussed already in 1982 by Alchourrón and Makinson [3, p. 35]. In the AGM paper three years later

[1, p. 525] it was shown to hold for maxichoice contraction (i.e. AGM contraction such that $\gamma(K \perp p)$ has at most one element).[4] As we saw in Section 2.5, maxichoice contraction has utterly implausible properties. In the framework of descriptor revision, decomposition does not take us to maxichoice contraction, but it is nevertheless a problematic property. Sometimes when we have to give up $p\&q$ we are equally (un)willing to give up p as to give up q, and the solution can then be to give up both. It may nevertheless be the case that $q \in K \div p$ and $p \in K \div q$, and then decomposition cannot hold. Ways to avoid the decomposition postulate will be discussed in Sections 9.5 and 10.4–10.6.

Conjunctive adjunction does not seem to have appeared previously in the literature. Provided that decomposition holds, it is reasonable to assume that conjunctive adjunction holds as well. To see this, note that if $p \notin K \div q$, then either (1) $K \div p = K \div q$ or (2) the belief set not containing p that is closest at hand for the agent is closer at hand than $K \div q$. In both cases, decomposition gives us reason to conclude that $K \div (p\&q) = K \div p$.

Conjunctive composition also seems to be a new postulate, but it holds for AGM contraction.[5] It says that if revocation by two sentences yields the same outcome, then these sentences are also exchangeable in revocations by conjunctions.

One of the postulates of Theorem 9.2, namely uniformity, is conspicuously absent from Theorem 9.6. However, that is only because it follows from the other postulates. The following observation shows that centrolinear revocation satisfies uniformity, and at the same time it introduces another property that holds for this operation.

Observation 9.7 *Let $\div$ be a sentential operation on a belief set K, with $\mathbb{X}$ as its outcome set. Then:*

(1) If $\div$ satisfies conjunctive adjunction, then it satisfies:

If $p \notin K \div q$ and $q \notin K \div p$, then $K \div p = K \div q$. (reciprocity)

(2) If $\div$ satisfies closure, relative success, persistence, and conjunctive adjunction, then it satisfies:

If it holds for all q that $K \div q \vdash p$ if and only if $K \div q \vdash p'$, then $K \div p = K \div p'$. (uniformity)

Reciprocity is closely related to the postulates with the same name for descriptor revision and sentential revision that we have discussed above (Sections 5.2 and 8.2). The variant for revocation does not appear in the AGM literature, but there we find instead the following postulate:

[4] The decomposition postulate should be distinguished from the decomposition principle that was discussed in Section 1.2.

[5] This is perhaps most easily seen in sphere systems. Let $[K \div p] = [K \div p']$. There are three cases. (1) If $K \div q$ is situated in a more central sphere than $[K \div p]$, then $[K \div (p\&q)] = [K \div (p'\&q] = [K \div q]$. (2) If $K \div q$ is situated in the same sphere as $[K \div p]$, then $[K \div (p\&q)] = [K \div p] \cup [K \div q] = [K \div p'] \cup [K \div q] = [K \div (p'\&q)]$. (3) If $K \div q$ is situated in a more peripheral sphere than $[K \div p]$, then $[K \div (p\&q)] = [K \div p] = [K \div p'] = [K \div (p'\&q)]$.

If $p \rightarrow q \in K \div q$ and $q \rightarrow p \in K \div p$, then $K \div p = K \div q$.
(implicative reciprocity)

Implicative reciprocity holds for full-blown AGM contraction, i.e. transitively relational partial meet contraction. This was shown by Hans Rott who also proved that in the presence of the basic AGM postulates, implicative reciprocity is equivalent via the Levi identity with the reciprocity condition for the corresponding sentential revision (If $p \in K * q$ and $q \in K * p$ then $K * p = K * q$).[6] In the presence of some of the basic AGM postulates, our reciprocity condition for revocation implies implicative reciprocity, but contrary to the latter it does not follow from the full set of AGM postulates:

Observation 9.8 *Let K be a belief set and $\div$ an operation on K that satisfies closure, success, extensionality and reciprocity. Then $\div$ satisfies implicative reciprocity.*

Observation 9.9 *Let K be a belief set and $\div$ a transitively relational partial meet contraction on K. It does not hold in general that $\div$ satisfies reciprocity.*

Although our reciprocity postulate may be unsuitable for the AGM framework it has some support in common patterns of belief change.

On several occasions I have heard Iminathi speak enthusiastically about the French novelist and poet Victor Hugo. She seems to know everything about him and his literary works. This makes me believe that she has read all his major works in the original French (r), and therefore also that she is proficient in French (f).

Case (i): I overhear her saying: "Thanks for sending me the newspaper clipping but unfortunately I don't read French." I contract my belief in f. In doing so I also lose my belief in r.

Case (ii): I overhear her saying: "I can't answer your question about *Les Misérables* since I have only read it in English." I contract my belief in r. In doing so I also lose my belief in f.

In this example we have $f \notin K \div r$ and $r \notin K \div f$. This suggests that giving up f and giving up r are processes with one and the same outcome, in other words that $K \div f = K \div r$.

Two of the basic AGM postulates, namely closure and vacuity, are among those used in Theorem 9.6 to characterize centrolinear revocation. A third AGM postulate, namely extensionality, holds for all centrolinear contractions although it is not used in the axiomatization. (It follows from uniformity.) The other three, namely success, inclusion, and recovery, do not hold in general for this operation. For success to hold, the same condition has to be satisfied that we used in Observation 9.3, namely $\bigcap \mathbb{X} = \mathrm{Cn}(\varnothing)$.

[6]Hans Rott showed that implicative reciprocity is equivalent to the conjunction of his conditions $\div$7c and $\div$8c [213, p. 54]. He also showed that $\div$7c corresponds exactly via the Levi identity to $*$7c and $\div$8c similarly to $*$8c. [213, pp. 50, 67–68] A proof that reciprocity for sentential revision is equivalent with the conjunction of $*$7c and $*$8c can be found in [73, p. 54] or [99, p. 274].

9.3 Entrenchment

In Section 5.5, relations of epistemic proximity were introduced as a generalization of epistemic entrenchment, and we also saw that an entrenchment relation $\lessdot$ can be derived from a relation $\succeq$ of epistemic proximity in the following way:

$p \lessdot q$ if and only if $\neg\mathfrak{B}p \succeq \neg\mathfrak{B}q$.

The strict part of $\lessdot$ is denoted $\lessdot$ and its symmetrical part $\dot{\simeq}$.

This definition does not refer to contraction but to the descriptors characteristic of revocation, namely descriptors of the form $\neg\mathfrak{B}p$. Consequently, entrenchment is defined not only for contraction but also, more generally, for revocation. As we saw in Observation 5.37, the entrenchment relation obtainable via the above definition satisfies all the standard properties of epistemic entrenchment, namely

If $p \lessdot q$ and $q \lessdot r$, then $p \lessdot r$.	(transitivity)
If $p \vdash q$, then $p \lessdot q$.	(dominance)
Either $p \lessdot p\&q$ or $q \lessdot p\&q$.	(conjunctiveness)
$p \notin \{r \mid \bot \lessdot r\}$ if and only if $p \lessdot q$ for all q.	(minimality)
If $q \lessdot p$ for all q, then $\vdash p$.	(maximality)

The set $\{r \mid \bot \lessdot r\}$ mentioned in the definition of minimality is identical with K. To see why this is so, let $\lessdot$ be derived from a belief set ordering $\leqq$, and let K be the $\leqq$-minimal set. Then $r \notin K$ holds if and only if $\neg\mathfrak{B}r \succeq \Psi$ for all descriptors Ψ. Focusing on revocation outcomes we can revise this condition to $\neg\mathfrak{B}r \succeq \neg\mathfrak{B}q$ for all $q \in \mathcal{L}$. Since $K \Vdash \neg\mathfrak{B}_\bot$ we have $\neg\mathfrak{B}_\bot \succeq \neg\mathfrak{B}q$ for all q, thus our condition that $\neg\mathfrak{B}r \succeq \neg\mathfrak{B}q$ for all q is equivalent to $\neg\mathfrak{B}r \simeq \neg\mathfrak{B}_\bot$. Thus due to Definition 5.36, $r \notin K$ if and only if $r \dot{\simeq} \bot$. Equivalently, $r \in K$ if and only if $r \not\dot{\simeq} \bot$. Due to dominance this means that $r \in K$ if and only if $\bot \lessdot r$. Thus $K = \{r \mid \bot \lessdot r\}$.[7]

In Section 5.5 we investigated the relationships among a belief set ordering $\leqq$, the centrolinear descriptor $\circ$ that is derivable from it, and the relation $\succeq$ of epistemic proximity that is based on it. We found that from each of these three constructs we can derive the other two. In Section 8.4 we performed a similar investigation on the part $\overset{\star}{\leqq}$ of $\leqq$ that is relevant for sentential revision (its additive restriction), the sentential revision $*$ that is derivable from $\circ$, and the believability relation $\overset{\star}{\succeq}$ that is derivable from $\succeq$. We found that although both $\overset{\star}{\succeq}$ and $*$ can be obtained from $\overset{\star}{\leqq}$, neither $\overset{\star}{\leqq}$ nor $*$ can be obtained from $\overset{\star}{\succeq}$, and neither $\overset{\star}{\leqq}$ nor $\overset{\star}{\succeq}$ can be obtained from $*$. Let us now proceed with a corresponding investigation for revocation.

In order to do so we need to restrict $\leqq$ to the elements of $\mathbb{X}$ that can be outcomes of a revocation. The following observation and definition show how this can be done:

---

[7]The same definition of K is obtainable in the AGM model. It follows from minimality (in the original version that was presented in Section 1.5) that $r \in K$ if and only if it is not the case that $r \lessdot s$ for all s, i.e. if and only if there is some s with $s \lessdot r$. It can straightforwardly be shown that $\bot \lessdot r$ holds if and only if there is some s with $s \lessdot r$. (For the non-trivial direction, let $s \lessdot r$, use dominance to obtain $\bot \lessdot s$ and then transitivity to obtain $\bot \lessdot r$.).

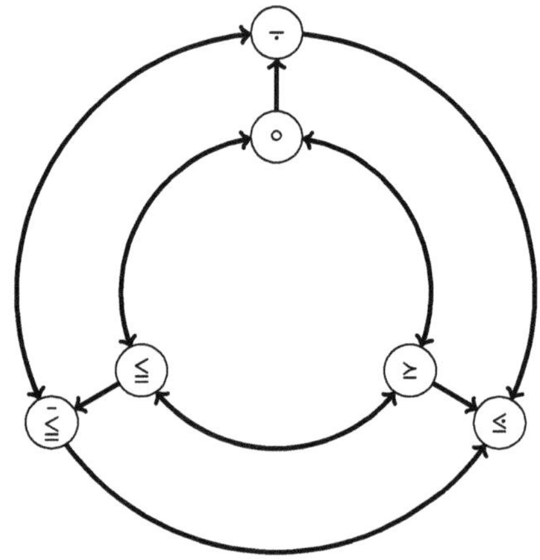

Fig. 9.1 Derivability diagram for revocation.

Observation 9.10 *Let $\leqq$ be a belief set ordering with the domain $\mathbb{X}$ and let K be its minimal element. Let $\div$ be the revocation on K that is based on $\leqq$. A belief set $Z \in \mathbb{X}$ is an element of the outcome set of $\div$ if and only if:*
$\bigcap\{Y \in \mathbb{X} \mid Y < Z\} \nsubseteq Z$.

Definition 9.11 ([126]) *Let $\leqq$ be an ordering on the set $\mathbb{X}$ of belief sets. Then the ordering $\bar{\leqq}$ such that*

$Z \bar{\leqq} W$ *if and only if*

$Z \leqq W$, $\bigcap\{Y \in \mathbb{X} \mid Y < Z\} \nsubseteq Z$, *and* $\bigcap\{Y \in \mathbb{X} \mid Y < W\} \nsubseteq W$

is the subtractive restriction *of* $\leqq$.

The strict part of $\bar{\leqq}$ is denoted $\bar{<}$.

A belief set ordering $\leqq$ is subtractively restricted *if and only if it is its own subtractive restriction.*

The following four observations provide us with the (un)derivabilities required. They are summarized in Fig. 9.1.

Observation 9.12 ([126]) *(1) Let $\circ$ be a centrolinear revision on a belief set K and $\div$ its derived operation of revocation. It does not hold in general that $\circ$ is derivable from $\div$ (not even if $\div$ is a contraction*[8]*).*

(2) Let $\leqq$ be a belief set ordering and $\bar{\leqq}$ its subtractive restriction. It does not hold in general that $\leqq$ is derivable from $\bar{\leqq}$ (not even if the revocation obtainable from $\leqq$ is a contraction).

(3) Let $\succeq$ be a relation of epistemic proximity and $\leq$ its corresponding entrenchment relation. It does not hold in general that $\succeq$ is derivable from $\leq$ (not even

[8] $\div$ is a contraction if and only if all elements of its outcome set are subsets of K.

if $\succeq$ is derivable from a belief set ordering whose centrolinear revocation is a contraction).

Observation 9.13 ([126]) *Let K be a belief set, $\leqq$ a descriptor-wellfounded belief set ordering with K as its minimal element, and $\circ = t^{b\circ}(\leqq)$. Furthermore, let $\bar{\leqq}$ be the subtractive restriction of $\leqq$ and $\div$ the operation of revocation that is derivable from $\circ$. Then:*

(1) $\div$ is derivable from $\bar{\leqq}$ as follows: (i) If the domain of $\bar{\leqq}$ contains some element not containing p, then $K \div p$ is the $\bar{\leqq}$-minimal element not containing p, and (ii) otherwise, $K \div p = K$.

(2) $\bar{\leqq}$ is derivable from $\div$ as follows: $\bar{\leqq}$ is the relation on the outcome set of $\div$ such that $X \bar{\leqq} Y$ if and only if there are p and q such that $X = K \div p$, $Y = K \div q$, $K \div p \nvdash p$, $K \div q \nvdash q$, and $K \div p = K \div (p\&q)$.

Observation 9.14 ([126]) *Let K be a belief set, $\leqq$ a descriptor-wellfounded belief set ordering with K as its minimal element, $\circ = t^{b\circ}(\leqq)$ and $\succeq = t^{bd}(\leqq)$. Furthermore, let $\div$ be the operation of revocation that is derivable from $\circ$ and $\leq$ the entrenchment relation derivable from $\succeq$. Then:*

(1) $\leq$ is derivable from $\div$ as follows: $p \leq q$ if and only if $K \div p = K \div (p\&q)$.

(2) $\div$ is not derivable from $\leq$, not even under the assumption that $\div$ is a contraction.

Observation 9.15 ([126]) *Let $\leqq$ be a descriptor-wellfounded belief set ordering and $\succeq = t^{bd}(\leqq)$. Furthermore, let $\bar{\leqq}$ be the subtractive restriction of $\leqq$ and $\leq$ the entrenchment relation derivable from $\succeq$. Then:*

(1) $\leq$ is derivable from $\bar{\leqq}$ as follows: $p \leq q$ if and only if, if there are any sets not containing q in the domain of $\bar{\leqq}$, then they are all preceded by some set not containing p.[9]

(2) $\bar{\leqq}$ is not derivable from $\leq$.

In summary, somewhat more extensive relations of interdefinability prevail among $\div$, $\bar{\leqq}$, and $\leq$ than among $*$, $\overset{\star}{\leqq}$, and $\overset{\star}{\succeq}$. The interdefinability between $\leqq$ and $\circ$ is retained between $\bar{\leqq}$ and $\div$, and from either of these we can derive $\leq$. However, neither $\bar{\leqq}$ nor $\div$ can be derived from $\leq$. It is particularly notable that the operation of revocation ($\div$) cannot be obtained from the relation of entrenchment ($\leq$). This should be compared to the AGM framework where each entrenchment relation gives rise to a unique transitively relational partial meet contraction, according to the following construction [69, 71]:

---

[9]Note that this holds if q is irrevocable, i.e. all sets in the domain of $\bar{\leqq}$ contain q.

$q \in K \div p$ if and only if $q \in K$ and either $p \lessdot p \vee q$ or $\vdash p$.

Perhaps not surprisingly, the operations obtained from an entrenchment relation in this way are not in general centrolinear revocations:

Observation 9.16 *Let $\lesssim$ be a relation on sentences that satisfies transitivity, dominance, conjunctiveness, minimality, and maximality with respect to a belief set K. Let $\div$ be the operation on K such that for all $p, q \in \mathcal{L}$: $q \in K \div p$ if and only if $q \in K$ and either $p \lessdot p \vee q$ or $\vdash p$. It does not hold in general that $\div$ is a centrolinear revocation.*

It follows from part 2 of Observation 9.15 that a given entrenchment relation can be obtainable from more than one belief set ordering. The following theorem and observation provide an exact characterization of these belief set orderings.

Theorem 9.17 ([126]) *Let $\lesssim$ satisfy transitivity, dominance, and conjunctiveness. Then $\lesssim$ is the entrenchment relation derivable from some belief set ordering $\leqq$ if and only if $\leqq$ is constructible by assigning to each equivalence class in $\lesssim$ exactly one logically closed set E_p such that*

$$\{s \mid p \lessdot s\} \subseteq E_p \subseteq \{s \mid (\forall t)(s \vee t \not\lesssim p)\}$$

for some p in that equivalence class, and letting $E_p \leqq E_q$ if and only if $p \lesssim q$.[10]

Observation 9.18 ([126]) *Let $\lesssim$ satisfy transitivity, dominance, and conjunctiveness. Then the sets $\{s \mid p \lessdot s\}$ and $\{s \mid (\forall t)(s \vee t \not\lesssim p)\}$ are both logically closed.*

Let $\lesssim$ be an entrenchment relation. For each sentence p, Theorem 9.17 provides us with a lower and an upper limit for the outcome $K \div p$ of any centrolinear revocation $\div$ that can be associated with $\lesssim$. This means that $K \div p$ has to be a superset of the lower limit and a subset of the upper limit. The lower limit can be further characterized as follows:

Definition 9.19 ([126]) *A belief set ordering $\leqq$ is* shrinking *if and only if it holds for all X and Y in its domain that if $X < Y$ then $Y \subset X$.*

Theorem 9.20 ([126]) *Let K be a belief set and $\leqq$ a subtractively restricted belief set ordering with K as its minimal element. Let $\div$ be the revocation on K and $\lesssim$ the entrenchment relation that are derivable from $\leqq$. Then the following four conditions are equivalent:*

(1) $\leqq$ is shrinking,

[10] As shown in Lemma 9.1 (p. 212), a logically closed set is a subset of $\{s \mid (\forall t)(s \vee t \not\lesssim p)\}$ if and only if it is a subset of $\{s \mid p \not\lesssim s\}$. Therefore, this theorem can alternatively be formulated with $\{s \mid p \not\lesssim s\}$ instead of $\{s \mid (\forall t)(s \vee t \not\lesssim p)\}$ as the upper limit. The present formulation is preferred because $\{s \mid (\forall t)(s \vee t \not\lesssim p)\}$ is logically closed, and therefore it is an unambiguous upper bound that E_p can coincide with.

(2) $\div$ *satisfies:* $K \div p \subseteq K \div q$ *or* $K \div q \subseteq K \div p$,

(3) $\div$ *is derivable from* $\leqslant$ *as follows:* $K \div p = \{s \mid p < s\}$, *and*

(4) $\leqq$ *is derivable from* $\leqslant$ *as follows:* $\leqq$ *is the relation on the sets obtainable as* $\{s \mid p < s\}$ *for some* $p \in \mathcal{L}$, *such that* $X \leqq Y$ *if and only if* $Y \subseteq X$.

Corollary to Theorems 9.17 and 9.20 *A relation* $\leqslant$ *on sentences is the entrenchment relation derivable from some belief set ordering if and only if it is the entrenchment relation derivable from some shrinking belief set ordering.*

It follows that if we construct $\div$ from a given entrenchment relation $\leqslant$ via the lower limit of belief set orderings compatible with $\leqslant$, then $\div$ will be a contraction. But more than that: It will be a centrolinear contraction with the rather special property that as we move away from K on the line defined by $\leqq$, each belief set is a proper subset of its predecessor. To see that such an operation is not plausible, note that it has the property:

If $q \in K \div p$, then $K \div (p\&q) = K \div p$.

Let q denote that Elvis Presley was born in 1935 and p that he died in 1977. I currently have both these beliefs. If I can in some way be induced to give up my belief that he died in 1977, then I will still retain my belief that he was born in 1935, i.e. $q \in K \div p$. However, if I am made to give up my belief that “Elvis Presley was born in 1935 and died in 1977”, then I will in all probability retain my belief about his death year and suspend my belief about his year of birth, i.e. $K \div (p\&q) = K \div q \neq K \div p$.

We can conclude that the minimal centrolinear revocation that is compatible with a given entrenchment relation is an implausible operation. It does not deserve a privileged or “canonical” status among the revocations that are compatible with the entrenchment relation. The question remains open whether there is some other construction that provides us with a more plausible candidate for a revocation with such a special status.

9.4 Multiple Revocation and Entrenchment

Multiple contraction is a generalized form of contraction that takes sets of sentences rather than single sentences as inputs. As mentioned in Section 4.3 there are two major types of multiple contraction, choice contraction and package contraction. The success condition of choice contraction is that at least one of the input sentences is absent from the outcome, whereas that of package contraction is that all of them are absent.[11] Analogous revocations can be defined as follows:

Definition 9.21 ([126]) *Let* $\circ$ *be a descriptor revision on the belief set* K.

(1) The operation $\div_{\vee}$ *such that*

[11] Previous studies of multiple contraction include [55, 64, 81, 123, 133, 168, 185, 198, 261, 262]. On iterated multiple contraction see [260].

$K \div_{\forall} A = K \circ \{\neg\mathfrak{B}a \mid a \in A\}$

for all sets $A \subseteq \mathcal{L}$ *is the* (multiple) package revocation *derivable from* $\circ$.

(2) The operation $\div_{\exists}$ *such that*

$K \div_{\exists} \{a_1, \ldots, a_n\} = K \circ \{\neg\mathfrak{B}a_1 \vee \ldots \vee \neg\mathfrak{B}a_n\}$

for all finite sets $\{a_1, \ldots, a_n\} \subseteq \mathcal{L}$ *is the* (multiple) choice revocation *derivable from* $\circ$.

The success condition of choice revocation, $\neg\mathfrak{B}a_1 \vee \ldots \vee \neg\mathfrak{B}a_n$, is equivalent with the simpler condition $\neg\mathfrak{B}(a_1 \& \ldots \& a_n)$. Therefore choice revocation is definable in terms of common (single-sentence) revocation.[12]

The definition of entrenchment relations can be extended to cover the inputs of multiple revocation. In what follows we will focus on package revocation.

Definition 9.22 *Let* $\succeq$ *be a relation of epistemic proximity. The relation* $\bar{\bar{\lesssim}}$ *on sets of sentences, such that*

$A \bar{\bar{\lesssim}} B$ *if and only if* $\{\neg\mathfrak{B}a \mid a \in A\} \succeq \{\neg\mathfrak{B}b \mid b \in B\}$

is the relation of multiple (epistemic) entrenchment *that is based on* $\succeq$.

The strict part of $\bar{\bar{\lesssim}}$ *is denoted* $\bar{\bar{<}}$.

We need to adjust the definition of a subtractive restriction of a belief set ordering.[13]

Observation 9.23 *Let* $\leqq$ *be a belief set ordering with the domain* $\mathbb{X}$, *and let* K *be its minimal element. Let* $\div_{\forall}$ *be the (multiple) package revocation on* K *that is based on* $\leqq$. *A belief set* $Z \in \mathbb{X}$ *is an element of the outcome set of* $\div_{\forall}$ *if and only if:*

$Y \not\subseteq Z$ *for all* $Y < Z$.

Definition 9.24 *Let* $\leqq$ *be an ordering on the set* $\mathbb{X}$ *of belief sets. Then the ordering* $\bar{\bar{\leqq}}$ *such that*

$Z \bar{\bar{\leqq}} W$ *if and only if* $Z \leqq W$, $Y \not\subseteq Z$ *for all* $Y < Z$, *and* $Y \not\subseteq W$ *for all* $Y < W$

is the multiply subtractive restriction *of* $\leqq$. *Its strict part is denoted* $\bar{\bar{<}}$.

[12] This applies only to choice revocation by finite sets. Choice revocation by infinite sets can be defined by introducing infinite disjunctions into the language. Cf. Section 8.5 on the same issue for choice revision.

[13] To see why this is necessary, let the language have only the two atoms p and q, and let $\leqq$ be completely characterized by

$\text{Cn}(\{p\&q\}) < \text{Cn}(\{p\}) < \text{Cn}(\{q\}) < \text{Cn}(\{p \leftrightarrow q\}) < \text{Cn}(\{p \vee q\}) < \text{Cn}(\{\varnothing\})$

Neither $\text{Cn}(\{p \vee q\})$ nor $\text{Cn}(\varnothing)$ is an element of the outcome set of the single-sentence revocation based on $\leqq$, and neither is retained in $\bar{\leqq}$. However, both are elements of the outcome set of the (multiple) package revocation based on $\leqq$, since $K \div_{\forall} \{p, q, p \leftrightarrow q\} = \text{Cn}(\{p \vee q\})$ and $K \div_{\forall} \{p \vee q, p \leftrightarrow q\} = \text{Cn}(\varnothing)$.

We can now repeat the investigations of interdefinabilities that we performed in Section 9.3, replacing the single-sentence constructions $\div$, $\bar{\leqq}$, and $\lesssim$ by their multiple versions $\div_{\vee}$, $\bar{\bar{\leqq}}$, and $\bar{\bar{\lesssim}}$. The results are summarized in the following four observations.

Observation 9.25 *(1) Let $\circ$ be a centrolinear descriptor revision on a belief set K and $\div_{\vee}$ its derived operation of (multiple) package revocation. It does not hold in general that $\circ$ is derivable from $\div_{\vee}$ (not even if $\div_{\vee}$ is a package contraction).*

(2) Let $\leqq$ be a belief set ordering and $\bar{\bar{\leqq}}$ its multiply subtractive restriction. It does not hold in general that $\leqq$ is derivable from $\bar{\bar{\leqq}}$ (not even if the package revocation obtainable from $\leqq$ is a package contraction).

(3) Let $\succeq$ be a proximity relation and $\bar{\bar{\lesssim}}$ its corresponding multiple entrenchment relation. It does not hold in general that $\succeq$ is derivable from $\bar{\bar{\lesssim}}$ (not even if $\succeq$ is derivable from a belief set ordering whose centrolinear package revocation is a package contraction).

Definition 9.26 ([116, p. 429]) *Let $X, Y \subseteq \mathcal{L}$. The* conjunctive product $X \overset{\&}{\times} Y$ *is defined as follows:*

$$X \overset{\&}{\times} Y = \{x \& y \mid x \in X \text{ and } y \in Y\}$$

Observation 9.27 *Let K be a belief set, $\leqq$ a descriptor-wellfounded belief set ordering with K as its minimal element, and $\circ = t^{bo}(\leqq)$. Furthermore, let $\bar{\bar{\leqq}}$ be the multiply subtractive restriction of $\leqq$, and $\div_{\vee}$ the operation of package revocation that is derived from $\circ$. Then:*

(1) $\div_{\vee}$ is derivable from $\bar{\bar{\leqq}}$ as follows: (i) If the domain of $\bar{\bar{\leqq}}$ has some element not containing any element of A, then $K \div_{\vee} A$ is the $\bar{\bar{\leqq}}$-minimal element not containing any element of A, and (ii) otherwise, $K \div_{\vee} A = K$.

(2) $\bar{\bar{\leqq}}$ is derivable from $\div_{\vee}$ as follows: $\bar{\bar{\leqq}}$ is the relation on the outcome set of $\div_{\vee}$ such that $X \mathrel{\bar{\bar{\leqq}}} Y$ if and only if there are A and B such that $X = K \div_{\vee} A$, $Y = K \div_{\vee} B$, $K \div_{\vee} A \nvdash_{\exists} A$, $K \div_{\vee} B \nvdash_{\exists} B$, and $K \div_{\vee} A = K \div_{\vee} (A \overset{\&}{\times} B)$.

Observation 9.28 *Let K be a belief set, $\leqq$ a descriptor-wellfounded belief set ordering with K as its minimal element, $\circ = t^{bo}(\leqq)$ and $\succeq = t^{bd}(\leqq)$. Furthermore, let $\div_{\vee}$ be the operation of multiple revocation derived from $\circ$ and $\bar{\bar{\lesssim}}$ the multiple entrenchment relation derived from $\succeq$. Then:*

(1) $\bar{\bar{\lesssim}}$ is derivable from $\div_{\vee}$ as follows: $A \mathrel{\bar{\bar{\lesssim}}} B$ if and only if $K \div_{\vee} A = K \div_{\vee} (A \overset{\&}{\times} B)$.

(2) $\div_{\vee}$ is derivable from $\bar{\bar{\lesssim}}$ as follows: $q \in K \div_{\vee} A$ if and only if $A \mathrel{\bar{\bar{<}}} A \cup \{q\}$.

Observation 9.29 *Let $\leqq$ be a descriptor-wellfounded belief set ordering and $\succeq = t^{bd}(\leqq)$. Furthermore, let $\bar{\bar{\leqq}}$ be the multiply subtractive restriction of $\leqq$ and $\bar{\bar{\lesssim}}$ the multiple entrenchment relation derived from $\succeq$. Then:*

(1) $\overset{=}{\lesssim}$ *is derivable from* $\overset{=}{\leqq}$ *as follows:* $A\overset{=}{\lesssim}B$ *if and only if: if there are in the domain of* $\overset{=}{\leqq}$ *any sets not containing any element of* B*, then they are all preceded by some set not containing any element of* A.

(2) $\overset{=}{\leqq}$ *is derivable from* $\overset{=}{\lesssim}$ *as follows: For any element* X *of the domain of* $\overset{=}{\leqq}$, *let* $\mathfrak{b}(X)$ *be some set such that* $X \cap \mathfrak{b}(X) = \varnothing$ *but* $Z \cap \mathfrak{b}(X) \neq \varnothing$ *for all* Z *with* $Z\overset{=}{<}X$*. Then:* $X\overset{=}{\leqq}Y$ *if and only if* $\mathfrak{b}(X)\overset{=}{\lesssim}\mathfrak{b}(Y)$.

Multiple revocation ($\underset{\cdot}{-}_{\forall}$) has an intermediate position between general descriptor revision ($\circ$) and single-sentence revocation ($\underset{\cdot}{-}$), in the sense that $\underset{\cdot}{-}_{\forall}$ is a special case of $\circ$ and $\underset{\cdot}{-}$ a special case of $\underset{\cdot}{-}_{\forall}$. In the same way, $\overset{=}{\leqq}$ is intermediate between $\leqq$ and $\bar{\leqq}$, and $\overset{=}{\lesssim}$ is intermediate between $\succeq$ and $\lesssim$. However, these derivabilities are all one-directional.

Observation 9.30 *Let* $\leqq$ *be a belief set ordering with* K *as its minimal element. Let* $\overset{=}{\leqq}$ *be the multiply subtractive restriction of* $\leqq$ *and* $\bar{\leqq}$ *its (single-sentence) subtractive restriction. Furthermore, let* $\underset{\cdot}{-}_{\forall}$ *be the (multiple) package revocation and* $\underset{\cdot}{-}$ *the (single-sentence) revocation that are derivable from* $t^{bo}(\leqq)$*, and let* $\overset{=}{\lesssim}$ *be the multiple and* $\lesssim$ *the single-sentence entrenchment relations that are based on* $t^{bd}(\succeq)$*. Then:*

(1) $\bar{\leqq}$ *is derivable from* $\overset{=}{\leqq}$.

(2) $\underset{\cdot}{-}$ *is derivable from* $\underset{\cdot}{-}_{\forall}$.

(3) $\lesssim$ *is derivable from* $\overset{=}{\lesssim}$.

(4) *It does not hold in general that* $\overset{=}{\leqq}$ *is derivable from* $\bar{\leqq}$.

(5) *It does not hold in general that* $\underset{\cdot}{-}_{\forall}$ *is derivable from* $\underset{\cdot}{-}$.

(6) *It does not hold in general that* $\overset{=}{\lesssim}$ *is derivable from* $\lesssim$.

The derivability results obtained in this and the preceding section are summarized in Fig. 9.2. The innermost circle refers to general descriptor revision. This operation is fully interderivable both with an ordering on belief sets and with a proximity relation on descriptors. In the intermediate circle these three formal entities are restricted to the framework of (multiple) package revocation. Full interdefinability is retained between the operation of package revocation and the appropriately restricted versions of the two types of relations. Finally, the outermost circle refers to the further restriction of the three formal entities to single-sentence revocation. As we saw in Section 9.3, some of the interderivabilities are lost on this level.

9.5 A Revisionary Account of Giving up Beliefs

We saw in Theorem 9.6 that centrolinear revocation satisfies the property

Either $K \div (p\&q) = K \div p$ or $K \div (p\&q) = K \div q$. (decomposition)

We also noted that this is not a plausible property for revocation (or contraction). The following example corroborates its problematic nature:

> Pauline and Quentin are my next-door neighbours. Based on what I have seen, I believe both that Pauline is a safe and careful driver (p) and that Quentin is a safe and careful driver (q).
>
> *Case i:* I see their car passing the zebra crossing outside my children's school at very high speed. I catch a glimpse of the driver's face. It is a woman with long black hair, and it might very well be Pauline. I therefore give up my belief that Pauline is a safe driver (p), but I still believe that Quentin is a safe driver (q).
>
> *Case ii:* I see the car driven in the same way. I catch a glimpse of the driver's face. It is a man with a beard, and it might very well be Quentin. I give up q but retain p.
>
> *Case iii:* I see the car driven in the same way, but I cannot see the driver. I give up both p and q, suspending my judgment on who is the reckless driver [127].

In the first case, the outcome can be represented as $K \div p$, and we have $q \in K \div p$. Similarly, in the second case the operation $K \div q$ is performed, and we have $p \in K \div q$. In the third case we perform the operation $K \div (p\&q)$, and the facts are such that $p \notin K \div (p\&q)$ and $q \notin K \div (p\&q)$. The example represents what seems to be a common and quite reasonable pattern of giving up beliefs. Sometimes there are two (or more) sentences such that we need to remove (at least) one of them in order to

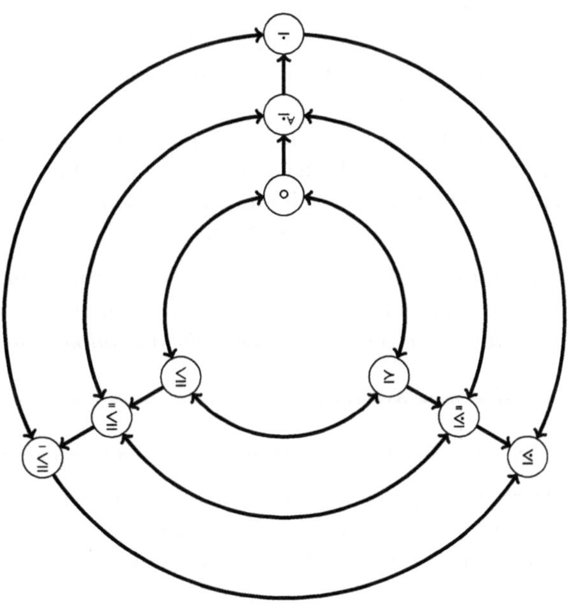

Fig. 9.2 Derivability diagram summarizing the results for package and single-sentence revocation. The inner circle represents general descriptor revision, the middle circle (multiple) package revocation and the outer circle single-sentence revocation.

perform a contraction. In the case of a tie between them, i.e. when the two potential removals are equally plausible, the most sensible solution is often to remove both of them. However, this pattern is incompatible with the postulate of decomposition, What has gone wrong?

A quite reasonable answer is that the problem arises only because we have decided to represent the example in terms of revocation (or contraction). That decision has (mis)led us to limit our attention to the information that was lost. The reason why I lost my belief in $p\&q$ in case (iii) was that I received and incorporated a piece of new information, namely that I have seen Pauline's and Quentin's car driven in a careless way, without seeing the driver. We can represent this information as c. Arguably, the belief change taking place in case (iii) should be represented as a revision by c, and the outcome should be a belief set $K * c$ that contains c but contains neither p nor q. Clearly, c is not included in the outcome in either case (i) or case (ii). Therefore, $K * c$ cannot be equal to either of them. This can readily be represented as a centrolinear revision in which $K * c$ is closest to K among the belief sets containing c, but it contains neither p nor q.

This can be called a *revisionary account of how we give up beliefs* [134]. It has much that speaks in its favour. However, we sometimes refer to someone as giving up a belief without specifying the incoming information that causes her to do so. ("She lost her belief in his innocence.") Arguably, this way of talking should be representable in the formal language. This can be achieved with an extension of the revisionary account. There are many potential new pieces of information that would make me give up my belief in $p\&q$. Let us make a list of them, calling them $c_1, c_2, \dots$. The process of giving up $p\&q$ can then be said to consist in performing the change which is common to all the revisions $K * c_1$, $K * c_2, \dots$. Consequently, the outcome of an idealized process of giving up $p\&q$ can be set equal to the intersection of all these sets, i.e. $K \div (p\&q) = \bigcap\{K * c_1, K * c_2, \dots\}$. (We can expect this set to contain some sentence that is not in K. Therefore this is a revocation and not a contraction.[14]) Alternatively, we can assign to each sentence p another sentence $\mathfrak{r}(p)$ that represents the minimal p-confuting information that is common to all pieces of information that would make the agent give up p. Tentatively, we can identify $\mathfrak{r}(p)$ with the sentence "There is sufficient reason not to believe that p." We can then let $K \div p = K * \mathfrak{r}(p)$. In formal terms:

Definition 9.31 *Let $*$ be a sentential revision with the outcome set $\mathbb{X}$ and let the* reject function $\mathfrak{r}$ *be a function from and to sentences, such that $p\&\mathfrak{r}(p) \notin X$ for all $p \in \mathcal{L}$ and $X \in \mathbb{X}$. Then the operation $\div$ on K such that*

$$K \div p = K * \mathfrak{r}(p)$$

for all $p \in \mathcal{L}$ is the revisionary revocation *based on $*$ and* $\mathfrak{r}$.

Clearly, $\mathfrak{r}(p\&q)$ need not be equal to $\mathfrak{r}(p)$ or $\mathfrak{r}(q)$, and therefore we have no reason to assume that $K * \mathfrak{r}(p\&q)$ should be equal to either $K * \mathfrak{r}(p)$ or $K * \mathfrak{r}(q)$. Consequently, the decomposition postulate will not hold.

[14] If the set $\{c_1, c_2 \dots\}$ is finite then we have $c_1 \vee c_2 \vee \dots \vee c_n \in K \div (p\&q)$.

It can be argued against the revisionary account of giving up beliefs that it deviates too much from the intuitive notion of "giving up" a belief that it was constructed to represent. If we wish to be true to that notion, then we should arguably look for a process that refers primarily to the sentence to be removed and to the choice between different ways to remove it. It is therefore legitimate to ask: Is there some way to represent the (idealized) notion of contraction within the descriptor revision framework, without having to accept the highly implausible property of decomposition? That is the topic of the next chapter.

Chapter 10
Contraction

Contraction differs from revocation in satisfying the inclusion postulate ($K \div p \subseteq K$). In spite of the problems connected with that postulate, it would be unwise to entirely remove operations of contraction from our research agenda. Contraction represents an interesting idealization, namely that in which losses of beliefs are described with an exclusive focus on the beliefs that are lost and a corresponding disregard for the additions to the belief set that push them out. Furthermore, studies of contraction can help us connect descriptor revision with important results from the AGM tradition in which contraction is often the primary object of study.

In Sections 10.1 and 10.2 some major ways to construct operations of contraction in the descriptor framework are introduced, and the properties of these constructions are investigated. In Section 10.3 the relations between these constructions and AGM contraction are investigated. These relations turn out to be much less harmonious than those that we found in Sections 8.1 and 8.2 for sentential revision. Two impossibility theorems make it clear that AGM contraction cannot be reconstructed as descriptor revision. In Sections 10.4–10.6 three extensions of the descriptor revision framework are introduced. They provide us with operations of contraction capable of dealing with ties in ways that are more consonant with how this is done in the AGM framework. Unsurprisingly, these contraction methods all deviate from the simple one-step choice mechanism that is one of the major advantages of the descriptor framework.

10.1 How to Construct Contraction

There are several ways to construct operations of contraction in the framework of descriptor revision. One option is to derive contraction from sentential revision in the same way as in the AGM framework:

$$K \div p = K \cap (K * \neg p) = K \cap (K \circ \mathfrak{B}\neg p) \qquad \text{(Harper identity)}$$

S.O. Hansson, *Descriptor Revision*, Trends in Logic 46,
DOI 10.1007/978-3-319-53061-1_10

The operation $\div$ derived in this way will always satisfy inclusion. Furthermore, if $K \circ \mathfrak{B}\neg p$ is a consistent set that contains $\neg p$, then $\div$ will satisfy success. However, this construction has no guarantee that $K \div p$ will always be an element of the outcome set of $\circ$.

In much the same way, contraction can be obtained from revocation:

$$K \div p = K \cap (K \mp p) = K \cap (K \circ \neg\mathfrak{B}p) \qquad \text{(revocation cut)}$$

The operation of contraction obtained through revocation cut will always satisfy inclusion, and if revocation satisfies success, then so does the derived operation of contraction.[1] However, the same problem that we found for the Harper identity arises here as well: $K \div p$ will not in general be an element of the outcome set of $\circ$.

We can avoid these difficulties by instead incorporating the inclusion postulate into the success condition:

$$K \div p = K \circ (\{\neg\mathfrak{B}p\} \cup \{\neg\mathfrak{B}q \mid q \notin K\}) \qquad \text{(canonical contraction)}$$

As indicated by the name "canonical", this can be seen as the standard construction of contraction in the descriptor framework. However, for many purposes it can be replaced by a simpler, equivalent construction, namely an outcome set all of whose elements are subsets of the original belief set K. It will then satisfy:

$$K \div p = K \circ \neg\mathfrak{B}p, \text{ where } \bigcup \mathbb{X} \subseteq K. \qquad \text{(contractive outcome set)}$$

An outcome set $\mathbb{X}$ will be called *contractive* for a belief set K if and only if $K \in \mathbb{X}$ and $\bigcup \mathbb{X} \subseteq K$. The operations obtainable with a contractive outcome set can easily be shown to coincide with those obtainable with canonical contraction. Requiring the outcome set to be contractive may have drastic consequences for operations whose success conditions have other forms than $\neg\mathfrak{B}p$. However, that need not concern us here since this chapter is devoted exclusively to contraction.

10.2 Centrolinear and Maximal Contraction

Using contractive outcome sets we can obtain operations of contraction that are variants of the operations of revocation introduced in Sections 9.1 and 9.2.

Definition 10.1 *Let $\div$ be a sentential operation on a belief set K, and let $\mathbb{X}$, with $K \in \mathbb{X}$, be its outcome set. Then:*

(1) $\div$ is a monoselective contraction *if and only if it is a monoselective revocation and $\mathbb{X}$ satisfies:*

$\bigcup \mathbb{X} \subseteq K$ *(contractiveness).*

(2) $\div$ is a centrolinear contraction *if and only if it is a centrolinear revocation and $\mathbb{X}$ satisfies contractiveness.*

[1] We can say that the success property is *inherited* from the operation of recovation to the operation of contraction that is derived from it through revocation cut. Many other postulates are inherited through revocation cut in the same way. [124, pp. 964–965].

Theorem 10.2 *A sentential operation* $\div$ *on the consistent belief set* K *is a monoselective contraction on* K *if and only if it satisfies:*

$K \div p = \mathrm{Cn}(K \div p)$. *(closure)*

If it holds for all q *that* $K \div q \vdash p$ *if and only if* $K \div q \vdash p'$, *then* $K \div p = K \div p'$. *(uniformity)*

Either $K \div p \nvdash p$ *or* $K \div p = K$. *(relative success)*

If $K \div p \vdash p$, *then* $K \div q \vdash p$. *(persistence)*

$K \div p \subseteq K$. *(inclusion)*

Theorem 10.3 *A sentential operation* $\div$ *on the consistent belief set* K *is a centrolinear contraction on* K *if and only if it satisfies:*

$K \div p = \mathrm{Cn}(K \div p)$. *(closure)*

Either $K \div p \nvdash p$ *or* $K \div p = K$. *(relative success)*

If $K \div p \vdash p$, *then* $K \div q \vdash p$. *(persistence)*

If $p \notin K$, *then* $K \div p = K$. *(vacuity)*

Either $K \div (p\&q) = K \div p$ *or* $K \div (p\&q) = K \div q$. *(decomposition)*

If $p \notin K \div q$, *then* $K \div (p\&q) = K \div p$. *(conjunctive adjunction)*

If $K \div p = K \div p'$, *then* $K \div (p\&q) = K \div (p'\&q)$. *(conjunctive composition)*

$K \div p \subseteq K$. *(inclusion)*

Operations of contraction should not induce larger losses of previous beliefs than what is justified by the removal of the input sentence. This is what the AGM postulate of recovery is intended to achieve, but as we saw in Section 3.3 that postulate has undesirable side effects. In the present framework, the following postulate can be used to express the requirement that the outcome retains as much of the original belief set as it can without violating the success condition:

If $K \div p \subset K \div q$, then $p \in K \div q$. (maximality[2])

Monoselective contraction does not in general satisfy maximality, but it will do so if the choice function satisfies the following condition:

[2]This property was discussed in another context under the name "maximal preservativity" [89, pp. 605–606], and it also figured in [122, p. 6]. Other postulates with the purpose of avoiding too drastic losses in previous beliefs are discussed in [89, pp. 606–607], [90, pp. 857–858], and [122, pp. 6–8].

Definition 10.4 *Let C be a choice function for the set $\mathbb{X}$ of belief sets.*

It is maximizing[3] *if and only if it holds for all $\mathbb{Y} \subseteq \mathbb{X}$ that if $X \subset Y$ and $Y \in \mathbb{Y}$, then $X \notin C(\mathbb{Y})$.*

It is maximizing for negated atomic descriptors *if and only if it holds for all $\mathbb{Y} \subseteq \mathbb{X}$ that if $\mathbb{Y} = [\![\neg\mathfrak{B}p]\!]$ for some $p \in \mathcal{L}$, $X \subset Y$, and $Y \in \mathbb{Y}$, then $X \notin C(\mathbb{Y})$.*

Definition 10.5 *An operation $\div$ on a belief set K is a* maximal contraction *if and only if it is a monoselective contraction whose selection function $\widehat{C}$ is maximizing for negated atomic descriptors.*

As can be seen from the following observation, no such addition is needed for centrolinear contraction.

Observation 10.6 *Let $\div$ be a centrolinear contraction. Then it satisfies maximality.*

The maximality postulate is far from unquestionable. But before considering the arguments that can be raised against it, we are going to investigate the relationships between AGM contraction and the operations of contraction obtainable in the descriptor framework.

10.3 Two Impossibility Theorems

In Section 8.2 we proved an important connection between AGM revision and descriptor revision: All full-blown AGM revisions (transitively relational partial meet revisions) are reconstructible as centrolinear revisions. Thus, AGM revision is a special case of descriptor revision. Now, what about contraction? Is AGM contraction a special case of centrolinear contraction?

The following theorem shows that the answer is negative:

Theorem 10.7 ([134]) *Let K be a consistent belief set that is not equal to $\mathrm{Cn}(\varnothing)$ and let $\div$ be a partial meet contraction on K. Then the following two conditions are equivalent:*

(I) $\div$ is a transitively relational maxichoice contraction.

(II) $\div$ is (reconstructible as) a centrolinear contraction.

Importantly, the theorem applies to all AGM contractions, not only those that are transitively relational. No AGM contraction (partial meet contraction) can be a centrolinear contraction without also being both maxichoice and transitively relational. The former property is the most problematic one. As we saw in Section 2.5, maxichoice AGM contraction is an utterly implausible limiting case.

[3] This is an adoption of the maximizing property on relations that was referred to in Section 8.2.

It follows from Theorem 10.7 that no operation can satisfy the three conditions that (1) it is a centrolinear contraction, (2) it is a partial meet contraction, and (3) it is not a (transitively relational) maxichoice contraction. Due to this trilemma the theorem can be described as an impossibility theorem. The following theorem answers even better to that description.

Theorem 10.8 ([134]) *Let the language $\mathcal{L}$ consist of infinitely many logically independent atoms and their truth-functional combinations. Let K be a finite-based and consistent belief set that is not equal to* $\text{Cn}(\varnothing)$*, and let $\div$ be a partial meet contraction on K that satisfies finite-based outcome (i.e. $K \div p$ is finite-based for all p). Then $\div$ is not (reconstructible as) a centrolinear contraction.*

Theorems 10.7 and 10.8 show that AGM contraction is much less compatible with descriptor revision than what AGM revision is. This can be related to our findings in Chapters 2 and 3 that indicate more severe problems for AGM contraction than AGM revision in terms of intuitive plausibility.

10.4 Monomaximal Contraction

The reason why contraction seems to be more difficult to model in the descriptor framework than in the AGM framework is closely related to the treatment of ties. In this and the following two sections we are going to investigate extensions of the descriptor framework that allow for different ways to handle ties. We will have use for the following simple example. Let $K = \text{Cn}(\{p\&q\})$ and:

$$\mathbb{X} = \{\text{Cn}(\{p\&q\}), \text{Cn}(\{p\}), \text{Cn}(\{q\}), \text{Cn}(\{p \vee q\}), \text{Cn}(\varnothing)\}$$

If an operation $\div$ of contraction does not inflict unnecessary losses of information, then $K \div p$ should be equal to $\text{Cn}(\{q\})$ in this case. The other potential outcomes that satisfy the success condition for contraction by p, namely $\text{Cn}(\{p \vee q\})$ and $\text{Cn}(\varnothing)$, are both proper subsets of $\text{Cn}(\{q\})$. They would both induce larger losses than what $\text{Cn}(\{q\})$ does, without saving anything instead. For similar reasons, $K \div q$ should be equal to $\text{Cn}(\{p\})$.

In contrast, it is not obvious what $K \div (p\&q)$ should be. We need to consider the elements of $\mathbb{X}$ that do not contain $p\&q$, i.e. the set $\{\text{Cn}(\{p\}), \text{Cn}(\{q\}), \text{Cn}(\{p \vee q\}), \text{Cn}(\varnothing)\}$. This set has two inclusion-maximal elements, $\text{Cn}(\{p\})$ and $\text{Cn}(\{q\})$. The contraction outcome could be either of these. However, if there is a tie between them, then a choice of one of them would be arbitrary, and it may be more appropriate to settle for $\text{Cn}(\{p \vee q\})$ that can be interpreted as expressing hesitation between them.

These deliberations can be summarized by saying that $\div$ should be based on a choice function that satisfies the following criterion:

Definition 10.9 *Let C be a choice function over a set $\mathbb{X}$ of belief sets. C is* monomaximizing *if and only if it holds for all $\mathbb{Y} \subseteq \mathbb{X}$ that if $\bigcup \mathbb{Y} \subseteq Y \in \mathbb{Y}$, then $C(\mathbb{Y}) = \{Y\}$.*

Observation 10.10 *If a choice function is maximizing then it is monomaximizing.*

Definition 10.11 *A sentential operation on a belief set K is a* monomaximal contraction *if and only if it is a monoselective contraction that is based on a monomaximizing choice function $\widehat{C}$.*

When a set of potential outcomes has a single inclusion-maximal element, then a monomaximal contraction has that element as its outcome. In our example, if $\div$ is monomaximal, then it follows directly that $K \div p = \text{Cn}(\{q\})$ and $K \div q = \text{Cn}(\{p\})$. However, the monomaximizing property has nothing to say about $K \div (p\&q)$. (In contrast, the maximizing property tells us not only that $K \div p = \text{Cn}(\{q\})$ and $K \div q = \text{Cn}(\{p\})$, but also that $K \div (p\&q)$ is either $\text{Cn}(\{q\})$ or $\text{Cn}(\{p\})$.)

Monomaximal contraction can be axiomatically characterized as follows:

Theorem 10.12 *Let $\div$ be a sentential operation on the consistent belief set K. It is a monomaximal contraction if and only if it satisfies:*

$K \div p = \text{Cn}(K \div p)$. *(closure)*

If it holds for all q that $K \div q \vdash p$ if and only if $K \div q \vdash p'$, then $K \div p = K \div p'$. (uniformity)

Either $K \div p \nvdash p$ or $K \div p = K$. (relative success)

If $K \div p \vdash p$, then $K \div q \vdash p$. (persistence)

$K \div p \subseteq K$. *(inclusion)*

If it holds for all r that $K \div r \nvdash p$ if and only if $K \div r \subseteq K \div q$, then $K \div p = K \div q$. (unique maximum [117]*)*

Monomaximal contraction has the advantage of upholding the preference for larger outputs in cases when a unique maximum is available. This is illustrated in our example by the contractions by p and q. However, this operation also has the disadvantage that when ties arise, then the preference for larger outcomes is completely abandoned. This gives us reason to look for operations in which the preference for larger sets still has an impact when a maximal set cannot be chosen.

10.5 Perimaximal Contraction

Let us return to the example given at the beginning of the previous section, and again assume that there is a tie between $\text{Cn}(\{p\})$ and $\text{Cn}(\{q\})$, so that neither of them can be chosen to be the outcome of contraction by $p\&q$. One obvious (and indeed, highly traditional) way to express this is to use a selection function that is not monoselective.

In this case it would choose both $\text{Cn}(\{p\})$ and $\text{Cn}(\{q\})$. The actual outcome will then have to be obtained in a second step after the resources of the selection function have been exhausted.[4] This brings us close to AGM contraction, but with the important difference that here, the first step provides us with a set of potential outcomes rather than a set of (inclusion-maximal) remainders. We will assume that the choice function is maximizing so that all the belief sets chosen in the first step are inclusion-maximal among the elements of $\mathbb{X}$ that do not imply p.[5] When there is a tie, i.e. more than one element of $\mathbb{X}$ was chosen, then the outcome will not be one of these elements. Instead it will be a non-maximal element of $\mathbb{X}$ that represents hesitation among the chosen maximal elements.

In this section we will assume that the choice function only yields a tie in cases when the outcome set contains a belief set that represents hesitation among the elements that the tie refers to. Just like $\text{Cn}(\{p \vee q\})$ in the example, such a hesitation-representing belief set should be uniquely inclusion-maximal among the belief sets that are subsets of all the (maximal) belief sets that the hesitation concerns.[6] The resulting operation is called *perimaximal contraction* and formally defined as follows:

Definition 10.13 ([120], modified) *Let K be a belief set. A contraction $\div$ on K with the outcome set $\mathbb{X}$ is a* perimaximal contraction *if and only if there is a maximizing choice function C on $\mathbb{X}$ such that (i) if $p \notin \bigcap \mathbb{X}$, then*

$$K \div p = \bigcup \{Z \in \mathbb{X} \mid Z \subseteq \bigcap C([\![\neg\mathfrak{B}p]\!])\},$$

and (ii) otherwise, $K \div p = K$.

If the set of belief sets selected by C contains only one element, then (calling that element Y) we have

$$\begin{aligned} K \div p &= \bigcup \{Z \in \mathbb{X} \mid Z \subseteq \bigcap C([\![\neg\mathfrak{B}p]\!])\} \\ &= \bigcup \{Z \in \mathbb{X} \mid Z \subseteq \bigcap\{Y\}\} \\ &= \bigcup \{Z \in \mathbb{X} \mid Z \subseteq Y\} \\ &= Y && \text{(since } Y \in \mathbb{X}) \end{aligned}$$

and consequently the single (inclusion-maximal) belief set chosen by C in the first step is also the outcome of the contraction. If C chooses more than one element of $\mathbb{X}$,

---

[4] A closely related approach employs a blockage relation instead of a selection function in the first step. This option will not be further discussed here, but it was explored in [116].

[5] Alternatively this can be expressed in a way that accentuates the similarity with AGM contraction. We can define $K \perp_{\mathbb{X}} p$ as the set of sets X such that (i) $X \in \mathbb{X}$, (ii) $X \subseteq K$, (iii) $X \nvdash p$, and (iv) if $X' \in \mathbb{X}$ and $X \subset X' \subseteq K$, then $X' \vdash p$. (Standard AGM remainders are then definable as follows: $K \perp p = K \perp_{\wp(K)} p$.) We can use $C(K \perp_{\mathbb{X}} p)$ instead of $C([\![\neg\mathfrak{B}p]\!])$ in the definition of perimaximal contraction.

[6] This is a weaker assumption than that made in the AGM framework, namely that when there is hesitation among the elements of a set $\mathbb{Y}$, then their intersection $\bigcap \mathbb{Y}$ is an element of the outcome set and can be used to represent the hesitant belief state.

then the contraction outcome has to be an element of $\mathbb{X}$ that is uniquely inclusion-maximal among those elements of $\mathbb{X}$ that are subsets of all the C-chosen belief sets. It represents a state of hesitation between these potential outcomes.

As already indicated, C can only make choices for which such an inclusion-maximal set is available within $\mathbb{X}$. Let us consider two examples of how this works. The first is the example already presented, with $K = \text{Cn}(\{p\&q\})$,

$$\mathbb{X} = \{\text{Cn}(\{p\&q\}), \text{Cn}(\{p\}), \text{Cn}(\{q\}), \text{Cn}(\{p \vee q\}), \text{Cn}(\varnothing)\}$$

and consequently:

$$[\![\neg\mathfrak{B}(p\&q)]\!]_{\mathbb{X}} = \{\text{Cn}(\{p\}), \text{Cn}(\{q\}), \text{Cn}(\{p \vee q\}), \text{Cn}(\varnothing)\}$$

Perimaximal contraction has three options for contraction by $p\&q$ in this case:

$$C([\![\neg\mathfrak{B}(p\&q)]\!]_{\mathbb{X}}) = \{\text{Cn}(\{p\})\} \text{ and } K \div (p\&q) = \text{Cn}(\{p\})$$

$$C([\![\neg\mathfrak{B}(p\&q)]\!]_{\mathbb{X}}) = \{\text{Cn}(\{q\})\} \text{ and } K \div (p\&q) = \text{Cn}(\{q\})$$

$$C([\![\neg\mathfrak{B}(p\&q)]\!]_{\mathbb{X}}) = \{\text{Cn}(\{p\}), \text{Cn}(\{q\})\} \text{ and } K \div (p\&q) = \text{Cn}(\{p \vee q\})$$

In the second example we still have $K = \text{Cn}(\{p\&q\})$ but the outcome set is:

$$\mathbb{X}' = \{\text{Cn}(\{p\&q\}), \text{Cn}(\{p\}), \text{Cn}(\{q\}), \text{Cn}(\{p \vee q \vee r\}), \text{Cn}(\{p \vee q \vee s\}), \text{Cn}(\varnothing)\}$$

In this case as well, the inclusion-maximal elements of the outcome set not implying $p\&q$ are $\text{Cn}(\{p\})$ and $\text{Cn}(\{q\})$. The difference is that if a choice function C selects both of them, then there is no outcome available for perimaximal contraction, since

$$\begin{aligned}&\bigcup\{Z \in \mathbb{X}' \mid Z \subseteq \bigcap\{\text{Cn}(\{p\}), \text{Cn}(\{q\})\}\}\\ &= \text{Cn}(\{p \vee q \vee r\}) \cup \text{Cn}(\{p \vee q \vee s\}) \cup \text{Cn}(\varnothing),\end{aligned}$$

and this is not an element of $\mathbb{X}'$. (It is not even logically closed.) Therefore, with the outcome set $\mathbb{X}'$ there are only two options available for the contraction outcome $K \div (p\&q)$:

$$C([\![\neg\mathfrak{B}(p\&q)]\!]_{\mathbb{X}'}) = \{\text{Cn}(\{p\})\} \text{ and } K \div (p\&q) = \text{Cn}(\{p\})$$
$$C([\![\neg\mathfrak{B}(p\&q)]\!]_{\mathbb{X}'}) = \{\text{Cn}(\{q\})\} \text{ and } K \div (p\&q) = \text{Cn}(\{q\})$$

More generally, if the outcome set contains no belief set that represents hesitation among a particular set of potential maximal outcomes, then that set of maximal outcomes cannot to be chosen by the choice function.

Perimaximal contraction can be axiomatically characterized as follows:

Theorem 10.14 *Let K be a consistent belief set and $\div$ an operation on K with the outcome set $\mathbb{X}$. Then $\div$ is a perimaximal contraction if and only if it satisfies:*

$K \div p = \text{Cn}(K \div p)$. *(closure)*

If it holds for all q that $K \div q \vdash p$ if and only if $K \div q \vdash p'$, then $K \div p = K \div p'$. (uniformity)

Either $K \div p \nvdash p$ *or* $K \div p = K$. *(relative success)*

If $K \div p \vdash p$, *then* $K \div q \vdash p$. *(persistence)*

$K \div p \subseteq K$. *(inclusion)*

If $K \div q \nvdash p$ *and* $K \div q \nsubseteq K \div p$, *then there is some* r *such that* $K \div p \subseteq K \div r \nvdash p$ *and that there is no* s *with* $(K \div q) \cup (K \div r) \subseteq K \div s \nvdash p$. *(preservativity)*

Preservativity is intermediate in strength between maximality and unique maximum:

Observation 10.15 *Let* K *be a consistent belief set and* $\div$ *an operation on* K. *Then:*

(1) If $\div$ *satisfies maximality and persistence, then it satisfies preservativity.*

(2) If $\div$ *satisfies preservativity and persistence, then it satisfies unique maximum.*

It follows from Observation 10.15 that each operation on the following list is a generalization of the one that precedes it:

maximal contraction
perimaximal contraction
monomaximal contraction
monoselective contraction.

10.6 Bootstrap Contraction

For our last construction of contraction we are going to reconsider the weakest of the three postulates that we compared in Observation 10.15, namely:

If it holds for all r that $K \div r \nvdash p$ if and only if $K \div r \subseteq K \div q$, then $K \div p = K \div q$. (unique maximum)

Unique maximum has the important effect of making the outcome of contraction by some sentences independent of the selection mechanism (for instance the choice function). These contraction outcomes can be determined based exclusively on the outcome set. To see how this works, consider again our example with $K = \text{Cn}(\{p\&q\})$ and

$$\mathbb{X} = \{\text{Cn}(\{p\&q\}), \text{Cn}(\{p\}), \text{Cn}(\{q\}), \text{Cn}(\{p \vee q\}), \text{Cn}(\varnothing)\}.$$

It can easily be verified that an element of $\mathbb{X}$ is a subset of $\text{Cn}(\{q\})$ if and only if it does not imply p. It follows that if an operation $\div$ of contraction satisfies unique maximum, then $K \div p = \text{Cn}(\{q\})$. For symmetrical reasons we also obtain $K \div q = \text{Cn}(\{p\})$. This follows without any reference to how the operation of contraction is constructed. Since unique maximum is a highly plausible property, these construction-independent assignments of contraction outcomes are intuitively quite satisfactory.

Unfortunately, this does not work for all sentences that we may wish to remove from a belief set. In our example, the set of elements of $\mathbb{X}$ that do not imply $p\&q$ is $\{\text{Cn}(\{p\}), \text{Cn}(\{q\}), \text{Cn}(\{p \vee q\}), \text{Cn}(\varnothing)\}$, and this is a set without a unique inclusion-maximal element. Therefore, unique maximum does not tell us how to contract by $p\&q$. However, with the help of two small tricks the construction-independent mode of contracting can be extended to all input sentences.

In this case there are three plausible outcomes of contraction by $p\&q$: It can be equal to either $\text{Cn}(\{p\})$, $\text{Cn}(\{q\})$, or $\text{Cn}(\{p \vee q\})$, depending on the comparative credibility of these three potential outcomes. Let us consider the case when $K \div (p\&q) = \text{Cn}(\{p\})$. We can express this by saying that $K \div (p\&q)$ is equal to $K \div q$, and this corresponds to a plausible way to talk about contraction. We can say that we give up $p\&q$ by giving up q. ("When I could no longer believe that both Pauline and Quentin are safe drivers, I gave up my belief that Quentin is a safe driver.") The first trick is to express this mode of speaking with a sentential selector (function from and to sentences) f representing the notion "... is contracted through contraction by ...". For this device to serve its purpose, f should take us to a sentence whose contraction outcome is determined by the outcome set (under the assumption that unique maximum holds). In this case we have $f(p\&q) = q$, and we have already seen that $K \div q$ is determined by the outcome set. Similarly, we can obtain $K \div (p\&q) = \text{Cn}(\{q\})$ by setting $f(p\&q) = p$ and then appealing to unique maximum in the same way.

However, the case when $K \div (p\&q) = \text{Cn}(\{p \vee q\})$ cannot be dealt with in this way. For that to be possible, there would have to be some sentence z such that $f(p\&q) = z$ and that each element of $\mathbb{X}$ implies z if and only if it is not a subset of $\text{Cn}(\{p \vee q\})$. It would follow from this that $z \in \text{Cn}(\{p\})$, $z \in \text{Cn}(\{q\})$, and $z \notin \text{Cn}(\{p \vee q\})$, which is clearly impossible. But this limitation of the method can be overcome with the second trick, which is to make use of (multiple) package contraction. For that purpose, we need to extend unique maximum to inputs that are sets containing more than one sentence:

If it holds for all D that $K \div D \nvdash_{\exists} A$ if and only if $K \div D \subseteq K \div B$, then $K \div A = K \div B$. (unique maximum, multiple version)[7]

The success condition for package contraction by $\{p, q\}$ is that the outcome contains neither p nor q. In our example, this is satisfied exactly by the subsets of $\text{Cn}(\{p \vee q\})$, and therefore unique maximum prescribes that $K \div \{p, q\} = \text{Cn}(\{p \vee q\})$. We can identify $K \div (p\&q)$ as $\text{Cn}(\{p \vee q\})$ by letting the sentential selector f assign $f(p\&q) = \{p, q\}$.

We now have a contraction method that is completely generalizable. Independently of what the outcome set looks like, each of its elements can be identified with the method we have just introduced.

[7] $A \vdash_{\exists} B$ holds if and only if A implies at least one element of B, i.e. $\text{Cn}(A) \cap B \neq \varnothing$.

Observation 10.16 *Let $\mathbb{X}$ be a set of belief sets and let $X \in \mathbb{X}$. Then there is a set B of sentences such that for each element Y of $\mathbb{X}$:*
$Y \subseteq X$ if and only if $Y \nvdash_{\exists} B$.

If we want $K \div p$ to be equal to the element X of $\mathbb{X}$, then we can achieve this by letting $f(p)$ be a set satisfying the condition of this observation, i.e.:

$Y \subseteq X$ if and only if $Y \nvdash_{\exists} f(p)$.

Then $K \div f(p) = X$ follows from unique maximum. In this way we can dispense completely with traditional selection mechanisms and instead perform all contractions with a sentential selector f. Obviously, we can perform not only single-sentence contractions (such as $K \div p$) in this way, but also multiple contractions (such as $K \div A$ where A is a set of sentences). This method is called "bootstrapping", and the formal definition is as follows:

Definition 10.17 ([117]) *Let K be a belief set and $\mathbb{X}$ a set of belief sets such that $K \in \mathbb{X}$ and $\bigcup \mathbb{X} \subseteq K$. Then:*

(1) A set $B \subseteq \mathcal{L}$ is bootstrapping *in $\mathbb{X}$ if and only if*
$\bigcup\{Y \in \mathbb{X} \mid Y \nvdash_{\exists} B\} \in \{Y \in \mathbb{X} \mid Y \nvdash_{\exists} B\}$.
(2) A function f from $\wp(\mathcal{L})$ to $\wp(\mathcal{L})$ is a bootstrapping selector *for $\mathbb{X}$ if and only if it holds for all $A \subseteq \mathcal{L}$ that $f(A)$ is bootstrapping in $\mathbb{X}$.*
(3) Let $\div$ be a contraction on K with $\mathbb{X}$ as its outcome set. Then a bootstrapping selector f for $\mathbb{X}$ bootstraps *$\div$ if and only if $K \div A = K \div f(A)$ for all $A \subseteq \mathcal{L}$.*

For each operation $\div$ of contraction there is a selector f that bootstraps it, i.e. such that $K \div A = K \div f(A)$ for all A. In this sense, all operations of contraction are bootstrappable. However, the following additional requirement is plausible:

Definition 10.18 *An operation $\div$ of contraction with the outcome set $\mathbb{X}$ is* preservingly bootstrappable *if and only if it is bootstrapped by a selector f such that:*

If A is bootstrapping in $\mathbb{X}$, then $f(A) = A$.

Observation 10.19 *An operation $\div$ of package contraction is preservingly bootstrappable if and only if it satisfies (the multiple version of) unique maximum.*

Almost all operations of contraction that have been seriously considered in the literature satisfy unique maximum, and they are therefore also all preservingly bootstrappable [117].

Bootstrapping can serve as a search heuristic that simplifies the search for a contraction outcome by restricting it to a limited number of readily available solutions. As an example of this, let d denote that the door of my fridge was properly closed when I left home a week ago, and e that the door of my freeze was properly closed at the same point in time. I know fairly well how my belief state would change after contraction by d. I also know how it would change after contraction by e or by both d and e. In other words, $K \div d$, $K \div e$, and $K \div \{d, e\}$ are readily available to me. It is much less clear to me what I would believe after contracting by $d\&e$. But if I can for some reason identify the contraction by $d\&e$ with one of the three more accessible ones just mentioned, then I can more easily arrive at the new belief state.

However, this only works if the bootstrapping actually reduces complexity. A bootstrapping function f that assigns an unmanageably large set $f(A)$ to a small contractee A will not be of much help. The type of cognitive unwieldiness that we need to avoid can be measured fairly accurately as the size of the smallest set that can be used for bootstrapping:

Definition 10.20 ([117], modified) *Let $\mathbb{X}$ be a set of belief sets. The* intricacy *of an element X of $\mathbb{X}$ is the lowest number of elements of any set B such that*

$$X = \bigcup\{Y \in \mathbb{X} \mid Y \nvdash_{\exists} B\} \in \{Y \in \mathbb{X} \mid Y \nvdash_{\exists} B\}$$

The intricacy of $\mathbb{X}$ is equal to the highest intricacy of any of its elements.

One of the mechanisms by which we keep down intricacy is to divide belief sets into independent compartments. In practice, most belief changes only affect a small part of the belief set, leaving the rest of it essentially unchanged [114, 139, 253]. It is often difficult to deal with input sentences that combine information from different compartments. Suppose that in a quiz you encounter the following question:

One of the following statements is false. Which?
s_1: Dar es Salaam is the capital of Tanzania.
s_2: Cucumbers are fruits.

If you previously had about equally strong beliefs in s_1 and s_2, then you should expectedly suspend your beliefs in both of them. This may be doable, but if the number of statements is much larger than two, then it is doubtful whether they can be suspended in the coordinated fashion demanded by the operation.[8] Our capacity to keep issues open in this way is limited. One option in such situations is not to give up any belief at all, which means of course that the success postulate will not be satisfied. The justification commonly given for violations of success is that some non-tautologous beliefs may be too strongly held to be contractible.[9] Another justification should be added, namely that some contractions cannot be performed because they are too complex (have too high intricacy) to be cognitively manageable. Bootstrap contraction has the advantage of making the complexity explicit that obstructs such contractions.

To conclude this chapter, it should again be emphasized that the operations introduced in it are all based on one and the same idealization, namely that we study removals of beliefs while disregarding the new information that caused them to be removed. This is a traditional idealization in the belief change literature, but as we saw in Chapter 9, it is not adequate for all purposes.

---

[8]This was expressed in Makinson's preface paradox [173]. On that paradox, see also [42, 43, 150, 156, 183, 208, 209, 226, 254, 255].

[9]See for instance [213, p. 54] and [51].

Part IV
Epilogue

Chapter 11
Looking Back – and Ahead

It is time to sum up what we have achieved and what remains to be done.

11.1 Looking Back

Our critical examination of the traditional account concluded in Section 3.8 with a list of ten desiderata for an alternative framework. Let us now reconsider the list and see to what extent they have been achieved.

1. Selection functions should operate directly on plausible outcomes, i.e. on potential belief sets, rather than on cognitively unmanageable objects such as remainders or possible worlds.

In Chapter 4 we introduced choice functions satisfying this criterion, in other words choice functions that select among the potential outcomes of the operation. This construction and its relational variant have been used throughout Chapters 5–10. Hopefully we have shown that the construction is viable.

2. Operations of revision and contraction that do not satisfy the success postulates should be available.

Such operations have been developed in Chapters 8–10 as special cases of descriptor revision.

3. Other types of belief change than contraction and revision should be available, including operations whose success conditions are not preserved under intersection.

Throughout Chapters 4–7 we have worked with a general account of belief change that covers operations with all success conditions that can be described with belief descriptors. One operation with a success condition that is not preserved under intersection, namely the operation of making up one's mind, was specifically investigated in Section 8.5.

S.O. Hansson, *Descriptor Revision*, Trends in Logic 46,
DOI 10.1007/978-3-319-53061-1_11

4. The postulate of finite-based outcome should be satisfiable.

This is easily achievable by letting the outcome set consist of finite-based belief sets. However, most of our formal results do not require this assumption, and therefore it has seldom been mentioned.

5. The recovery postulate for contraction should not hold in general.

Recovery does not hold for any of the revocation and contraction operations that we have introduced in Chapters 9 and 10.

6. Contraction-like operations that do not satisfy the inclusion postulate should be available.

Chapter 9 provided a whole series of such operations that have been axiomatically characterized with plausible postulates.

7. The operation of revision should not be required to satisfy the expansion property.

None of the operations of (sentential) revision that we developed in Chapter 8 has this property.

8. Conditional sentences satisfying the Ramsey test should be includible in the belief sets.

This was achieved in Chapter 7. Moreover, it was achieved not only with traditional Ramsey test conditionals but also with an extended set of expressions, Ramsey descriptors, that allow for much richer descriptions of how changes in a belief state bring about other such changes.

9. Modal sentences and sentences expressing autoepistemic beliefs should be includible in belief sets.

For autoepistemic beliefs this was achieved in Sections 7.1 and 7.3, and for modal beliefs in Section 7.6.

10. Operations of change should be iterable.

Global and therefore iterable descriptor revision was introduced in Chapter 6 and further studied in Section 8.3.

11.2 Looking Ahead

As so often, the answers to research questions give rise to new questions. Many of the constructions introduced on the foregoing pages are in need of much more thorough investigations. Furthermore, quite a few extensions of the framework remain to investigate. A few examples:

1. The indeterministic variants of the various types of descriptor revision remain to investigate.
2. Several issues concerning blockage revision remain open. Neither its local nor its global version has been axiomatically characterized. Furthermore, a wider range of properties of the relation $\rightharpoondown$ should be investigated.
3. Weakened versions of the axioms for epistemic proximity should be studied. There may be interesting connections with the weakened versions of epistemic entrenchment that have been explored by Hans Rott [217, 218].
4. Sentential centrolinear revision has only been axiomatically characterized under restrictions that it would be of interest to relax.
5. The formal properties of choice revision remain to investigate.
6. The logic of the descriptor-based modal operations introduced in Section 7.6 is an open issue, and so are their connections with the various modal update logics that have been proposed previously.
7. Just like entrenchment, believability relations can be generalized to a multiple version. It would be particularly interesting to see how such relations can connect with choice revision.
8. In Section 9.3 we left it open whether the set of centrolinear revocations that are compatible with a given entrenchment relation has an element that deserves a privileged or "canonical" status.
9. Multiple entrenchment is in need of a full investigation. This should also cover its relations to multiple operations in the AGM framework.
10. The formal properties of the revisionary account of contraction (Section 9.5) remain to investigate.

There are also larger issues that can be attacked with the tools developed here. As was mentioned in Section 7.1, descriptor revision is a suitable framework for developing more realistic accounts of autoepistemic beliefs. We can model belief states in which some but not all true autoepistemic sentences of a certain formal type (such as Ramsey conditionals) are believed by the agent. We can also include false dynamic beliefs in belief sets to represent an agent's incorrect beliefs about how she will react to possible future information. The presence of such incorrect beliefs should not necessarily make the belief set inconsistent.

We can also extend the basic structure introduced in Definition 4.1 to represent an agent's overall state of mind rather than only the part of her state of mind that constitutes her belief state. Such an enlarged model can incorporate representations of for instance beliefs, preferences, value assignments, emotions, and desires. A unified change operator similar to our $\circledcirc$ or $\circ$ can be used to study the dynamic relationships among different components of mental states, for instance the effects of changes in belief on value judgments and vice versa.

Appendices

S.O. Hansson, *Descriptor Revision*, Trends in Logic 46,
DOI 10.1007/978-3-319-53061-1

Proofs

Proofs for Section 2.5

Lemma 2.1 ([99, p. 38]) *Let K be a logically closed set, and let $X \in K \perp p$. Then $X = \mathrm{Cn}(X)$.*

Proof of Lemma 2.1 Suppose to the contrary that $X \neq \mathrm{Cn}(X)$. Then $X \subset \mathrm{Cn}(X) \subseteq K$ and $p \notin \mathrm{Cn}(X)$, contrary to $X \in K \perp p$. □

Lemma 2.2 ([99, p. 47]) *Let K be a logically closed set such that $p\&q \in K$. If $X \in K \perp (p\&q)$, then exactly one of p, q, and $p \leftrightarrow q$ is an element of X.*

Proof of Lemma 2.2 We are first going to show that at least one of p, q and $p \leftrightarrow q$ is an element of X. Suppose that $p \notin X$ and $q \notin X$. We have to show that $p \leftrightarrow q \in X$.

It follows from $X \in K \perp (p\&q)$ and $p \notin X$ that $X \cup \{p\} \vdash p\&q$. By the deduction property we obtain $X \vdash p \rightarrow p\&q$, that is truth-functionally equivalent to $X \vdash p \rightarrow q$.

Similarly, it follows from $X \in K \perp (p\&q)$ and $q \notin X$ that $X \cup \{q\} \vdash p\&q$. By the deduction property we obtain $X \vdash q \rightarrow p\&q$, that is equivalent to $X \vdash q \rightarrow p$.

It follows from $X \vdash p \rightarrow q$ and $X \vdash q \rightarrow p$ that $X \vdash p \leftrightarrow q$. Due to Lemma 2.1, X is logically closed, thus, $p \leftrightarrow q \in X$.

It remains to be proved that no more than one of the three is an element of X. If more than one of p, q and $p \leftrightarrow q$ is an element of X, then one of the three sets $\{p, q\}, \{p, p \leftrightarrow q\}$, and $\{q, p \leftrightarrow q\}$, is a subset of X. Since each of these sets implies $p\&q$, this contradicts $X \in K \perp (p\&q)$. The contradiction concludes the proof. □

Proof of Observation 2.2 ([99, p. 47]) It follows from the existence of a set X with $X \in K \perp p$ that p is not a tautology. Since p is logically equivalent to $(p \vee q)\&(p \vee \neg q)$, it follows from Lemma 2.2 that either $p \vee q$, $p \vee \neg q$, or $p \vee q \leftrightarrow p \vee \neg q$ is an element of X. Since $p \vee q \leftrightarrow p \vee \neg q$ is logically equivalent to p, it follows from $X \in K \perp p$ that $p \vee q \leftrightarrow p \vee \neg q$ is not an element of X. We can therefore conclude that either $p \vee q \in X$ or $p \vee \neg q \in X$. □

S.O. Hansson, *Descriptor Revision*, Trends in Logic 46,
DOI 10.1007/978-3-319-53061-1

Postulate [2] *(The upper bound property) If* $X \subseteq A$ *and* $p \notin \mathrm{Cn}(X)$*, then there is some* X' *such that* $X \subseteq X' \in A \perp p$*.*

Comment on the postulate: The upper bound property follows from compactness and Zorn's lemma that is equivalent with the axiom of choice.

Proof of Observation 2.4 (1) follows from (3) and (2) from (4). Note that $p \in W \in \mathcal{L} \perp_{\perp}$ if and only if $W \in \mathcal{L} \perp \neg p$. Therefore we only have to prove (3) and (4).

Part 3 [109]: Let $X \in K \perp p$. Suppose for contradiction that X is finite-based. Then $X = \mathrm{Cn}(\{x\})$ for some sentence x. Let d be an atom that is not a subformula of x or p. (More precisely, let d be an atom such that there is some x' with $\vdash x \leftrightarrow x'$ and some p' with $\vdash p \leftrightarrow p'$ such that d is not a subformula either of x' or of p'.) It follows from $X \in K \perp p$ that $X \nvdash p$, thus $X \cup \{\neg p\}$ is logically consistent. Due to Observation 2.2, either $p \vee d \in X$ or $p \vee \neg d \in X$. Equivalently, either $\neg p \rightarrow d \in X$ or $\neg p \rightarrow \neg d \in X$, thus either $X \cup \{\neg p\} \vdash d$ or $X \cup \{\neg p\} \vdash \neg d$, which is impossible due to the atomic structure assumed. This concludes the proof.

Part 4 [109]: Let D be the set of atoms in the language that are not subformulas of p. (More precisely, $d \in D$ if and only if there is some p' such that $\vdash p' \leftrightarrow p$ and d is not a subformula in p'.) For each subset Z of D we can form the set $Z^+ = \{z \vee p \mid z \in Z\} \cup \{\neg z \vee p \mid z \in D \setminus Z\}$. It follows from the upper bound property that there is at least one set V such that $Z^+ \subseteq V \in K \perp p$. This construction provides us with an infinite number of distinct elements of $K \perp p$. □

Proof for Section 3.1

Proof of Observation 3.1 In the proofs of parts 1 and 2 we assume that the language consists of infinitely many logically independent atoms and their truth-functional combinations.

Part 1: [109] Let $p \in K \setminus \mathrm{Cn}(\varnothing)$ and let γ be such that $\gamma(K \perp p)$ has exactly one element. Then it follows from Observation 2.4 that $K \div p$ is not finite-based.

Part 2: Let $K = \mathrm{Cn}(\{\neg p\})$ and let S be an infinite set of logical atoms that are not subformulas of p. (More precisely, for each $s \in S$ there is some p' such that $\vdash p \leftrightarrow p'$ and s is not a subformula of p'.) Then $\{\neg p \vee s \mid s \in S\}$ is a subset of K that does not imply $\neg p$. It follows from the upper bound property that there is some X such that $\{\neg p \vee s \mid s \in S\} \subseteq X \in K \perp \neg p$. Let γ be a selection function such that $\gamma(K \perp \neg p) = \{X\}$ and let $*$ be the partial meet revision based on γ. Then $\{\neg p \vee s \mid s \in S\} \cup \{p\} \subseteq K * p$, thus $S \subseteq K * p$ and consequently $K * p$ is not finite-based.

Part 3: The case when $p \notin K$ is straight-forward. For the other case, let $p \in K$ and suppose for contradiction that $K \div p$ is finite-based. Due to the recovery postulate that holds for partial meet contraction [1], $K = \mathrm{Cn}((K \div p) \cup \{p\})$, so that K is finite-based, contrary to the assumption. □

Proofs for Section 3.7

Lemma 3.1 *Let K be a belief set and p and q sentences such that:*

$$\mathrm{Cn}(\varnothing) \subset \mathrm{Cn}(\{q\}) \subset \mathrm{Cn}(\{p\}) \subseteq K.$$

Then $K \perp p$ has more than one element.

Proof of Lemma 3.1 Since $q \nvdash p$ and $q \in K$ it follows from the upper bound property that there is some X_1 such that $q \in X_1 \in K \perp p$. Furthermore, since $\nvdash q$ and $\vdash p \rightarrow q$ we have $\nvdash p \vee q$, or equivalently $q \rightarrow p \nvdash p$. Since $q \rightarrow p \in K$ it follows from the upper bound property that there is some X_2 such that $q \rightarrow p \in X_2 \in K \perp p$. Suppose that $X_1 = X_2$. Then $q \in X_1$ and $q \rightarrow p \in X_1$ yield $X_1 \vdash p$, contrary to $X_1 \in K \perp p$. □

Proof of Observation 3.2 First step, proof that $\vdash p_1 \leftrightarrow p_2$: Since $\mathcal{L}$ is infinite there is a sentence q such that $\mathrm{Cn}(\varnothing) \subset \mathrm{Cn}(\{q\}) \subset \mathrm{Cn}(\{p_2\})$. It follows from Lemma 3.1 that $K_2 \perp p_2$ has more than one element, hence $\bigcup(K_2 \perp p_2) \vdash p_2$, and since $\bigcup(K_2 \perp p_2) = \bigcup(K_1 \perp p_1) \subseteq K_1$ we have $K_1 \vdash p_2$. Since $K_1 \perp p_1 = K_2 \perp p_2$ no element of $K_1 \perp p_1$ includes p_2. It follows from the upper bound property that $\vdash p_2 \rightarrow p_1$. The proof that $\vdash p_1 \rightarrow p_2$ is symmetrical, and we can conclude that $\vdash p_1 \leftrightarrow p_2$.

Second step, proof that $K_1 = K_2$: We now have $K_1 \perp p_1 = K_2 \perp p_1$. Suppose that $K_1 \neq K_2$. Without loss of generality, we may then assume that there is some $z \in K_2 \setminus K_1$. It follows from $z \notin K_1$ and $p_1 \in K_1$ that $p_1 \rightarrow z \notin K_1$, hence $p_1 \rightarrow z \notin \bigcup(K_1 \perp p_1) = \bigcup(K_2 \perp p_1)$. Due to the upper bound property it follows from $p_1 \rightarrow z \in K_2 \setminus \bigcup(K_2 \perp p_1)$ that $p_1 \rightarrow z \vdash p_1$, hence $\vdash p_1$, contrary to the conditions. □

Proof for Section 4.1

Proof of Observation 4.3 *Part 1*: Suppose to the contrary that all three conditions are satisfied. Due to changeability there are $\mathcal{K}$ and $\imath_1$ such that $\mathcal{K} \circledcirc \imath_1 \neq \mathcal{K}$. It follows from successive access that there are $\imath_2, \dots, \imath_n$ such that $\mathcal{K} \circledcirc \imath_1 \circledcirc \imath_2 \circledcirc \dots \circledcirc \imath_n = \mathcal{K}$, contrary to non-reversion.

Part 2: Let $\mathcal{K} \neq \mathcal{K}'$ and let retainability and direct access be satisfied. It follows from direct access that there is some $\imath$ with $\mathcal{K} \circledcirc \imath = \mathcal{K}'$ and from retainability that there is some $\imath'$ with $\mathcal{K}' = \mathcal{K}' \circledcirc \imath'$, thus $\mathcal{K} \circledcirc \imath = \mathcal{K}' \circledcirc \imath'$, contrary to non-convergence.

Part 3: Let retainability and non-convergence hold. Suppose that non-reversion is not satisfied. We can then assume, without loss of generality, that there are $\mathcal{K}$ and $\imath_1, \dots, \imath_{m+1}$ such that $\mathcal{K} \circledcirc \imath_1 \circledcirc \dots \circledcirc \imath_m \neq \mathcal{K}$ but $\mathcal{K} \circledcirc \imath_1 \circledcirc \dots \circledcirc \imath_m \circledcirc \imath_{m+1} = \mathcal{K}$. Due to retainability there is some $\imath_0$ such that $\mathcal{K} = \mathcal{K} \circledcirc \imath_0$. We then have $(\mathcal{K} \circledcirc \imath_1 \circledcirc \dots \circledcirc \imath_m) \neq \mathcal{K}$ and $(\mathcal{K} \circledcirc \imath_1 \circledcirc \dots \circledcirc \imath_m) \circledcirc \imath_{m+1} = \mathcal{K} \circledcirc \imath_0$, contrary to non-convergence. Contradiction. □

Proof for Section 4.2

Proof of Observation 4.5 *Part 1*: Since $\mathcal{L}$ is logically infinite there is an infinite set of logically independent, consistent sentences $\{p_1, p_2, \dots\}$. Due to direct believability there is for each p_k some $\imath_k \in \mathbb{I}$ such that $p_k \in \mathfrak{s}(\mathcal{K} \circledcirc \imath_k)$. Due to finite direct access set, the number of such sets $\mathcal{K} \circledcirc \imath_k$ is finite. Then at least one of them must have a support set $\mathfrak{s}(\mathcal{K} \circledcirc \imath_k)$ that contains an infinite subset of $\{p_1, p_2, \dots\}$, contrary to finite-based outcome.

Part 2: Since $\mathcal{L}$ is logically infinite there is an infinite set of logically independent, consistent sentences $\{p_1, p_2, \dots\}$. Due to successive believability there is for each p_k some finite set $\{\imath_{k_1}, \dots, \imath_{k_m}\} \subseteq \mathbb{I}$ such that $p_k \in \mathfrak{s}(\mathcal{K} \circledcirc \imath_{k_1} \circledcirc \dots \circledcirc \imath_{k_m})$. Due to finite successive access set, the number of such sets $\mathcal{K} \circledcirc \imath_{k_1} \circledcirc \dots \circledcirc \imath_{k_m}$ is finite. Then at least one of them must have a support set $\mathfrak{s}(\mathcal{K} \circledcirc \imath_{k_1} \circledcirc \dots \circledcirc \imath_{k_m})$ that contains an infinite subset of $\{p_1, p_2, \dots\}$, contrary to finite-based outcome. □

Proofs for Section 4.4

Proof of Observation 4.8 (1) and (2) follow directly from Definition 4.6. The proof of (3) is straight-forward. For the left-to-right direction, let $\Psi' = \Psi \cup \Xi$. □

Proof of Observation 4.11 We need to show that for all belief sets X:

(1) If $X \Vdash \Psi$ or $X \Vdash \Xi$, then $X \Vdash \Psi \veebar \Xi$, and

(2) If $X \Vdash \Psi \veebar \Xi$, then either $X \Vdash \Psi$ or $X \Vdash \Xi$.

(1) follows directly. For (2), let $X \Vdash \Psi \veebar \Xi$. We need to show that if $X \nVdash \Psi$ then $X \Vdash \Xi$. Let $X \nVdash \Psi$. Then there is some $\alpha' \in \Psi$ such that $X \nVdash \alpha'$, and consequently $X \Vdash \neg\alpha'$. It follows from $X \Vdash \Psi \veebar \Xi$ that $X \Vdash \{\alpha' \vee \beta \mid \beta \in \Xi\}$. Combining this with $X \Vdash \neg\alpha'$ we obtain $X \Vdash \{\beta \mid \beta \in \Xi\}$, i.e., $X \Vdash \Xi$, as desired. □

Lemma 4.1 *For all descriptors Ψ there is some descriptor Ψ' such that $\Psi \dashv\Vdash \Psi'$ and that all elements of Ψ' have one of the forms*

$$\neg\mathfrak{B}p_1 \vee \dots \vee \neg\mathfrak{B}p_m \vee \mathfrak{B}q_1 \vee \dots \vee \mathfrak{B}q_n \textit{ and } \neg\mathfrak{B}p_1 \vee \dots \vee \neg\mathfrak{B}p_m,$$

with $m \geq 1$ and $n \geq 1$.

Proof of Lemma 4.1 It follows from the conjunctive normal form theorem that each element α of Ψ is equivalent with a descriptor that has the form

$\alpha_1 \& \dots \& \alpha_v$,

where each α_u has one of the three forms

$\neg\mathfrak{B}p_1 \vee \dots \vee \neg\mathfrak{B}p_m \vee \mathfrak{B}q_1 \vee \dots \vee \mathfrak{B}q_n$,

$\neg\mathfrak{B}p_1 \vee \dots \vee \neg\mathfrak{B}p_m$, or

$\mathfrak{B}q_1 \vee \dots \vee \mathfrak{B}q_n$,

with $m \geq 1$ and $n \geq 1$. If α_u has the form $\mathfrak{B}q_1 \vee \dots \vee \mathfrak{B}q_n$, then we can replace it by the equivalent $\neg\mathfrak{B}\top \vee \mathfrak{B}q_1 \vee \dots \vee \mathfrak{B}q_n$. Therefore we can assume that each α_u has one of the two forms stated in the lemma. Furthermore, since $\alpha_1 \& \dots \& \alpha_v$ and $\{\alpha_1, \dots, \alpha_v\}$ are satisfied by the same belief sets, we can replace Ψ by $(\Psi \setminus \{\alpha\}) \cup \{\alpha_1, \dots, \alpha_v\}$, and similarly for other elements of Ψ. □

Proof of Observation 4.12 Let the object language consist of the infinite set $\{a_0, a_1, \dots\}$ of atoms and their truth-functional combinations. Let $X = \text{Cn}(\{a_0\})$. The descriptor Π_X (see Definition 4.14) is satisfied by X and by no other belief set. Suppose for contradiction that there is a descriptor $\neg\!\!\!\neg\, \Pi_X$ that is satisfied by all belief sets except X. Due to Lemma 4.1 we can assume that each element α of $\neg\!\!\!\neg\, \Pi_X$ has one of the forms:

$$\neg\mathfrak{B}p_1 \vee \dots \vee \neg\mathfrak{B}p_m \vee \mathfrak{B}q_1 \vee \dots \vee \mathfrak{B}q_n \text{ or } \neg\mathfrak{B}p_1 \vee \dots \vee \neg\mathfrak{B}p_m$$

with $\{p_1, \dots, p_m, q_1, \dots, q_n\} \subseteq \mathcal{L}$ and $m \geq 1$ and $n \geq 1$. If $a_0 \nvdash p_k$ for some p_k or $a_0 \vdash q_k$ for some q_k, then $X \Vdash \alpha$. It follows from $\alpha \in \neg\!\!\!\neg\, \Pi_X$ that $Y \Vdash \alpha$ for all belief sets Y such that $Y \neq X$, thus $Y \Vdash \alpha$ for all belief sets Y, thus $\Psi \dashv\Vdash \Psi \setminus \{\alpha\}$ for all descriptors Ψ. We can therefore assume that there is no such α in $\neg\!\!\!\neg\, \Pi_X$, in other words that $a_0 \vdash p_k$ for each p_k and $a_0 \nvdash q_k$ for each q_k.

Let a_t be any atom other than a_0. Consider the belief set $\text{Cn}(\{a_0, a_t\})$. Since $a_0 \vdash p_k$ for each p_k, $\text{Cn}(\{a_0, a_t\})$ does not satisfy $\neg\mathfrak{B}p_1 \vee \dots \vee \neg\mathfrak{B}p_m$, and therefore it has to satisfy $\mathfrak{B}q_1 \vee \dots \vee \mathfrak{B}q_n$. Thus there must be some q_k such that $a_0 \nvdash q_k$ but $a_0 \& a_t \vdash q_k$. For this to be the case, q_k must contain the atom a_t.

It follows that each atom a_t except a_0 has to be present in $\{q_1, \dots q_n\}$. But this is impossible since the number of such atoms is infinite and $\{q_1, \dots q_n\}$ is a finite set of sentences. □

Definition A.1 ([124]) *For any finite set* $\mathbb{Y} = \{Y_1, \dots Y_n\}$ *of belief sets:* $\Pi_{\mathbb{Y}} = \Pi_{Y_1} \underline{\vee} \cdots \underline{\vee}\, \Pi_{Y_n}$.

Lemma 4.2 *Let* $\mathbb{Y}$ *be a finite set of belief sets. Then it holds for all belief sets* X *that* $X \in \mathbb{Y}$ *iff* $X \Vdash \Pi_{\mathbb{Y}}$.

Proof of Lemma 4.2 From Observation 4.11. □

Proof of Observation 4.16 *Part 1*: From Lemma 4.2.

Part 2: This follows from the proof of Observation 4.12 but can also be proved as follows: The set of possible worlds ($\mathcal{L} \perp \perp$) has cardinality $2^{\aleph_0}$. Since possible worlds are belief sets, the set of belief sets has at least cardinality $2^{\aleph_0}$. The set of sets of belief sets is its power set and therefore has higher cardinality than $2^{\aleph_0}$. Descriptors are sets of sentences in a denumerable language and therefore the cardinality of the set of descriptors cannot be higher than $2^{\aleph_0}$. □

Proof of Observation 4.18 Let Ψ consist of elements of the forms (i), (ii), and (iii). Let $X \Vdash \Psi$ for all $X \in \mathbb{X}$. Then it holds for each $\alpha \in \Psi$ that $X \Vdash \alpha$ for all $X \in \mathbb{X}$. We are going to show that $\bigcap \mathbb{X} \Vdash \alpha$. There are three cases:

Case (i): α has the form $\mathfrak{B}p$: Then $p \in X$ for all $X \in \mathbb{X}$, thus $p \in \bigcap \mathbb{X}$, thus $\bigcap \mathbb{X} \Vdash \alpha$.

Case (ii): α has the form $\neg\mathfrak{B}p$: Then $p \notin X$ for all $X \in \mathbb{X}$, thus $p \notin \bigcap \mathbb{X}$, thus $\bigcap \mathbb{X} \Vdash \alpha$.

Case (iii): α has the form shown in (iii) in the observation: If $q \notin X$ for some $X \in \mathbb{X}$ then $q \notin \bigcap \mathbb{X}$ and we are done. If $q \in X$ for all $X \in \mathbb{X}$ then for each $X \in \mathbb{X}$ there is some $p_m \in \{p_1, \dots, p_n\}$ such that $p_m \in X$ and consequently $q \& (p_1 \vee \dots \vee p_n) \in X$.

Thus $q\&(p_1 \vee \ldots \vee p_n) \in \bigcap \mathbb{X}$. We have $q \vdash p_1 \vee \ldots \vee p_n \rightarrow p_k$ for some p_k, thus $p_k \in \bigcap \mathbb{X}$ and $\bigcap \mathbb{X} \Vdash \alpha$.

We can conclude that $\bigcap \mathbb{X} \Vdash \alpha$ for each $\alpha \in \Psi$, thus $\bigcap \mathbb{X} \Vdash \Psi$. □

Proof for Section 4.5

Proof of Observation 4.21 For each $\imath \in \mathbb{I}$, use the descriptor $\Pi_{\mathcal{K}\circledcirc\imath} = \{\mathfrak{B}x \mid x \in \mathfrak{s}(\mathcal{K} \circledcirc \imath)\} \cup \{\neg\mathfrak{B}x \mid x \notin \mathfrak{s}(\mathcal{K} \circledcirc \imath)\}$. □

Proofs for Section 5.1

Proof of Theorem 5.3 *From (I) to (II)*: Left to the reader.

From (II) to (I): Let $\mathbb{X} = \{X \mid (\exists\Psi)(X = K \circ \Psi)\}$ and let $\widehat{C}$ be a monoselective choice function on $\mathbb{X}$ such that $\widehat{C}(\llbracket\Psi\rrbracket) = K \circ \Psi$ whenever Ψ is satisfiable within $\mathbb{X}$. To verify the construction we need to show that (1) $\widehat{C}$ is well-defined, i.e. it is indeed a function, (2) all elements of $\mathbb{X}$ are logically closed, (3) $K \in \mathbb{X}$, (4) $\widehat{C}(\llbracket\Psi\rrbracket) \in \llbracket\Psi\rrbracket$ if $\llbracket\Psi\rrbracket \neq \varnothing$, (5) if there is some $X \in \mathbb{X}$ with $X \Vdash \Psi$, then $K \circ \Psi = \widehat{C}(\llbracket\Psi\rrbracket)$, and (6) if $X \nVdash \Psi$ for all $X \in \mathbb{X}$, then $K \circ \Psi = K$.

(1): It follows directly from uniformity that if $\llbracket\Psi\rrbracket = \llbracket\Psi'\rrbracket$, then $K \circ \Psi = K \circ \Psi'$.

(2) follows from closure.

(3): It follows from relative success that $K \circ \Pi_K = K$.

(4) follows from regularity. (5) and (6) follow from the construction, using regularity and relative success. □

Proof of Observation 5.5 Left to the reader. Use Observation 4.11 for part 3. □

Proof of Observation 5.7 Left to the reader. □

Proofs for Section 5.2

Lemma 5.1 *Let $\circ$ be a descriptor revision on a belief set K. If it satisfies relative success, regularity, and cumulativity, then it satisfies:*

(1) If $K \circ \Xi \Vdash \Psi$ iff $K \circ \Xi \Vdash \Psi'$ for all Ξ, then $K \circ \Psi = K \circ \Psi'$. (uniformity), and

(2) If $K \circ (\Psi \veebar \Xi) \Vdash \Psi$, then $K \circ (\Psi \veebar \Xi) = K \circ \Psi$. (disjunctive implication)

Proof of Lemma 5.1 *Part 1*: Let Ψ and Ψ' be descriptors such that $K \circ \Xi \Vdash \Psi$ iff $K \circ \Xi \Vdash \Psi'$ for all Ξ.

Case 1, there is no Ξ such that $K \circ \Xi \Vdash \Psi$: Then by supposition there is no Ξ such that $K \circ \Xi \Vdash \Psi'$. It follows from relative success that $K \circ \Psi = K$ and $K \circ \Psi' = K$.

Case 2, there is some Ξ' such that $K \circ \Xi' \Vdash \Psi$: Then by the assumption $K \circ \Xi' \Vdash \Psi'$. Due to regularity it follows from $K \circ \Xi' \Vdash \Psi$ that $K \circ \Psi \Vdash \Psi$ and from $K \circ \Xi' \Vdash \Psi'$ that $K \circ \Psi' \Vdash \Psi'$. Due to our assumption (substituting Ψ for Ξ) we can conclude from $K \circ \Psi \Vdash \Psi$ that $K \circ \Psi \Vdash \Psi'$. Similarly (substituting Ψ' for Ξ) we can conclude from $K \circ \Psi' \Vdash \Psi'$ that $K \circ \Psi' \Vdash \Psi$. Applying cumulativity to $K \circ \Psi \Vdash \Psi'$ we obtain $K \circ \Psi = K \circ (\Psi \cup \Psi')$, and applying the same postulate to $K \circ \Psi' \Vdash \Psi$ we obtain $K \circ \Psi' = K \circ (\Psi \cup \Psi')$. Thus $K \circ \Psi = K \circ \Psi'$ in this case as well.

Part 2: Let $K \circ (\Psi \veebar \Xi) \Vdash \Psi$. Cumulativity yields $K \circ (\Psi \veebar \Xi) = K \circ (\Psi \cup (\Psi \veebar \Xi))$, and uniformity (that follows from Part 1) yields $K \circ (\Psi \cup (\Psi \veebar \Xi)) = K \circ \Psi$. □

Lemma 5.2 (modified from [124]) *Let $\leqq$ be a relation on a set $\mathbb{X}$ of belief sets. Then the following three conditions are equivalent:*

(A) For all descriptors Ψ that are satisfiable within $\mathbb{X}$ there is a unique $\leqq$-minimal Ψ-satisfying element X of $\mathbb{X}$, i.e. a unique element X such that $X \leqq Y$ for all $Y \in \mathbb{X}$ with $Y \Vdash \Psi$.

(B) $\leqq$ is antisymmetric, complete, transitive and descriptor-wellfounded.

(C) $\leqq$ is antisymmetric and descriptor-wellfounded.

Proof of Lemma 5.2 *From (A) to (B)*:

Completeness: Let $X, Y \in \mathbb{X}$. Apply (A) to the descriptor $\Pi_{\{X,Y\}}$, as defined in Definition A.1 (p. 169).

Antisymmetry: Suppose to the contrary that $X \leqq Y \leqq X$ and $X \neq Y$. Since both $X \leqq X$ and $Y \leqq Y$ hold due to completeness it follows that both X and Y are $\leqq$-minimal elements for the descriptor $\Pi_{\{X,Y\}}$. This contradicts (A).

Transitivity: Let $X \leqq Y \leqq Z$ and suppose to the contrary that $X \nleqq Z$. Since $\leqq$ is complete it is reflexive, thus $X \neq Z$. It also follows from completeness and $X \nleqq Z$ that $Z \leqq X$.

If $X = Y$ then $Y \leqq Z$ would yield $X \leqq Z$, contrary to what we have assumed. Thus $X \neq Y$. If $Y = Z$, then $X \nleqq Z$ would yield $X \nleqq Y$, also contradicting our assumptions. Thus $Y \neq Z$. We therefore have the cycle $X \leqq Y \leqq Z \leqq X$ of three distinct elements. Due to antisymmetry, $X < Y < Z < X$, which means that there is no unique $\leqq$-minimal element for $\Pi_{\{X,Y,Z\}}$, contrary to (A). We can conclude that $X \leqq Z$.

Descriptor-wellfoundedness follows directly from (A).

From (C) to (A): Let Ψ be a descriptor that is satisfiable within $\mathbb{X}$. Since $\leqq$ is descriptor-wellfounded there is some $\leqq$-minimal Ψ-satisfying element X of $\mathbb{X}$. Suppose that there is some other such element Y. Then $X \leqq Y$ and $Y \leqq X$, and antisymmetry yields $X = Y$. This proves the uniqueness of X. □

Proof of Theorem 5.10 The equivalence of (I) and (II) follows from Lemma 5.2. The direction from (II) to (III) is left to the reader. For the direction from (III) to (I) we define the set $\mathbb{X} = \{X \mid (\exists\Psi)(X = K \circ \Psi)\}$ and the relation $\leqq$ on $\mathbb{X}$ such that for all Ψ and Ξ:

$K \circ \Psi \leqq K \circ \Xi$ if and only if $K \circ \Psi = K \circ (\Psi \veebar \Xi)$.

We have to prove (1) that $\mathbb{X}$ is a set of belief sets, (2) that it contains K, (3) that if there is some $X \in \mathbb{X}$ with $X \Vdash \Psi$, then $K \circ \Psi$ is the unique $\leqq$-minimal element of $\mathbb{X}$ that satisfies Ψ, and (4) if $X \nVdash \Psi$ for all $X \in \mathbb{X}$, then $K \circ \Psi = K$.

(1) follows from closure.

(2): It follows from relative success that $K \circ \Pi_K = K$.

(3): Let $K \circ \Xi \Vdash \Psi$. It follows from regularity that $K \circ \Psi \Vdash \Psi$. To prove the unique $\leqq$-minimality of $K \circ \Psi$ among Ψ-satisfying elements of $\mathbb{X}$, we first prove minimality and then uniqueness.

For minimality, suppose to the contrary that $K \circ \Xi \Vdash \Psi$ and $K \circ \Psi \nleqq K \circ \Xi$, i.e. $K \circ \Psi \neq K \circ (\Psi \veebar \Xi)$. It follows from disjunctive implication (Lemma 5.1) that $K \circ (\Psi \veebar \Xi) \nVdash \Psi$.

It follows from $K \circ \Psi \Vdash \Psi$ that $K \circ \Psi \Vdash \Psi \veebar \Xi$, and regularity yields $K \circ (\Psi \veebar \Xi) \Vdash \Psi \veebar \Xi$, thus due to Observation 4.11 either $K \circ (\Psi \veebar \Xi) \Vdash \Psi$ or $K \circ (\Psi \veebar \Xi) \Vdash \Xi$. Thus $K \circ (\Psi \veebar \Xi) \Vdash \Xi$, and disjunctive implication (Lemma 5.1) yields $K \circ \Xi = K \circ (\Psi \veebar \Xi)$, thus $K \circ (\Psi \veebar \Xi) \Vdash \Psi$, contrary to what was just shown.

For uniqueness, suppose to the contrary that there is some $X \in \mathbb{X}$ such that $X \Vdash \Psi$ and $X \leqq K \circ \Psi \neq X$. It follows from our definition of $\mathbb{X}$ that $X = K \circ \Xi$ for some Ξ. Due to our definition of $\leqq$, $K \circ \Xi \leqq K \circ \Psi$ yields $K \circ \Xi = K \circ (\Psi \veebar \Xi)$. It follows from $X \Vdash \Psi$, i.e. $K \circ (\Psi \veebar \Xi) \Vdash \Psi$, and disjunctive implication (Lemma 5.1) that $K \circ (\Psi \veebar \Xi) = K \circ \Psi$. But we already have $K \circ \Psi \neq X$ and $X = K \circ \Xi = K \circ (\Psi \veebar \Xi)$. Contradiction. Since this holds for all $X \in \mathbb{X}$ with $X \Vdash \Psi$ we can conclude that $K \circ \Psi$ is the unique $\leqq$-minimal Ψ-satisfying element of $\mathbb{X}$.

(4) Let $K \circ \Xi \nVdash \Psi$ for all $K \circ \Xi \in \mathbb{X}$. Then $K \circ \Psi \nVdash \Psi$, and relative success yields $K \circ \Psi = K$. □

Proof of Observation 5.11 This was proved as part of Lemma 5.1. □

Proof of Observation 5.12 *From cumulativity to reciprocity*: Let $K \circ \Psi \Vdash \Xi$ and $K \circ \Xi \Vdash \Psi$. Then cumulativity yields $K \circ \Psi = K \circ (\Psi \cup \Xi) = K \circ \Xi$.

From reciprocity to cumulativity: Let $K \circ \Psi \Vdash \Xi$. There are two cases.

Case (i), $K \circ \Psi \nVdash \Psi$: Regularity yields $K \circ (\Psi \cup \Xi) \nVdash \Psi$, thus $K \circ (\Psi \cup \Xi) \nVdash \Psi \cup \Xi$. Relative success yields $K \circ \Psi = K = K \circ (\Psi \cup \Xi)$.

Case (ii), $K \circ \Psi \Vdash \Psi$: Then $K \circ \Psi \Vdash \Psi \cup \Xi$. Regularity yields $K \circ (\Psi \cup \Xi) \Vdash \Psi \cup \Xi$. We thus have $K \circ \Psi \Vdash \Psi \cup \Xi$ and $K \circ (\Psi \cup \Xi) \Vdash \Psi$, and reciprocity yields $K \circ \Psi = K \circ (\Psi \cup \Xi)$. □

Proof of Observation 5.13 Let $K^{+} = K \circ \mathfrak{B}\top$ and let $K^{+} \Vdash \Psi$, i.e. $K \circ \mathfrak{B}\top \Vdash \Psi$. Cumulativity yields $K \circ \mathfrak{B}\top = K \circ (\{\mathfrak{B}\top\} \cup \Psi)$. We also have $K \circ \Psi \Vdash \mathfrak{B}\top$, and cumulativity yields $K \circ \Psi = K \circ (\Psi \cup \{\mathfrak{B}\top\})$. Thus $K \circ \Psi = K \circ \mathfrak{B}\top = K^{+}$. □

Lemma 5.3 *Let $\circ$ be the linear descriptor revision on a belief set K that is based on the relation $\leqq$ on its outcome set $\mathbb{X}$. Then $K \circ \Psi \leqq K \circ \Xi$ if and only if $K \circ \Psi = K \circ (\Psi \veebar \Xi)$.*

Proof of Lemma 5.3 Left to the reader. □

Proof of Observation 5.14 For one direction of the proof, let confirmation be satisfied. We have to show that K is the $\leqq$-minimal element of $\mathbb{X}$. Let $X \in \mathbb{X}$. It follows from regularity (Theorem 5.10) that $X = K \circ \Pi_X$. Due to uniformity (Lemma 5.1, p. 170), $K \circ \mathfrak{B}\top = K \circ (\mathfrak{B}\top \veebar \Pi_X)$, and it follows from Lemma 5.3 that $K \circ \mathfrak{B}\top \leqq K \circ \Pi_X$. Since $K \Vdash \mathfrak{B}\top$, confirmation yields $K \circ \mathfrak{B}\top = K$, and we can conclude that $K \leqq K \circ \Pi_X$, i.e. $K \leqq X$.

The other direction of the proof is left to the reader. □

Proofs for Section 5.4

Proof of Observation 5.16 For one direction, let $\rightharpoondown$ satisfy irreflexivity within $\mathbb{X} \setminus \{K\}$. It follows directly from $K \in \mathbb{X}$ and Definition 5.15 that the outcome set is a subset of $\mathbb{X}$. We also have to show that each element of $\mathbb{X}$ is an element of the outcome set. For each $X \in \mathbb{X} \setminus \{K\}$, $\{X\}$ is the set of Π_X-satisfying elements of $\mathbb{X}$, and due to the irreflexivity of $\rightharpoondown$, X is the unique unblocked element of $\{X\}$, thus $K \circ \Pi_X = X$. Furthermore, $K = K \circ \neg\mathfrak{B}\top$ (where $\top$ is a tautology) due to clause (ii) of Definition 5.15. Thus all elements of $\mathbb{X}$ are elements of the outcome set.

For the other direction, assume that $X \rightharpoondown X$ for some $X \in \mathbb{X} \setminus \{K\}$. Then $K \circ \Psi \neq X$ for all Ψ, thus X is not in the outcome set although it is in $\mathbb{X}$. □

Proof of Observation 5.17 Left to the reader. □

Proof of Observation 5.18 *Part 1*: Let $\mathbb{X} = \{K, X, Y\}$ and $\rightharpoondown = \{\langle K, X\rangle, \langle K, Y\rangle, \langle X, Y\rangle, \langle Y, X\rangle\}$. Then $K \circ \Pi_X \Vdash \Pi_X \veebar \Pi_Y$ but $K \circ (\Pi_X \veebar \Pi_Y) \nVdash \Pi_X \veebar \Pi_Y$.

Part 2: Let $\mathbb{X} = \{K, X, Y\}$ and $\rightharpoondown = \{\langle K, X\rangle, \langle X, Y\rangle, \langle Y, K\rangle\}$. Then $K \circ (\Pi_K \veebar \Pi_X \veebar \Pi_Y) = K$ and $K \circ ((\Pi_K \veebar \Pi_X \veebar \Pi_Y) \cup (\Pi_K \veebar \Pi_Y)) = K \circ (\Pi_K \veebar \Pi_Y) = Y$, thus $K \circ (\Pi_K \veebar \Pi_X \veebar \Pi_Y) \Vdash \Pi_K \veebar \Pi_Y$ but $K \circ (\Pi_K \veebar \Pi_X \veebar \Pi_Y) \neq K \circ ((\Pi_K \veebar \Pi_X \veebar \Pi_Y) \cup (\Pi_K \veebar \Pi_Y))$.

Part 3: Let $\mathbb{X} = \{K, X, Y\}$ and $\rightharpoondown = \{\langle K, X\rangle, \langle X, Y\rangle, \langle Y, K\rangle\}$. Then $K \circ (\Pi_K \veebar \Pi_X \veebar \Pi_Y) \Vdash \Pi_K \veebar \Pi_Y$ and $K \circ (\Pi_K \veebar \Pi_Y) \Vdash \Pi_K \veebar \Pi_X \veebar \Pi_Y$ but $K \circ (\Pi_K \veebar \Pi_X \veebar \Pi_Y) \neq K \circ (\Pi_K \veebar \Pi_Y)$.

Part 4: Let $\mathbb{X} = \{K, X\}$ and let $\rightharpoondown = \{\langle X, K\rangle\}$. Then $K \Vdash \Pi_K \veebar \Pi_X$ but $K \circ (\Pi_K \veebar \Pi_X) = X$. □

Proof of Observation 5.19 *Part 1, peripheral cumulativity*: Let $K \circ \Psi \neq K \neq K \circ (\Psi \cup \Xi)$ and $K \circ \Psi \Vdash \Xi$. It follows from Definition 5.15 and $K \neq K \circ \Psi$ that $K \circ \Psi \Vdash \Psi$, thus $K \circ \Psi \Vdash \Psi \cup \Xi$, i.e. $K \circ \Psi \in [\![\Psi \cup \Xi]\!]$. It also follows from $K \neq K \circ \Psi$ that $K \circ \Psi$ is unblocked within $[\![\Psi]\!]$, and since $[\![\Psi \cup \Xi]\!] \subseteq [\![\Psi]\!]$ it is then also unblocked within $[\![\Psi \cup \Xi]\!]$. Thus $K \circ \Psi$ is $\Psi \cup \Xi$-satisfying and unblocked within $[\![\Psi \cup \Xi]\!]$. Since $K \neq K \circ (\Psi \cup \Xi)$ there is exactly one belief set with that property, namely $K \circ (\Psi \cup \Xi)$. It follows that $K \circ \Psi = K \circ (\Psi \cup \Xi)$.

Part 1, peripheral disjunctive identity: Since $K \neq K \circ \Psi$, all elements of $[\![\Psi]\!] \setminus \{K \circ \Psi\}$ are blocked within $[\![\Psi]\!]$, and similarly all elements of $[\![\Xi]\!] \setminus \{K \circ \Psi\}$ are blocked within $[\![\Xi]\!]$. Since $[\![\Psi \veebar \Xi]\!] = [\![\Psi]\!] \cup [\![\Xi]\!]$, all elements of $[\![\Psi \veebar \Xi]\!] \setminus \{K \circ \Psi\}$ are blocked within $[\![\Psi \veebar \Xi]\!]$. Since $K \circ \Psi$ is unblocked both within $[\![\Psi]\!]$ and within $[\![\Xi]\!]$, it is unblocked within $[\![\Psi \veebar \Xi]\!]$. Thus $K \circ \Psi$ is the only unblocked element within $[\![\Psi \veebar \Xi]\!]$, thus $K \circ \Psi = K \circ (\Psi \veebar \Xi)$.

Part 2, peripheral cumulativity: Let $\mathbb{X} = \{K, X, Y, Z\}$ and let $\circ$ be based on a monoselective choice function $\widehat{C}$ such that $\widehat{C}(\{X, Y, Z\}) = \{Y\}$ and $\widehat{C}(\{Y, Z\}) = \{Z\}$. Then $K \circ (\Pi_X \veebar \Pi_Y \veebar \Pi_Z) = Y$ and $Y \Vdash \Pi_Y \veebar \Pi_Z$ but $K \circ ((\Pi_X \veebar \Pi_Y \veebar \Pi_Z) \cup (\Pi_Y \veebar \Pi_Z)) = K \circ (\Pi_Y \veebar \Pi_Z) = Z$.

Part 2, peripheral disjunctive identity: Let $\mathbb{X} = \{K, X, Y, Z\}$ and let $\circ$ be based on a monoselective choice function $\widehat{C}$ such that $\widehat{C}(\{X, Y\}) = \widehat{C}(\{Y, Z\}) = Y$ and $\widehat{C}(\{X, Y, Z\}) = X$. Then $K \circ (\Pi_X \veebar \Pi_Y) = Y$, $K \circ (\Pi_Y \veebar \Pi_Z) = Y$, and $K \circ ((\Pi_X \veebar \Pi_Y) \veebar (\Pi_Y \veebar \Pi_Z)) = X$. □

Proof of Theorem 5.20 *From regularity to peripheral non-occlusion*: Let $K \notin [\![\Psi]\!]$ and $X \in [\![\Psi]\!]$. Due to Observation 5.16 and our assumption that $\rightharpoondown$ is irreflexive there is some Ξ with $X = K \circ \Xi$. Then $K \circ \Xi \Vdash \Psi$ and regularity yields $K \circ \Psi \Vdash \Psi$. From this and $K \circ \Psi \neq K$ it follows according to Definition 5.15 that $K \circ \Psi$ is an unblocked element within $[\![\Psi]\!]$.

From regularity to peripheral weak connectedness: Let $X, Y \in \mathbb{X}$ and $X \neq Y \neq K \neq X$. Due to Observation 5.16 and our assumption that $\rightharpoondown$ is irreflexive there is some Ξ with $X = K \circ \Xi$. Thus $K \circ \Xi \Vdash \Pi_X \veebar \Pi_Y$. Regularity yields $K \circ (\Pi_X \veebar \Pi_Y) \Vdash \Pi_X \veebar \Pi_Y$, thus $K \circ (\Pi_X \veebar \Pi_Y) \in \{X, Y\}$, from which it follows that either $X \rightharpoondown Y$ or $Y \rightharpoondown X$.

From peripheral non-occlusion and peripheral weak connectedness to regularity: Let $K \circ \Xi \Vdash \Psi$.

First case, $K \nVdash \Psi$: Due to peripheral non-occlusion it follows from $K \circ \Xi \Vdash \Psi$ and $K \nVdash \Psi$ that $[\![\Psi]\!]$ has at least one non-blocked element. It follows from peripheral weak connectedness that it has at most one such element. Due to clause (i) of Definition 5.15, that element is equal to $K \circ \Psi$, thus $K \circ \Psi \Vdash \Psi$.

Second case, $K \Vdash \Psi$: According to Definition 5.15, $K \circ \Psi$ is either an element of $[\![\Psi]\!]$ or equal to K. In both cases, $K \circ \Psi \Vdash \Psi$. □

Proof of Observation 5.21 For one direction, let $\circ$ be a monoselective descriptor revision. Then $\circ$ satisfies regularity, and we can conclude from Theorem 5.20 that $\rightharpoondown$ satisfies peripheral non-occlusion and peripheral weak connectedness.

For the other direction, let $\rightharpoondown$ satisfy peripheral non-occlusion and peripheral weak connectedness, and let $\circ$ be the operation of descriptor revision generated from $\rightharpoondown$. It follows from Theorem 5.20 that $\circ$ satisfies regularity and from Observation 5.17 that it satisfies closure, relative closure, and uniformity. We can conclude from Theorem 5.3 that it is a monoselective descriptor revision. □

Proof of Theorem 5.22 *From cumulativity to peripheral non-occlusion*: Suppose to the contrary that $K \notin [\![\Psi]\!] \neq \varnothing$ and $[\![\Psi]\!]$ has no unblocked element. Let $X \in [\![\Psi]\!]$. Then $K \circ \Psi = K$, $K \circ \Psi \Vdash \Pi_K \veebar \Pi_X$, and $K \circ (\Psi \cup (\Pi_K \veebar \Pi_X)) = K \circ \Pi_X = X$, contrary to cumulativity.

From cumulativity to peripheral weak connectedness: Suppose to the contrary that there are $X, Y \in \mathbb{X}$ such that $X \neq Y \neq K \neq X$ and $X \not\rightharpoondown Y \not\rightharpoondown X$. Then $K \circ (\Pi_X \veebar \Pi_Y) = K$, thus $K \circ (\Pi_X \veebar \Pi_Y) \Vdash \Pi_K \veebar \Pi_Y$, but $K \circ ((\Pi_X \veebar \Pi_Y) \cup (\Pi_K \veebar \Pi_Y)) = K \circ \Pi_Y = Y$, contrary to cumulativity.

From cumulativity to top adjacency: We will assume that cumulativity holds but top adjacency does not hold, and show that this leads to a contradiction. Let $X, Y \in \mathbb{X} \setminus \{K\}$ and $X \rightharpoondown Y \rightharpoondown K$. Since top adjacency does not hold, if $K \rightharpoondown X$ then $K \not\rightharpoondown Y$, and furthermore, if $K \not\rightharpoondown X$ then $X \not\rightharpoondown K$. We therefore have the following two cases:

Case 1, $K \rightharpoondown X$ and $K \not\rightharpoondown Y$: Then there is no unblocked element within $\{K, X, Y\}$ but there is a unique unblocked element within $\{K, Y\}$, namely Y. It follows that $K \circ (\Pi_K \veebar \Pi_X \veebar \Pi_Y) = K$, thus $K \circ (\Pi_K \veebar \Pi_X \veebar \Pi_Y) \Vdash \Pi_K \veebar \Pi_Y$, but $K \circ ((\Pi_K \veebar \Pi_X \veebar \Pi_Y) \cup (\Pi_K \veebar \Pi_Y)) = Y$, contrary to cumulativity.

Case 2, $K \not\rightharpoondown X$ and $X \not\rightharpoondown K$: We have $X \rightharpoondown Y$ and by applying peripheral non-occlusion (that we have just proved) to $\{X, Y\}$ we obtain $Y \not\rightharpoondown X$. Thus $\{K, X, Y\}$ has a unique unblocked element namely X, whereas $\{K, X\}$ has the unblocked elements K and X. It follows that $K \circ (\Pi_K \veebar \Pi_X \veebar \Pi_Y) = X$, thus $K \circ (\Pi_K \veebar \Pi_X \veebar \Pi_Y) \Vdash \Pi_K \veebar \Pi_X$, but $K \circ ((\Pi_K \veebar \Pi_X \veebar \Pi_Y) \cup (\Pi_K \veebar \Pi_X)) = K$, contrary to cumulativity.

In combination, the two cases show that it is impossible for cumulativity to hold without top adjacency also holding.

From peripheral non-occlusion, peripheral weak connectedness, and top adjacency to cumulativity: Let $K \circ \Psi \Vdash \Xi$.

Case 1, $[\![\Psi]\!] = \varnothing$: Then $[\![\Psi \cup \Xi]\!] = \varnothing$, and we have $K \circ \Psi = K$ and $K \circ (\Psi \cup \Xi) = K$.

Case 2, $K \notin [\![\Psi]\!] \neq \varnothing$: It follows from peripheral non-occlusion and peripheral weak connectedness that $[\![\Psi]\!]$ has exactly one unblocked element, and due to Definition 5.15 that element is equal to $K \circ \Psi$. It follows that $K \circ \Psi \Vdash \Psi$ and we already have $K \circ \Psi \Vdash \Xi$, so $K \circ \Psi \Vdash \Psi \cup \Xi$. Since $K \circ \Psi$ is unblocked within $[\![\Psi]\!]$, it is also unblocked within its subset $[\![\Psi \cup \Xi]\!]$. It follows from $K \notin [\![\Psi]\!]$ and $[\![\Psi \cup \Xi]\!] \subseteq [\![\Psi]\!]$ that $K \notin [\![\Psi \cup \Xi]\!]$. It follows from peripheral non-occlusion and peripheral weak connectedness that $[\![\Psi \cup \Xi]\!]$ has exactly one unblocked element, and then $K \circ \Psi$ is that element. Due to Definition 5.15, $K \circ \Psi = K \circ (\Psi \cup \Xi)$.

Case 3, $K \in [\![\Psi]\!]$ and $K \neq K \circ \Psi$: It follows from Definition 5.15 that $K \circ \Psi$ is the unique unblocked element of $[\![\Psi]\!]$. Since $K \circ \Psi \Vdash \Xi$ we have $K \circ \Psi \in [\![\Psi \cup \Xi]\!]$. Since $K \circ \Psi$ is unblocked within $[\![\Psi]\!]$, it is also unblocked within its subset $[\![\Psi \cup \Xi]\!]$.

Case 3A, $K \notin [\![\Psi \cup \Xi]\!]$: It follows from peripheral non-occlusion and peripheral weak connectedness that $[\![\Psi \cup \Xi]\!]$ has exactly one unblocked element, and then $K \circ \Psi$ is that element. Due to Definition 5.15, $K \circ \Psi = K \circ (\Psi \cup \Xi)$.

Case 3B, $K \in [\![\Psi \cup \Xi]\!]$: Since $K \circ \Psi$ is the unique unblocked element of $[\![\Psi]\!]$, there is some $X \in [\![\Psi]\!]$ with $X \rightharpoondown K$. Since $K \circ \Psi$ is unblocked within $[\![\Psi]\!]$ we also have $X \not\rightharpoondown K \circ \Psi$ and $K \not\rightharpoondown K \circ \Psi$. Due to peripheral weak connectedness, $K \circ \Psi \rightharpoondown X$. We conclude from top adjacency that $K \circ \Psi \rightharpoondown K$.

Next, let $Z \in [\![\Psi \cup \Xi]\!] \setminus \{K, K \circ \Psi\}$. Since $K \circ \Psi$ is unblocked within $[\![\Psi \cup \Xi]\!]$ it follows from peripheral weak connectedness that $K \circ \Psi \rightharpoondown Z$. Thus $K \circ \Psi$ is the only unblocked element within $[\![\Psi \cup \Xi]\!]$, thus $K \circ \Psi = K \circ (\Psi \cup \Xi)$.

Case 4, $K \in [\![\Psi]\!]$ and $K = K \circ \Psi$: It follows from $K \circ \Psi \Vdash \Xi$ that $K \in [\![\Psi \cup \Xi]\!]$.

Case 4A, *K is unblocked within* $[\![\Psi \cup \Xi]\!]$: If K is the only unblocked element within $[\![\Psi \cup \Xi]\!]$, then $K = K \circ (\Psi \cup \Xi)$ due to clause (i) of Definition 5.15. If it is one of at least two unblocked elements within $[\![\Psi \cup \Xi]\!]$, then $K = K \circ (\Psi \cup \Xi)$ due to clause (ii) of the same definition.

Case 4B, *K is blocked within* $[\![\Psi \cup \Xi]\!]$: Due to peripheral non-occlusion and peripheral weak connectedness there is some $Y \in [\![\Psi \cup \Xi]\!] \setminus \{K\}$ such that $Y \rightharpoondown X \not\rightharpoondown Y$ for all $X \in [\![\Psi \cup \Xi]\!] \setminus \{K, Y\}$. It follows that $K \circ (\Psi \cup \Xi)$ is either Y or K. We are going to show that it is not Y. Suppose that it is. Then clearly $K \not\rightharpoondown Y$.

Case 4Ba, $Y \not\rightharpoondown K$: Since K is blocked within $[\![\Psi \cup \Xi]\!]$ there is then some $X \in [\![\Psi \cup \Xi]\!] \setminus \{K, Y\}$ such that $X \rightharpoondown K$. We then have $Y \rightharpoondown X \not\rightharpoondown Y$, $K \not\rightharpoondown Y \not\rightharpoondown K$ and $X \rightharpoondown K$, contrary to top adjacency.

Case 4Bb, $Y \rightharpoondown K$: Due to peripheral non-occlusion and peripheral weak connectivity there is some $Z \in [\![\Psi]\!] \setminus \{K\}$ such that $Z \rightharpoondown V \not\rightharpoondown Z$ for all $V \in [\![\Psi]\!] \setminus \{K, Z\}$. Since K is blocked (within $[\![\Psi \cup \Xi]\!]$ and therefore also) within $[\![\Psi]\!]$ and $K \circ \Psi \neq Z$, Z is blocked within $[\![\Psi]\!]$, thus $K \rightharpoondown Z$. Since by assumption $K \not\rightharpoondown Y$, $Y \neq Z$. We then have $Z \rightharpoondown Y \not\rightharpoondown Z$, $K \not\rightharpoondown Y \rightharpoondown K$ and $K \rightharpoondown Z$, contrary to top adjacency.

Thus, in neither subcase is $K \circ (\Psi \cup \Xi)$ equal to Y. We can conclude that $K \circ (\Psi \cup \Xi) = K$, thus $K \circ \Psi = K \circ (\Psi \cup \Xi)$ in case 4B as well. □

Proof of Observation 5.23 *From linear revision to blockage revision*: Let $\rightharpoondown$ be the strict part of $\leqq$.

From blockage revision to linear revision: It follows from Theorem 5.22 that $\rightharpoondown$ satisfies cumulativity, from Theorem 5.20 that it satisfies regularity, and from Observation 5.17 that it satisfies closure and relative success. It then follows from Theorem 5.10 that it is a linear revision. □

Proof of Observation 5.24 *From cumulativity to regularity*: Directly from Theorems 5.20 and 5.22.

From cumulativity to reciprocity: Let $K \circ \Psi \Vdash \Xi$ and $K \circ \Xi \Vdash \Psi$. Then cumulativity yields $K \circ \Psi = K \circ (\Psi \cup \Xi) = K \circ \Xi$.

From regularity and reciprocity to cumulativity: Let $K \circ \Psi \Vdash \Xi$. There are two cases.

Case (i), $K \circ \Psi \nVdash \Psi$: Regularity yields $K \circ (\Psi \cup \Xi) \nVdash \Psi$, thus $K \circ (\Psi \cup \Xi) \nVdash \Psi \cup \Xi$. It follows from Definition 5.15 that $K \circ \Psi = K = K \circ (\Psi \cup \Xi)$.

Case (ii), $K \circ \Psi \Vdash \Psi$: Then $K \circ \Psi \Vdash \Psi \cup \Xi$. Regularity yields $K \circ (\Psi \cup \Xi) \Vdash \Psi \cup \Xi$. We thus have $K \circ \Psi \Vdash \Psi \cup \Xi$ and $K \circ (\Psi \cup \Xi) \Vdash \Psi$, and reciprocity yields $K \circ \Psi = K \circ (\Psi \cup \Xi)$. □

Proof of Observation 5.25 *Part 1*: Let $\mathbb{X} = \{K, X, Y\}$ and $\rightharpoondown = \{\langle X, Y\rangle, \langle Y, K\rangle\}$. It follows from Theorem 5.20 that regularity is satisfied. We have $K \circ (\Pi_K \veebar \Pi_X \veebar \Pi_Y) = X$ and $K \circ (\Pi_K \veebar \Pi_X) = K$, thus $K \circ (\Pi_K \veebar \Pi_X \veebar \Pi_Y) \Vdash \Pi_K \veebar \Pi_X$ and $K \circ (\Pi_K \veebar \Pi_X) \Vdash \Pi_K \veebar \Pi_X \veebar \Pi_Y$ but $K \circ (\Pi_K \veebar \Pi_X \veebar \Pi_Y) \neq K \circ (\Pi_K \veebar \Pi_X)$, which shows that reciprocity does not hold.

Part 2: Let $\mathbb{X} = \{K, X, Y\}$ and $\rightharpoondown = \{\langle K, X\rangle, \langle K, Y\rangle\}$. In order to show that reciprocity holds it is sufficient to show that there are no Ψ and Ξ such that either (1) $K \circ \Psi = K$, $K \Vdash \Xi$, $K \circ \Xi = X$, and $X \Vdash \Psi$, (2) $K \circ \Psi = K$, $K \Vdash \Xi$, $K \circ \Xi = Y$, and $Y \Vdash \Psi$, or (3) $K \circ \Psi = X$, $X \Vdash \Xi$, $K \circ \Xi = Y$, and $Y \Vdash \Psi$.

Suppose that (1) holds. Due to our construction of $\circ$ it follows from $K \circ \Xi = X$ that $X \Vdash \Xi$, $K \nVdash \Xi$, and $Y \nVdash \Xi$. But we also have $K \Vdash \Xi$, thus (1) does not hold. A symmetrical proof shows that (2) does not hold. Suppose that (3) holds. It then follows from $K \circ \Psi = X$ that $X \Vdash \Psi$, $K \nVdash \Psi$, and $Y \nVdash \Psi$, but we also have $Y \Vdash \Psi$, so that this case is impossible as well. Thus there are no Ψ and Ξ that satisfy either (1), (2), or (3), thus reciprocity holds.

It follows from $K \circ \Pi_X = X$, $X \Vdash \Pi_X \veebar \Pi_Y$, and $K \circ (\Pi_X \veebar \Pi_Y) = K$ that regularity does not hold. □

Proof of Theorem 5.26 *From confirmation to near-superiority*: Let $X \rightharpoondown K$. Confirmation yields $K \circ (\Pi_K \veebar \Pi_X) = K$, which is not the case if $X \rightharpoondown K \not\rightharpoondown X$. Thus $K \rightharpoondown X$.

If also follows from confirmation that $K \circ (\Pi_K \veebar \Pi_X \veebar \Pi_Y) = K$. Since $X \rightharpoondown K$, $K \circ (\Pi_K \veebar \Pi_X \veebar \Pi_Y)$ cannot follow from clause (i) of Definition 5.15, so it must be based on clause (ii). Thus Y must be blocked by either K or X.

From near-superiority to confirmation: Let $K \Vdash \Psi$. If $[\![\Psi]\!]$ does not have exactly one unblocked element, then clause (ii) of Definition 5.15 yields $K \circ \Psi = K$. It remains to treat the case when $[\![\Psi]\!]$ has exactly one unblocked element. Suppose that element is not K. Then there is some $X \in [\![\Psi]\!]$ with $X \rightharpoondown K$. It follows from near-superiority that all elements of $[\![\Psi]\!]$ are blocked (either by K or by X). This contradicts the assumption that there is some unblocked element of $[\![\Psi]\!]$ which is not K. We conclude that it is K and that therefore $K \circ \Psi = K$ in this case as well. □

Proof of Observation 5.27 *From (I) to (III)*: Let $\rightharpoondown$ be the strict part of $\leqq$.

From (III) to (II): Left to the reader.

From (II) to (I): It follows from Theorem 5.26 that $\rightharpoondown$ satisfies confirmation, and we know from Observation 5.23 that it is a linear revision. It follows that it is a centrolinear revision. □

Proof of Observation 5.28 (1) Directly by substitution. (2) Let $\rightharpoondown$ be transitive and irreflexive, and suppose for contradiction that it is not asymmetric. Then there are Y and Z such that $Y \rightharpoondown Z$ and $Z \rightharpoondown Y$. Transitivity yields $Y \rightharpoondown Y$, contrary to irreflexivity. □

Proof of Observation 5.29 Let $\mathbb{X} = \{K, X, Y\}$ and $\rightharpoondown = \{\langle X, Y\rangle, \langle X, K\rangle, \langle Y, K\rangle\}$. It follows from Theorem 5.22 that $\circ$ satisfies cumulativity and from $K \circ (\Pi_K \veebar \Pi_X) = X$ that it does not satisfy confirmation. □

Proof of Observation 5.30 *Part 1*: Let $\mathbb{X} = \{K, X, Y\}$ and $\rightharpoondown = \{\langle K, X\rangle, \langle K, Y\rangle\}$. It follows from Theorem 5.26 that $\circ$ satisfies confirmation and from $K \circ \Pi_X = X$, $X \Vdash \Pi_X \veebar \Pi_Y$, and $K \circ (\Pi_X \veebar \Pi_Y) = K$ that it does not satisfy regularity.

Part 2: Let $\mathbb{X} = \{K, X, Y, Z, V\}$ and $\rightharpoondown = \{\langle K, X\rangle, \langle K, Y\rangle, \langle K, Z\rangle, \langle K, V\rangle, \langle X, Z\rangle, \langle Z, Y\rangle, \langle Y, V\rangle, \langle V, X\rangle\}$. It follows from Theorem 5.26 that $\circ$ satisfies confirmation. However, $K \circ (\Pi_X \veebar \Pi_Y \veebar \Pi_Z) = X$ and $K \circ (\Pi_X \veebar \Pi_Y \veebar \Pi_V) = Y$. Since $K \circ (\Pi_X \veebar \Pi_Y \veebar \Pi_Z) \Vdash \Pi_X \veebar \Pi_Y \veebar \Pi_V$ and $K \circ (\Pi_X \veebar \Pi_Y \veebar \Pi_V) \Vdash \Pi_X \veebar \Pi_Y \veebar \Pi_Z$, it follows that reciprocity is not satisfied. □

Proof of Observation 5.31 *Parts 1 and 2* are left to the reader.

Part 3: Let $\mathbb{X} = \{K, X, Y\}$ and $\rightharpoondown = \{\langle X, Y\rangle, \langle X, K\rangle, \langle Y, K\rangle\}$.

Part 4: Let $\mathbb{X} = \{K, X, Y\}$ and $\rightharpoondown = \{\langle X, Y\rangle, \langle Y, X\rangle, \langle Y, K\rangle, \langle K, Y\rangle\}$. □

Proofs for Section 5.5

Lemma 5.4 *For any maxispecified descriptors Ψ and Ψ' : $\Psi \simeq \Psi'$ if and only if $\Psi \dashv\Vdash \Psi'$.*

Proof of Lemma 5.4 For one direction, let $\Psi \dashv\Vdash \Psi'$. Then $\Psi \simeq \Psi'$ follows from counter-dominance. For the other direction, let $\Psi \simeq \Psi'$ and let X be the unique belief set with $X \Vdash \Psi$ and X' the unique belief set with $X' \Vdash \Psi'$. Then $\Psi \dashv\Vdash \Pi_X$ and $\Psi' \dashv\Vdash \Pi_{X'}$. Suppose for contradiction that $X \neq X'$. Without loss of generality we can then assume that there is some $p \in X \setminus X'$. We then have $\Pi_X \Vdash \mathfrak{B}p$ and $\Pi_{X'} \Vdash \neg\mathfrak{B}p$, thus $\Pi_X \nVdash \Pi_{X'}$, thus $\Psi \nVdash \Psi'$. Contradiction. □

Lemma 5.5 ([126]) *(1)* $\widehat{\Psi} \Vdash \Psi$*, and*
(2) $\widehat{\Psi} \simeq \Psi$.

Proof of Lemma 5.5 *Part 1*: When $\Psi \Vdash \bot$, $\widehat{\Psi} \Vdash \Psi$ holds trivially. When $\Psi \nVdash \bot$, note that $\Psi \cup \{\beta(p)\} \succ \bot$ holds for all p due to $\Psi \cup \{\beta(p)\} \simeq \Psi$ and transitivity. Due to the coupling postulate and the compactness of the logic there is some belief set X such that for all sentences p, $X \Vdash \Psi \cup \{\beta(p)\}$, and due to the exhaustive construction of $\widehat{\Psi}$ there can only be one such set. It is the set specified by $\widehat{\Psi}$, and it satisfies Ψ. We can conclude that $\widehat{\Psi} \Vdash \Psi$.

Part 2: Due to part 1, $\widehat{\Psi} \Vdash \Psi$. Since $\widehat{\Psi}$ is maxispecified it follows from $\widehat{\Psi} \Vdash \Psi$ that $\widehat{\Psi} \cup \Psi \dashv\Vdash \widehat{\Psi}$, thus due to counter-dominance, $\widehat{\Psi} \cup \Psi \simeq \widehat{\Psi}$. Now suppose that $\widehat{\Psi} \cup \Psi \not\simeq \Psi$. Due to the axiom of choice and the construction in Definition 5.34 there is then some Δ and some p such that $\Psi \subseteq \Delta \subset \Delta \cup \{\beta(p)\} \subseteq \widehat{\Psi} \cup \Psi$ and $\Psi \simeq \Delta$ and $\Psi \not\simeq \Delta \cup \{\beta(p)\}$. But due to counter-dominance, amplification and the construction in Definition 5.34 $\Psi \cup \{\beta(p)\} \simeq \Psi$, thus by transitivity and coupling, $\Delta \cup (\Psi \cup \{\beta(p)\}) \simeq \Psi$, or equivalently $\Delta \cup \{\beta(p)\} \simeq \Psi$. This contradiction shows that $\widehat{\Psi} \cup \Psi \simeq \Psi$, thus by transitivity and $\widehat{\Psi} \cup \Psi \simeq \widehat{\Psi}$ we have $\widehat{\Psi} \simeq \Psi$, as desired. □

Lemma 5.6 *Let* $\succeq$ *satisfy transitivity, counter-dominance, coupling, and amplification. Then it satisfies:*
(1) either $\Psi \succeq (\Psi \veebar \Xi)$ *or* $\Xi \succeq (\Psi \veebar \Xi)$ *(*disjunctiveness*),*
(2) $(\Psi \veebar \Xi) \simeq \Psi$ *if and only if* $\Psi \succeq \Xi$*, and*
(3) $\Psi \succeq \Xi$ *or* $\Xi \succeq \Psi$ *(*completeness*).*

Proof of Lemma 5.6 *Part 1*:
$\widehat{\Psi \veebar \Xi} \Vdash \Psi \veebar \Xi$ (Lemma 5.5, Part 1)
Either $\widehat{\Psi \veebar \Xi} \Vdash \Psi$ or $\widehat{\Psi \veebar \Xi} \Vdash \Xi$ (Observation 4.11, since $\widehat{\Psi \veebar \Xi}$ is maxispecified)
Either $\Psi \succeq \widehat{\Psi \veebar \Xi}$ or $\Xi \succeq \widehat{\Psi \veebar \Xi}$ (counter-dominance)
Either $\Psi \succeq (\Psi \veebar \Xi)$ or $\Xi \succeq (\Psi \veebar \Xi)$ (transitivity and Lemma 5.5, Part 2)

Part 2: For one direction, let $(\Psi \veebar \Xi) \simeq \Psi$. Since $\Xi \Vdash \Psi \veebar \Xi$, counter-dominance yields $(\Psi \veebar \Xi) \succeq \Xi$, and with transitivity we can derive $\Psi \succeq \Xi$.

For the other direction, let $\Psi \succeq \Xi$. Suppose for contradiction that $(\Psi \veebar \Xi) \not\simeq \Psi$. Then due to counter-dominance, $(\Psi \veebar \Xi) \succ \Psi$. Part 1 yields $\Xi \succeq (\Psi \veebar \Xi)$, and transitivity yields $\Xi \succ \Psi$, contrary to the assumption. We can conclude that $(\Psi \veebar \Xi) \simeq \Psi$.

Part 3: Due to part 1, either $\Psi \succeq (\Psi \veebar \Xi)$ or $\Xi \succeq (\Psi \veebar \Xi)$. In the former case we use counter-dominance to obtain $(\Psi \veebar \Xi) \succeq \Xi$ and then transitivity to obtain $\Psi \succeq \Xi$. In the latter case $\Xi \succeq \Psi$ is obtained in the same way. □

Proof of Observation 5.33 *Part 1*: Counter-dominance yields $\Psi \simeq \Psi'$ and $\Xi \simeq \Xi'$. The rest follows from transitivity.

Part 2: Already proved in Lemma 5.6. □

Proof of Observation 5.35 *Part 1* was proved in Lemma 5.5.

Part 2, left to right: Leaving out the straight-forward case when $\Psi \Vdash \bot$, let $\widehat{\Psi} \simeq \Psi \cup \Xi$ and $\Psi \nVdash \bot$. It follows from Part 1 and transitivity that $\widehat{\Psi} \simeq \widehat{\Psi \cup \Xi}$. Since $\widehat{\Psi}$ is maxispecified, there is exactly one belief set X with $X \Vdash \widehat{\Psi}$, and it follows from Lemma 5.4 that $X \Vdash \widehat{\Psi \cup \Xi}$. It follows that $X \Vdash \Xi$, thus $\widehat{\Psi} \Vdash \Xi$.

Part 2, right-to-left: Let $\widehat{\Psi} \Vdash \Xi$. Since $\widehat{\Psi} \Vdash \Psi$ due to Lemma 5.5, we then have $\widehat{\Psi} \Vdash \Psi \cup \Xi$. Counter-dominance yields $(\Psi \cup \Xi) \succeq \widehat{\Psi}$. Due to $\Psi \cup \Xi \Vdash \Psi$ and counter-dominance we also have $\Psi \succeq (\Psi \cup \Xi)$. Part 1 and transitivity yield $\widehat{\Psi} \succeq (\Psi \cup \Xi)$. We can conclude that $\widehat{\Psi} \simeq (\Psi \cup \Xi)$.

Part 3: This was proved as Lemma 5.4. □

Lemma 5.7 ([71]) *Let $\lesssim$ be a relation on sentences that satisfies transitivity, dominance, and conjunctiveness. Then it satisfies completeness.*

Proof of Lemma 5.7 Due to conjunctiveness, either $p \lesssim p\&q$ or $q \lesssim p\&q$. In the former case we use dominance to obtain $p\&q \lesssim q$ and transitivity to obtain $p \lesssim q$. In the latter case $q \lesssim p$ is obtained in the same way. □

Proof of Observation 5.37 *Transitivity*: Directly from the definition of $\lesssim$.

Dominance:

$p \vdash q$

$\mathfrak{B}p \Vdash \mathfrak{B}q$

$\neg\mathfrak{B}q \Vdash \neg\mathfrak{B}p$

$\neg\mathfrak{B}p \succeq \neg\mathfrak{B}q$ (counter-dominance)

$p \lesssim q$ (Definition of $\lesssim$)

Conjunctiveness:

$\neg\mathfrak{B}p \succeq (\neg\mathfrak{B}p \vee \neg\mathfrak{B}q)$ or $\neg\mathfrak{B}q \succeq (\neg\mathfrak{B}p \vee \neg\mathfrak{B}q)$ (Lemma 5.6)

$\neg\mathfrak{B}p \succeq \neg\mathfrak{B}(p\&q)$ or $\neg\mathfrak{B}q \succeq \neg\mathfrak{B}(p\&q)$ (intersubstitutivity, Observation 5.33)

$p \lesssim (p\&q)$ or $q \lesssim (p\&q)$ (Definition of $\lesssim$)

Minimality:

$p \notin \{r \mid \bot < r\}$

iff $\bot \nless p$

iff $p \lesssim \bot$ (Lemma 5.7)

iff $p \lesssim q$ for all q ($\bot \lesssim q$ due to dominance, transitivity)

Maximality:

$q \lesssim p$ for all q

$\neg\mathfrak{B}q \succeq \neg\mathfrak{B}p$ for all q

$\neg\mathfrak{B}\top \succeq \neg\mathfrak{B}p$

$\bot \succeq \neg\mathfrak{B}p$ ($\neg\mathfrak{B}\top \dashv\Vdash \bot$, counter-dominance, transitivity)

$\vdash p$ (absurdity avoidance) □

Lemma 5.8 *Let $\leqq$ be a complete, transitive, antisymmetric, and descriptor-wellfounded relation on a set $\mathbb{X}$ of belief sets. For each descriptor Ψ that is satisfiable within $\mathbb{X}$, let M_Ψ be the $\leqq$-minimal Ψ-satisfying element of $\mathbb{X}$. Then $t^{bd}(\leqq)$ is the relation $\succeq$ such that $\Xi \succeq \Psi$ if and only if either $M_\Xi \leqq M_\Psi$ or Ψ is unsatisfiable within $\mathbb{X}$.*

Proof of Lemma 5.8 Leaving out the case when Ψ is unsatisfiable within $\mathbb{X}$ we have:
$\Xi \succeq \Psi$
iff there is some X with $X \Vdash \Xi$ such that $X \leqq Y$ for all Y with $Y \Vdash \Psi$ $\quad (t^{bd})$
iff $M_\Xi \leqq Y$ for all Y with $Y \Vdash \Psi$
($\leqq$ is transitive and $M_\Xi \leqq X$ for all X with $X \Vdash \Xi$)
iff $M_\Xi \leqq M_\Psi$ $\quad$ ($\leqq$ is transitive and $M_\Psi \leqq Y$ for all Y with $Y \Vdash \Psi$) □

Lemma 5.9 *Let $\succeq$ be a relation on sentences that satisfies transitivity, counter-dominance, coupling, and amplification. Let let $t^{db}(\succeq) = \leqq$. Then $X \leqq Y$ if and only if $\Pi_X \succeq \Pi_Y$.*

Proof of Lemma 5.9 Let $t^{b\circ}(\leqq) = \circ$. We first note that if X is an element of the domain of $\leqq$, then there is some Ψ such that $X = K \circ \Psi$ and $K \circ \Psi \Vdash \Psi$. This follows since due to t^{db}, X is completely characterized by some $\widehat{\Xi}$, and then due to $t^{b\circ}$, $X = K \circ \widehat{\Xi}$ and $K \circ \widehat{\Xi} \Vdash \widehat{\Xi}$.

Next, let X and Y be elements of the domain of $\leqq$. As we have just seen, there are Ξ and Ψ such that $X = K \circ \Xi$, $K \circ \Xi \Vdash \Xi$, $Y = K \circ \Psi$, and $K \circ \Psi \Vdash \Psi$. We then have:
$X \leqq Y$
iff $K \circ \Xi \leqq K \circ \Psi$
iff there is some Φ with $K \circ \Psi \Vdash \Phi$ such that $\Sigma \succeq \Phi$ for all Σ with $K \circ \Xi \Vdash \Sigma$ $\quad (t^{db})$
iff there is some Φ with $\Pi_{K\circ\Psi} \Vdash \Phi$ such that $\Sigma \succeq \Phi$ for all Σ with $\Pi_{K\circ\Xi} \Vdash \Sigma$
iff there is some Φ with $\Pi_{K\circ\Psi} \Vdash \Phi$ such that $\Pi_{K\circ\Xi} \succeq \Phi$
(counter-dominance and the transitivity of $\succeq$)
iff $\Pi_{K\circ\Xi} \succeq \Pi_{K\circ\Psi}$ $\quad$ (counter-dominance and the transitivity of $\succeq$)
iff $\Pi_X \succeq \Pi_Y$ □

Proof of Theorem 5.39 *Part 1* follows from Lemma 5.8.

Part 2: Let $t^{bd}(\leqq) = \succeq$, $t^{db}(\succeq) = \leqq'$, and $t^{b\circ}(\leqq) = \circ$. Due to the definition of t^{db}, the domain of $\leqq'$ is the set of belief sets specified by some $\widehat{\Psi}$ such that Ψ is satisfiable within the domain of $\leqq$, i.e. $\leqq'$ has the same domain as $\leqq$. Furthermore:
$X \leqq' Y$
iff $\Pi_X \succeq \Pi_Y$ $\quad$ (Lemma 5.9 whose conditions are satisfied due to part 1)
iff $K \circ \Pi_X \leqq K \circ \Pi_Y$ $\quad$ (Lemma 5.8)
iff $X \leqq Y$

Part 3: Let $t^{b\circ}(\leqq) = \circ$. Let $t^{bd}(\leqq) = \succeq$ and $t^{d\circ}(\succeq) = \circ'$. Let $\circ_i$ be the part of $\circ$ that is based on clause (i) of the definition of $t^{b\circ}$ and $\circ'_i$ the part of $\circ'$ that is based on clause (i) of the definition of $t^{d\circ}$. Since $\leqq$ is descriptor-wellfounded, it follows from the definition of $t^{b\circ}$ that the domain of $\circ_i$ is the set of descriptors that are satisfiable within the domain of $\leqq$. It follows from $t^{d\circ}$ that the domain of $\circ'_i$ is the set of descriptors Ψ satisfying $\Psi \succ \perp$, and due to t^{bd} these are the descriptors that are satisfiable within the domain of $\leqq$. Thus $\circ_i$ and $\circ'_i$ have the same domain. Therefore, in order to prove $\circ = \circ'$ it is sufficient to show that $\circ_i = \circ'_i$. We have:

$q \in K \circ'_i \Psi$
iff $\Psi \cup \{\mathfrak{B}q\} \simeq \Psi \succ \perp$ $\quad (t^{d\circ})$
iff $\Psi \cup \{\mathfrak{B}q\} \simeq \Psi$ $\quad$ (domain of $\circ'_i$)
iff $\Psi \cup \{\mathfrak{B}q\} \succeq \Psi$ and $\Psi \succeq \Psi \cup \{\mathfrak{B}q\}$
iff $K \circ_i (\Psi \cup \{\mathfrak{B}q\}) \leqq K \circ_i \Psi$ and $K \circ_i \Psi \leqq K \circ_i (\Psi \cup \{\mathfrak{B}q\})$ $\quad$ (Lemma 5.8)
iff $K \circ_i (\Psi \cup \{\mathfrak{B}q\}) = K \circ_i \Psi$ $\quad$ (antisymmetry of $\leqq$)
iff $q \in K \circ_i \Psi$ $\quad (t^{b\circ})$

Part 4: Let $t^{b\circ}(\leqq) = \circ$ and $t^{\circ b}(\circ) = \leqq'$. Due to the definitions of $t^{b\circ}$ and $t^{\circ b}$, both $\leqq$ and $\leqq'$ have the domain $\{K \circ \Psi \mid K \circ \Psi \Vdash \Psi\}$. Within that domain:

$K \circ \Psi \leqq' K \circ \Xi$
iff $K \circ \Psi = K \circ (\Psi \veebar \Xi)$ $\quad (t^{\circ b})$
iff $K \circ \Psi \leqq K \circ \Xi$ $\quad$ ($t^{b\circ}$ and Observation 4.11)

Part 5: Let $t^{b\circ}(\leqq) = \circ$, and $t^{\circ d}(\circ) = \succeq$, and $t^{bd}(\leqq) = \succeq'$. Then:

$\Xi \succeq \Psi$
iff *either* $K \circ \Xi \Vdash \Xi$, $K \circ \Psi \Vdash \Psi$, and $K \circ \Xi = K \circ (\Xi \veebar \Psi)$ *or* $K \circ \Psi \nVdash \Psi$ $\quad (t^{\circ d})$
iff *either* Ξ and Ψ are both satisfiable within $\mathbb{X}$ and its $\leqq$-minimal $\Xi \veebar \Psi$-satisfying element satisfies Ξ, *or* Ψ is unsatisfiable within $\mathbb{X}$ $\quad (t^{b\circ})$
iff $\Xi \succeq' \Psi$ $\quad (t^{bd})$ □

Lemma 5.10 *Let $\succeq$ satisfy transitivity, counter-dominance, coupling, and amplification. Let $t^{db}(\succeq) = \leqq$. Then $\{p \mid \widehat{\Psi} \Vdash \mathfrak{B}p\}$ is the $\leqq$-minimal Ψ-satisfying element of the domain of $\leqq$.*

Proof of Lemma 5.10 Suppose not. Then there is some descriptor Ξ such that $\{p \mid \widehat{\Xi} \Vdash \mathfrak{B}p\} \Vdash \Psi$ and:

(1) $\{p \mid \widehat{\Xi} \Vdash \mathfrak{B}p\} < \{p \mid \widehat{\Psi} \Vdash \mathfrak{B}p\}$.

Due to the definition of t^{db} and the completeness of $\succeq$ (Lemma 5.6, p. 184), (1) is equivalent with:

(2) For all Σ such that $\{p \mid \widehat{\Xi} \Vdash \mathfrak{B}p\} \Vdash \Sigma$ there is some Φ with $\{p \mid \widehat{\Psi} \Vdash \mathfrak{B}p\} \Vdash \Phi$ and $\Sigma \succ \Phi$.

Due to counter-dominance and transitivity this is equivalent with:

(3) $\widehat{\Xi} \succ \widehat{\Psi}$

Since $\{p \mid \widehat{\Xi} \Vdash \mathfrak{B}p\} \Vdash \Psi$ we also have $\widehat{\Xi} \Vdash \Psi$. It follows from counter-dominance that $\Psi \succeq \widehat{\Xi}$ and from Lemma 5.5 (p. 184), part 2, that $\widehat{\Psi} \simeq \Psi$. Thus $\Psi \succeq \widehat{\Xi} \succ \widehat{\Psi} \simeq \Psi$. This contradicts transitivity and we can conclude from this contradiction that $\{p \mid \widehat{\Psi} \Vdash \mathfrak{B}p\}$ is indeed the $\leqq$-minimal Ψ-satisfying element of the domain of $\leqq$. □

Lemma 5.11 *Let $\succeq$ satisfy transitivity, counter-dominance, coupling, and amplification. Then $\widehat{\mathfrak{B}\top} \Vdash \mathfrak{B}p$ if and only if $\mathfrak{B}\top \simeq \mathfrak{B}p$.*

Proof of Lemma 5.11 For one direction, let $\widehat{\mathfrak{B}\top} \Vdash \mathfrak{B}p$. Counter-dominance yields $\mathfrak{B}p \succeq \widehat{\mathfrak{B}\top}$. It follows from Lemma 5.5 (p. 184) and transitivity that $\mathfrak{B}p \succeq \mathfrak{B}\top$. Since $\mathfrak{B}p \Vdash \mathfrak{B}\top$, counter-dominance yields $\mathfrak{B}\top \succeq \mathfrak{B}p$ and we can conclude that $\mathfrak{B}\top \simeq \mathfrak{B}p$.

For the other direction, let $\mathfrak{B}\top \simeq \mathfrak{B}p$. Due to Lemma 5.5 and transitivity, $\widehat{\mathfrak{B}\top} \simeq \mathfrak{B}p$. Due to coupling, $\widehat{\mathfrak{B}\top} \simeq \widehat{\mathfrak{B}\top} \cup \{\mathfrak{B}p\}$. Since $\widehat{\mathfrak{B}\top}$ is maxispecified, $\widehat{\mathfrak{B}\top} \Vdash \mathfrak{B}p$. □

Proof of Theorem 5.40 *Part 1*. Let $t^{db}(\succeq) = \leqq$. *Completeness*: From Observation 5.33, part 2 and Lemma 5.9 (p. 186).

Transitivity: Directly from the transitivity of $\succeq$, using Lemma 5.9.

Antisymmetry:

$X \leqq Y \leqq X$

$\Pi_X \simeq \Pi_Y$ (Lemma 5.9, p. 186)

$\Pi_X \dashv\Vdash \Pi_Y$ (Lemma 5.4, p. 183)

$X = Y$ (Definition 4.14)

Descriptor-wellfoundedness: From Lemma 5.10.

Part 2: Let $t^{db}(\succeq) = \leqq$, $t^{bd}(\leqq) = \succeq'$, and $t^{b\circ}(\leqq) = \circ$.

$\Psi \succeq' \Xi$ iff either $K \circ \Psi \leqq K \circ \Xi$ or Ξ is unsatisfiable within the domain of $\leqq$

(Lemma 5.8, p. 186, whose conditions are satisfied due to part 1)

iff $\Pi_{K\circ\Psi} \succeq \Pi_{K\circ\Xi}$ or $\Xi \simeq \perp$ (Lemma 5.9, p. 186[1])

iff $\widehat{\Psi} \succeq \widehat{\Xi}$ or $\Xi \simeq \perp$ (Lemma 5.10)

iff $\Psi \succeq \Xi$ or $\Xi \simeq \perp$ (Lemma 5.5, p. 184, and the transitivity of $\succeq$)

iff $\Psi \succeq \Xi$ ($\Psi \succeq \perp$ from counter-dominance, transitivity)

Part 3: Let $t^{db}(\succeq) = \leqq$, $t^{b\circ}(\leqq) = \circ'$, and $t^{d\circ}(\succeq) = \circ$. Let $\mathbb{X}$ be the domain of $\leqq$. Since $\mathfrak{B}\top$ holds in all belief sets, it holds in the minimal element of the domain of $\leqq$, and it follows from t^{db} that this element is $\{p \mid \widehat{\mathfrak{B}\top} \Vdash \mathfrak{B}p\}$, which according to Lemma 5.11 is identical to $\{p \mid \mathfrak{B}p \simeq \mathfrak{B}\top\}$.

$q \in K \circ' \Psi$

iff *either* $\widehat{\Psi} \Vdash \mathfrak{B}q$ and Ψ is satisfiable within $\mathbb{X}$ *or* $q \in K$ and Ψ is unsatisfiable within $\mathbb{X}$ (Lemma 5.10)

iff *either* $\widehat{\Psi} \simeq \Psi \cup \{\mathfrak{B}q\}$ and $\Psi \succ \perp$ *or* $q \in K$ and $\Psi \simeq \perp$

(Observation 5.35, part 2)

iff *either* $\Psi \simeq \Psi \cup \{\mathfrak{B}q\}$ and $\Psi \succ \perp$ *or* $q \in K$ and $\Psi \simeq \perp$

(Lemma 5.5, p. 184, and the transitivity of $\succeq$)

iff $q \in K \circ \Psi$ ($t^{d\circ}$)

---

[1]To see that Ξ is unsatisfiable within the domain of $\leqq$ if and only if $\Xi \simeq \perp$, let Ξ be unsatisfiable within the domain of $\leqq$. Then due to t^{db} it is not the case that $\Xi \succ \perp$, and since $\perp \Vdash \Xi$ counter-dominance yields $\Xi \simeq \perp$. The other direction follows directly.

Part 4: Let $t^{d\circ}(\succeq) = \circ$ and $t^{\circ d}(\circ) = \succeq'$. Then:
$\Psi \succeq' \Xi$
iff *either* $K \circ \Psi \Vdash \Psi$, $K \circ \Xi \Vdash \Xi$, and $K \circ (\Psi \veebar \Xi) = K \circ \Psi$ *or* $K \circ \Xi \nVdash \Xi$ $(t^{\circ d})$
iff *either* $K \circ \Psi \Vdash \Psi$, $K \circ \Xi \Vdash \Xi$, and for all q: $(\Psi \veebar \Xi) \simeq (\Psi \veebar \Xi) \cup \{\mathfrak{B}q\} \succ \bot$ iff $\Psi \simeq \Psi \cup \{\mathfrak{B}q\} \succ \bot$ *or* $\Xi \simeq \bot$ $(t^{d\circ})$
iff *either* $K \circ \Psi \Vdash \Psi$, $K \circ \Xi \Vdash \Xi$, and $\widehat{\Psi \veebar \Xi} \simeq \widehat{\Psi} \succ \bot$ *or* $\Xi \simeq \bot$ (Definition 5.34)
iff *either* $K \circ \Psi \Vdash \Psi$, $K \circ \Xi \Vdash \Xi$, and $\Psi \veebar \Xi \simeq \Psi \succ \bot$ *or* $\Xi \simeq \bot$ (Lemma 5.5, p. 184)
iff $\Psi \succeq \Xi$ and $\Psi \succ \bot$ *or* $\Xi \simeq \bot$ (Lemma 5.6, p. 184, part 2)
iff $\Psi \succeq \Xi$

Part 5: Let $t^{d\circ}(\succeq) = \circ$, $t^{\circ b}(\circ) = \leqq$, and $t^{db}(\succeq) = \leqq'$. First note that due to $t^{\circ b}$ the domain of $\leqq$ is the set of belief sets X such that $X = K \circ \Xi$ for some Ξ, and due to $t^{d\circ}$ and t^{db} this is equal to the domain of $\leqq'$.

For any $K \circ \Xi$ and $K \circ \Psi$ in that domain we have:
$K \circ \Xi \leqq K \circ \Psi$
iff $K \circ \Xi = K \circ (\Xi \veebar \Psi)$ $(t^{\circ b})$
iff $\Xi \simeq (\Xi \veebar \Psi) \succ \bot$ $(t^{d\circ})$
iff $\Xi \succeq \Psi \succ \bot$ (Lemma 5.6, p. 184, part 2)
iff $\widehat{\Xi} \succeq \widehat{\Psi} \succ \bot$ (Lemma 5.5, p. 184, transitivity)
iff there is some Φ with $\{q \mid \widehat{\Psi} \Vdash \mathfrak{B}q\} \Vdash \Phi$ such that $\Sigma \succeq \Phi$ for all Σ with $\{q \mid \widehat{\Xi} \Vdash \mathfrak{B}q\} \Vdash \Sigma$.
(Due to counter-dominance $\Phi \succeq \widehat{\Psi}$ whenever $\{q \mid \widehat{\Psi} \Vdash \mathfrak{B}q\} \Vdash \Phi$, and $\Sigma \succeq \widehat{\Xi}$ whenever $\{q \mid \widehat{\Xi} \Vdash \mathfrak{B}q\} \Vdash \Sigma$.)
iff $\{q \mid \widehat{\Xi} \Vdash \mathfrak{B}q\} \leqq' \{q \mid \widehat{\Psi} \Vdash \mathfrak{B}q\}$ (t^{db})
iff $K \circ \Xi \leqq' K \circ \Psi$ (Lemma 5.10) □

Proofs for Section 6.1

Proof of Theorem 6.4 *From (A) to (B)*: For closure, relative success, local regularity, and cumulativity this follows as in Theorem 5.10, and for confirmation as in Observation 5.14.

From (B) to (A): For each $X \in \mathbb{X}$, let $a(X) = \{Y \mid (\exists\Psi)(Y = X \circ \Psi\}$ and let $\leqq_X$ be the relation on $a(X)$ such that for all Ψ and Ξ: $X \circ \Psi \leqq_X X \circ \Xi$ if and only if $X \circ \Psi = X \circ (\Psi \veebar \Xi)$. Then proceed as in the proof of Theorem 5.10. □

Proof of Theorem 6.5 *From (A^e) to (B^e)*: Closure, relative success, and cumulativity follow as in Theorem 5.10, and confirmation as in Observation 5.14. For global regularity, let $X \circ \Xi \Vdash \Psi$. We then have $X \circ \Xi \in a(X)$, and since $\circ$ is coextensive we have $a(X) = a(Y)$, thus Ψ is satisfiable within $a(Y)$, and it follows from Definition 6.3 that $Y \circ \Psi \Vdash \Psi$.

From (B^e) to (A^e): For each $X \in \mathbb{X}$, let $a(X) = \{Y \mid (\exists\Psi)(Y = X \circ \Psi\}$. It follows from confirmation for all $Y \in \mathbb{X}$ that $Y \circ \Pi_Y \Vdash \Pi_Y$. Due to global regularity, $X \circ \Pi_Y \Vdash \Pi_Y$, thus $X \circ \Pi_Y = Y$. Hence $a(X) = \mathbb{X}$ for all $X \in \mathbb{X}$. Let $\leqq_X$ be the relation on $\mathbb{X}$ such that for all Ψ and Ξ: $X \circ \Psi \leqq_X X \circ \Xi$ if and only if $X \circ \Psi = X \circ (\Psi \veebar \Xi)$. Then proceed as in the proof of Theorem 5.10. □

Proofs for Section 6.2

Proof of Theorem 6.6 *From (C^e) to (A^e)*: Let (C^e) hold. Due to Theorem 6.5 it is sufficient to show that (B^e) holds. Closure, relative success, and global regularity follow directly from (C^e).

For cumulativity, let $X \circ \Psi \Vdash \Xi$. There are two cases:

Case 1, $X \circ \Psi \nVdash \Psi$: Then Ψ is unsatisfiable within $\mathbb{X}$, and so is $\Psi \cup \Xi$. It follows from clause (ii) of (C^e) that $X \circ \Psi = X$ and $X \circ (\Psi \cup \Xi) = X$.

Case 2, $X \circ \Psi \Vdash \Psi$: Then $X \circ \Psi \Vdash \Psi \cup \Xi$. Suppose that $X \circ \Psi \neq X \circ (\Psi \cup \Xi)$. Then due to clause (i) of (C^e) there is some $Z \in \mathbb{X}$ such that $Z \neq X \circ \Psi$, $Z \Vdash \Psi \cup \Xi$ and $\delta(X, Z) \leq \delta(X, Y)$ whenever $Y \Vdash \Psi \cup \Xi$, thus $\delta(X, Z) \leq \delta(X, X \circ \Psi)$. Due to righthand uniqueness, $\delta(X, Z) < \delta(X, X \circ \Psi)$. Since $Z \Vdash \Psi$ this contradicts clause (i) of (C^e). We can conclude from this contradiction that $X \circ \Psi = X \circ (\Psi \cup \Xi)$.

For confirmation, note that due to selfcloseness $\delta(X, X) = 0$ and that it follows from non-negativity and righthand uniqueness that if $X \neq Y$ then $\delta(X, Y) > 0$.

From (A^e) to (C^e): Let (A^e) hold. It follows from Definition 6.3 that there is for each $X \in \mathbb{X}$ a relation $\leqq_X$ on $\mathbb{X}$ such that $X \leqq_X Y$ for all $Y \in \mathbb{X}$ and that (i) if Ψ is satisfiable within $\mathbb{X}$ then $X \circ \Psi$ is the unique $\leqq_X$-minimal element of $[\![\Psi]\!]_{\mathbb{X}}$ and otherwise, $X \circ \Psi = X$. It follows from Lemma 5.2 (p. 177) that $\leqq_X$ is antisymmetric, transitive, complete, and descriptor-wellfounded. Due to Cantor's representation theorem for countable sets [211, pp. 109–111], if $\mathbb{X}$ is countable then there is for each $X \in \mathbb{X}$ some real-valued measure δ_X such that for all $Y, Z \in \mathbb{X}$: $\delta_X(Y) \leq \delta_X(Z)$ if and only if $Y \leqq_X Z$. Let δ be the real-valued measure such that $\delta(X, Y) = \delta_X(Y) - \delta_X(X)$ for all $X, Y \in \mathbb{X}$. Then δ satisfies the conditions given in (C^e). □

Proof of Theorem 6.7 *From (C) to (A)*: Due to Theorem 6.4 it is sufficient to show that (B) holds. Closure, relative success, and local regularity follow directly from (C). For cumulativity, the proof of Theorem 6.6 can be used with the adjustment that references to satisfiability within $\mathbb{X}$ have to be replaced by references to satisfiability within $\{Y \in \mathbb{X} \mid \delta(X, Y) < 1\}$. For confirmation the same proof can be used as in Theorem 6.6.

From (A) to (C): For each $X \in \mathbb{X}$, there is due to (A) a relation $\leqq_X$ on a set $a(X)$ with $X \in a(X) \subseteq \mathbb{X}$, such that $X \leqq_X Y$ for all $Y \in a(X)$ and that:

(i) If Ψ is satisfiable within $a(X)$, then $X \circ \Psi$ is the unique $\leqq_X$-minimal element of $a(X)$ that satisfies Ψ, and
(ii) Otherwise, $X \circ \Psi = X$.

Due to Cantor's representation theorem for countable sets [211, pp. 109–111] there is for all $X \in \mathbb{X}$ some real-valued function δ_X such that for all $Y, Z \in \mathbb{X}$: $\delta_X(Y) \leq \delta_X(Z)$ if and only if $Y \leqq_X Z$. It follows from Lemma 5.2 (p. 177) that $\leqq_X$ is antisymmetric, transitive, complete, and descriptor-wellfounded. Let δ be the real-valued function on $\mathbb{X} \times \mathbb{X}$ such that:

(a) $\delta(X, Y) = \frac{\delta_X(Y) - \delta_X(X)}{\delta_X(Y) - \delta_X(X) + 1}$ when $Y \in a(X)$, and
(b) $\delta(X, Y) = 1$ when $Y \in \mathbb{X} \setminus a(X)$.

It can be straight-forwardly verified that δ satisfies the conditions in (C) (self-closeness, non-negativity, and righthand uniqueness) and that the operation that it gives rise to through (C) coincides with the operation $\circ$ given in (A). □

Proofs for Section 6.3

Lemma 6.1 *Let $\rightharpoondown_X$ be a transitive and irreflexive relation, and let $\bar{\rightharpoondown}_X$ be the relation such that $Y \bar{\rightharpoondown}_X Z$ iff either $Y \rightharpoondown_X Z$ or $Y = Z$. Then $\bar{\rightharpoondown}_X$ is transitive, reflexive, and antisymmetric.*

Proof of Lemma 6.1 Reflexivity follows from the reflexivity of identity. For transitivity, note that if $Y \bar{\rightharpoondown}_X Z$ & $Z \bar{\rightharpoondown}_X V$, then there are four possibilities, namely

$Y \rightharpoondown_X Z$ & $Z \rightharpoondown_X V$,
$Y = Z$ & $Z \rightharpoondown_X V$,
$Y \rightharpoondown_X Z$ & $Z = V$, and
$Y = Z$ & $Z = V$.

In the first case $Y \rightharpoondown_X V$ and thus $Y \bar{\rightharpoondown}_X V$ follow from the transitivity of $\rightharpoondown_X$, and in the following two cases $Y \bar{\rightharpoondown}_X V$ follows from substitution of identicals. In the fourth case $Y \bar{\rightharpoondown}_X V$ follows from the transitivity of identity. For antisymmetry, let $Y \bar{\rightharpoondown}_X Z$ and $Z \bar{\rightharpoondown}_X Y$. It follows from the definition of $\bar{\rightharpoondown}_X$ that either $Y = Z$ or $Y \rightharpoondown_X Z$ & $Z \rightharpoondown_X Y$. The latter is impossible due to the asymmetry of $\rightharpoondown_X$ that follows from Observation 5.28. Thus $Y = Z$. □

Lemma 6.2 *Let $\rightharpoondown$ be a ternary relation that satisfies asymmetry and weak connectedness, and $\bar{\rightharpoondown}$ the relation such that $X \bar{\rightharpoondown}_Y Z$ iff either $X \rightharpoondown_Y Z$ or $X = Z$. Then:*

(1) It holds for all X, Y, and Z that $X \bar{\rightharpoondown}_Y Z$ iff $Z \not\rightharpoondown_Y X$.

(2) If $\rightharpoondown$ satisfies negative transmission, then $\bar{\rightharpoondown}$ satisfies:

If $X_1 \bar{\rightharpoondown}_{X_2} X_3,\ X_2 \bar{\rightharpoondown}_{X_3} X_4, \ldots, X_{n-2} \bar{\rightharpoondown}_{X_{n-1}} X_n$, and $X_2 = X_{n-1}$, then $X_1 \bar{\rightharpoondown}_{X_2} X_n$.

Proof of Lemma 6.2 *Part 1*: First let $X \bar{\rightharpoondown}_Y Z$, i.e. either $X \rightharpoondown_Y Z$ or $X = Z$. In the former case $Z \not\rightharpoondown_Y X$ follows from the asymmetry of $\rightharpoondown$ and in the latter case from its irreflexivity that follows from its asymmetry. Next let $Z \not\rightharpoondown_Y X$. It follows from weak connectedness that either $X \rightharpoondown_Y Z$ or $X = Z$, i.e. $X \bar{\rightharpoondown}_Y Z$.

Part 2: Consider the following substitution instance of negative transmission:

If $X_n \not\rightharpoondown_{X_{n-1}} X_{n-2},\ X_{n-1} \not\rightharpoondown_{X_{n-2}} X_{n-3}, \ldots, X_4 \not\rightharpoondown_{X_3} X_2,\ X_3 \not\rightharpoondown_{X_2} X_1$ and $X_2 = X_{n-1}$, then $X_n \not\rightharpoondown_{X_{n-1}} X_1$.

According to part 1 it is equivalent with

If $X_{n-2} \bar{\rightharpoondown}_{X_{n-1}} X_n,\ X_{n-3} \bar{\rightharpoondown}_{X_{n-2}} X_{n-1}, \ldots, X_2 \bar{\rightharpoondown}_{X_3} X_4,\ X_1 \bar{\rightharpoondown}_{X_2} X_3$ and $X_2 = X_{n-1}$, then $X_1 \bar{\rightharpoondown}_{X_2} X_n$.

as desired. □

Proof of Observation 6.9 Let $\bar{\rightharpoondown}$ be the relation such that $X \bar{\rightharpoondown}_Y Z$ iff either $X \rightharpoondown_Y Z$ or $X = Z$. We are first going to show that $\bar{\rightharpoondown}$ is transitive. Let $Y_1 \bar{\rightharpoondown}_X Y_2$ and $Y_2 \bar{\rightharpoondown}_X Y_3$.

Due to the definition of $\bar{\rightharpoonup}$, we also have $X \bar{\rightharpoonup}_{Y_2} X$. We can apply Lemma 6.2, part 2, to $Y_1 \bar{\rightharpoonup}_X Y_2$, $X \bar{\rightharpoonup}_{Y_2} X$, and $Y_2 \bar{\rightharpoonup}_X Y_3$ and obtain $Y_1 \bar{\rightharpoonup}_X Y_3$, as desired.

Due to a standard theorem in preference logic (see for instance [101, p. 324]), it follows from the transitivity of $\bar{\rightharpoonup}_X$ that its strict part $\rightharpoonup_X$ is also transitive. □

Proof of Theorem 6.10 Directly from Observation 5.27. □

Proof of Theorem 6.11 *From* (C^s) *to* (D^s): Let $X \rightharpoonup_Y Z$ iff $\delta(Y, X) < \delta(Y, Z)$. Weak connectedness, asymmetry, and superiority follow directly. Negative transmission can be proved as follows:

$X_1 \not\rightharpoonup_{X_2} X_3,\ X_2 \not\rightharpoonup_{X_3} X_4, \dots, X_{n-2} \not\rightharpoonup_{X_{n-1}} X_n$, and $X_2 = X_{n-1}$

$\delta(X_2, X_3) \geq \delta(X_2, X_1),\ \delta(X_3, X_4) \geq \delta(X_3, X_2), \dots, \delta(X_{n-1}, X_n) \geq \delta(X_{n-1}, X_{n-2})$, and $X_2 = X_{n-1}$

$\delta(X_{n-1}, X_n) \geq \delta(X_{n-1}, X_{n-2}), \dots, \delta(X_3, X_4) \geq \delta(X_3, X_2), \delta(X_2, X_3) \geq \delta(X_2, X_1)$, and $X_2 = X_{n-1}$

$\delta(X_n, X_{n-1}) \geq \delta(X_{n-1}, X_{n-2}), \dots, \delta(X_4, X_3) \geq \delta(X_3, X_2), \delta(X_3, X_2) \geq \delta(X_2, X_1)$, and $X_2 = X_{n-1}$ (symmetry of δ)

$\delta(X_n, X_2) \geq \delta(X_2, X_1)$

$\delta(X_2, X_n) \geq \delta(X_2, X_1)$ (symmetry of δ)

$X_1 \not\rightharpoonup_{X_2} X_n$

From (D^s) *to* (C^s) *when* $\mathbb{X}$ *is countable*: We are going to construct a measure δ and then verify that it gives rise to the same global descriptor revision as the blockage revision.

For the construction, let $\bar{\rightharpoonup}$ be the relation on the elements of $\mathbb{X}$ such that $X \bar{\rightharpoonup}_Y Z$ iff either $X \rightharpoonup_Y Z$ or $X = Z$. It follows from Observation 6.9 that $\rightharpoonup$ is transitive, and then from Lemma 6.1 that for all X, $\bar{\rightharpoonup}_X$ is transitive, reflexive, and antisymmetric. For each $X \in \mathbb{X}$ we introduce the relation $\trianglelefteq_X$ on subsets of $\mathbb{X}$ containing X and at most one additional element, such that for all $Y, Z \in \mathbb{X}$:

$\{X, Y\} \trianglelefteq_X \{X, Z\}$ if and only if $Y \bar{\rightharpoonup}_X Z$

Since there is a one-to-one correspondence between the sets eligible as arguments of $\trianglelefteq_X$ and the elements of $\mathbb{X}$, it follows from the transitivity, reflexivity, and antisymmetry of $\bar{\rightharpoonup}_X$ that $\trianglelefteq_X$ is also transitive, reflexive, and antisymmetric. Let $\approx$ be the relation such that $\{X, Y\} \approx \{Z, V\}$ iff $X = Y$ and $Z = V$. Construct the union of $\approx$ and all the relations $\trianglelefteq_X$ for $X \in \mathbb{X}$. The transitive closure of this union is denoted $\trianglelefteq^*$. Its domain consists of all subsets of $\mathbb{X}$ with either one or two elements. Let $\underline{\triangle}^*$ be the symmetrical part of $\trianglelefteq^*$, i.e. $\{X, Y\} \underline{\triangle}^* \{Z, V\}$ iff $\{X, Y\} \trianglelefteq^* \{Z, V\} \trianglelefteq^* \{X, Y\}$. Since $\trianglelefteq^*$ is transitive and reflexive, $\underline{\triangle}^*$ is transitive, symmetric, and reflexive, i.e. an equivalence relation. For each set $\{X, Y\}$ of one or two elements of $\mathbb{X}$, let $\lfloor\{X, Y\}\rfloor$ be its equivalence class under $\underline{\triangle}^*$. Let $\trianglelefteq^\#$ be the relation on these equivalence classes such that $\lfloor\{X, Y\}\rfloor \trianglelefteq^\# \lfloor\{X, Z\}\rfloor$ if and only if $\{X, Y\} \trianglelefteq^* \{X, Z\}$.[2] It follows directly

[2]To show that $\trianglelefteq^\#$ is well-defined it is sufficient to show that if $\{X, Y\} \underline{\triangle}^* \{X', Y'\}$ and $\{X, Z\} \underline{\triangle}^* \{X', Z'\}$, then $\{X, Y\} \trianglelefteq^* \{X, Z\}$ iff $\{X', Y'\} \trianglelefteq^* \{X', Z'\}$. Let $\{X, Y\} \underline{\triangle}^* \{X', Y'\}$, $\{X, Z\} \underline{\triangle}^* \{X', Z'\}$, and $\{X, Y\} \trianglelefteq^* \{X, Z\}$. It follows from $\{X, Y\} \underline{\triangle}^* \{X', Y'\}$ that $\{X', Y'\} \trianglelefteq^* \{X, Y\}$ and from $\{X, Z\} \underline{\triangle}^* \{X', Z'\}$ that $\{X, Z\} \trianglelefteq^* \{X', Z'\}$. Using $\{X, Y\} \trianglelefteq^* \{X, Z\}$ and the transitivity of $\trianglelefteq^*$ we obtain $\{X', Y'\} \trianglelefteq^* \{X', Z'\}$.

that $\trianglelefteq^{\#}$ is transitive, reflexive, and antisymmetric, i.e. a partial ordering. ($\trianglelefteq^{\#}$ is the reduction of $\trianglelefteq^{*}$, cf. [211, pp. 30–31].)

Due to the order extension principle (that follows from the axiom of choice [145, p. 19] there is a linear ordering (transitive, complete, and antisymmetric relation) $\trianglelefteq^{\ddagger}$ that extends $\trianglelefteq^{\#}$. Due to Cantor's representation theorem for countable sets [211, pp. 109–111] there is a real-valued function δ' such that $\lfloor\{X, Y\}\rfloor \trianglelefteq^{\ddagger} \lfloor\{Z, V\}\rfloor$ iff $\delta'(\lfloor\{X, Y\}\rfloor) \leq \delta'(\lfloor\{Z, V\}\rfloor)$. To obtain a calibrated measure, let $\delta(\lfloor\{X, Y\}\rfloor) = \delta'(\lfloor\{X, Y\}\rfloor) - \delta'(\lfloor\{X, X\}\rfloor)$.[3] We can apply δ also to the elements of the equivalence classes, setting $\delta(\{X, Y\}) = \delta(\lfloor\{X, Y\}\rfloor)$. With this our construction of a distance measure is finished.

Verification: It follows from the construction that for all $X, Y \in \mathbb{X}$: $\delta(X, Y) = \delta(Y, X)$, $\delta(X, X) = 0$, and $\delta(X, Y) \geq 0$. It remains to show (A) that $\delta(X, Y) \neq \delta(X, Z)$ whenever $Y \neq Z$, and (B) that if Ψ is satisfiable within $\mathbb{X}$, then $\delta(X, X \circ \Psi) \leq \delta(X, Y)$ whenever $Y \Vdash \Psi$. The satisfaction of clause (ii) of (C^s) follows directly.

Part A: We will assume that $\delta(X, Y) = \delta(X, Z)$ and prove that $Y = Z$. If $X = Y$ then $\delta(X, Y) = \delta(X, X) = 0$, thus $\delta(X, Z) = 0$, thus $\delta'(X, Z) = \delta'(X, X)$, thus $\{X, Z\} \underline{\triangle}^{*} \{X, X\}$, thus $\{X, Z\} \trianglelefteq^{*} \{X, X\}$, thus $Z \bar{\rightharpoondown}_X X$, thus $X = Z$.

In the principal case when $X \neq Y$ it follows from $\delta(X, Y) = \delta(X, Z)$ that $\lfloor\{X, Y\}\rfloor \trianglelefteq^{\#} \lfloor\{X, Z\}\rfloor$ and $\lfloor\{X, Z\}\rfloor \trianglelefteq^{\#} \lfloor\{X, Y\}\rfloor$.

It follows from $\lfloor\{X, Y\}\rfloor \trianglelefteq^{\#} \lfloor\{X, Z\}\rfloor$ that there is a chain of links from $\{X, Y\}$ to $\{X, Z\}$, where each link has either of the forms

(a) $\ldots \{A, B\} \trianglelefteq_B \{B, C\},\ \{B, C\} \trianglelefteq_C \{C, D\} \ldots$ or

(b) $\ldots \{A, B\} \trianglelefteq_B \{B, C\},\ \{B, C\} \trianglelefteq_B \{B, D\} \ldots$

which means equivalently that either

(a') $\ldots A \bar{\rightharpoondown}_B C,\ B \bar{\rightharpoondown}_C D \ldots$ or

(b') $\ldots A \bar{\rightharpoondown}_B C,\ C \bar{\rightharpoondown}_B D \ldots$

In case (b') we can use $B \bar{\rightharpoondown}_C B$ that follows from the reflexivity of $\bar{\rightharpoondown}$ to expand the chain as follows:

(b") $\ldots A \bar{\rightharpoondown}_B C,\ B \bar{\rightharpoondown}_C B,\ C \bar{\rightharpoondown}_B D \ldots$

which means that in all cases of $\lfloor\{X, Y\}\rfloor \trianglelefteq^{\#} \lfloor\{X, Z\}\rfloor$ we have a chain of type (a'). More precisely, taking into account the beginnings and ends of these chains we have in all cases a chain of one of the four following types:

(1) $X \bar{\rightharpoondown}_Y S_1,\ Y \bar{\rightharpoondown}_{S_1} S_2, \ldots, S_{n-1} \bar{\rightharpoondown}_{S_n} X,\ S_n \bar{\rightharpoondown}_X Z$

(2) $X \bar{\rightharpoondown}_Y S_1,\ Y \bar{\rightharpoondown}_{S_1} S_2, \ldots, S_{n-1} \bar{\rightharpoondown}_{S_n} Z,\ S_n \bar{\rightharpoondown}_Z X$

(3) $Y \bar{\rightharpoondown}_X S_1,\ X \bar{\rightharpoondown}_{S_1} S_2, \ldots, S_{n-1} \bar{\rightharpoondown}_{S_n} X,\ S_n \bar{\rightharpoondown}_X Z$

(4) $Y \bar{\rightharpoondown}_X S_1,\ X \bar{\rightharpoondown}_{S_1} S_2, \ldots, S_{n-1} \bar{\rightharpoondown}_{S_n} Z,\ S_n \bar{\rightharpoondown}_Z X$

By adding $Y \bar{\rightharpoondown}_X Y$ at the beginning of all chains of types 1 and 2 and $Z \bar{\rightharpoondown}_X Z$ at the end of all chains of types 2 and 4 we will have in all cases a chain of type 3. It follows from negative transmission and Lemma 6.2, part 2, that $Y \bar{\rightharpoondown}_X Z$, i.e. either $Y \rightharpoondown_X Z$ or $Y = Z$

Similarly, it follows from $\lfloor\{X, Z\}\rfloor \trianglelefteq^{\#} \lfloor\{X, Y\}\rfloor$ that $Z \bar{\rightharpoondown}_X Y$, i.e. either $Z \rightharpoondown_X Y$ or $Y = Z$. Since $\rightharpoondown_X$ is asymmetric it follows that $Y = Z$.

---

[3]Note that $\{X, X\} \approx \{Y, Y\}$ yields $\delta'(\lfloor\{X, X\}\rfloor) = \delta'(\lfloor\{Y, Y\}\rfloor)$.

Part B can be proved by showing that for each belief set X and each descriptor Ψ that is satisfiable within $\mathbb{X}$, there is a unique Ψ-satisfying element of $\mathbb{X}$ that has the smallest distance to X of all Ψ-satisfying elements of $\mathbb{X}$. Given the result in part A we can show this by proving that $Y \rightharpoondown_X Z$ iff $\delta(X, Y) < \delta(X, Z)$.

For one direction, let $Y \rightharpoondown_X Z$. It follows from the construction that $\lfloor\{X, Y\}\rfloor \trianglelefteq^{\#} \lfloor\{X, Z\}\rfloor$, thus $\delta(X, Y) \leq \delta(X, Z)$. It follows from the asymmetry of $\rightharpoondown_X$ that $Y \neq Z$ and thus from part A that $\delta(X, Y) \neq \delta(X, Z)$, thus $\delta(X, Y) < \delta(X, Z)$.

For the other direction, let $\delta(X, Y) < \delta(X, Z)$. Then $Y \neq Z$. Due to the definition of δ we also have $\lfloor\{X, Y\}\rfloor \trianglelefteq^{\#} \lfloor\{X, Z\}\rfloor$, and then it follows as in part A that $Y \overline{\rightharpoondown}_X Z$. From this and $Y \neq Z$ we can conclude that $Y \rightharpoondown_X Z$, as desired. □

Proofs for Section 7.3

Proof of Observation 7.2
$\langle K, \circ\rangle \twoheadleftarrow \mathfrak{B}\top \Rightarrow \Xi$ iff $K \circ \mathfrak{B}\top \Vdash \Xi$ (Definition 7.1)
iff $K \Vdash \Xi$ (confirmation yields $K \circ \mathfrak{B}\top = K$) □

Proof of Observation 7.3 *Part 1*:
$\langle K, \circ\rangle \twoheadleftarrow \Psi \Rightarrow \Xi$ iff $K \circ \Psi \Vdash \Xi$ (Definition 7.1)
iff $K \circ \Psi = K \circ (\Psi \cup \Xi)$ (Theorem 5.10, cumulativity and regularity)
Part 2:
$\langle K, \circ\rangle \twoheadleftarrow \Psi \Rightarrow \Xi$ iff $K \circ \Psi = K \circ (\Psi \cup \Xi)$ (Part 1)
iff $K \circ (\Psi \veebar (\Psi \cup \Xi)) = K \circ (\Psi \cup \Xi)$
(uniformity, from Theorem 5.10 and Observation 5.11)
iff $\Psi \cup \Xi \succeq \Psi$ (Definition 5.38, $t^{\circ d}$)
Part 3:
$\Psi \succeq \Xi$ iff: either $K \circ \Psi = K \circ (\Psi \veebar \Xi)$ or Ξ is unsatisfiable within $\mathbb{X}$
(Definition 5.38, $t^{\circ d}$; note that Ψ is satisfiable within $\mathbb{X}$)
iff $K \circ \Psi = K \circ (\Psi \veebar \Xi)$
(Observation 5.11. When Ξ is unsatisfiable, $X \Vdash \Psi$ iff $X \Vdash \Psi \veebar \Xi$ for all $X \in \mathbb{X}$.)
iff $K \circ ((\Psi \veebar \Xi) \cup \Psi) = K \circ (\Psi \veebar \Xi)$ (Observation 5.11)
iff $\langle K, \circ\rangle \twoheadleftarrow (\Psi \veebar \Xi \Rightarrow \Psi)$ (Part 1) □

Proof of Theorem 7.4 Since (II) follows directly from (I), it remains to prove that (II) implies (III) and that (III) implies (I).

From (II) to (III): It follows from Theorem 5.10 that $\circ$ satisfies closure, relative success, regularity, and cumulativity. Left logical equivalence follows from uniformity that we have from Lemma 5.1 (p. 176). Unitarity follows from closure. (Let $Y = K \circ \Psi$.) Reflexivity follows from regularity since Ψ is by assumption satisfiable within $\mathbb{X}$, and the cumulativity of $\Rightarrow$ follows from that of $\circ$.

From (III) to (I): Let K be the set of sentences such that for all Ψ, $K \Vdash \Psi$ if and only if $\langle K, \circ\rangle \twoheadleftarrow \mathfrak{B}\top \Rightarrow \Psi$. It follows from unitarity that K is a belief set. Let $\circ$ be the descriptor revision such that for all Ψ and Ξ: $K \circ \Psi \Vdash \Xi$ if and only if $\langle K, \circ\rangle \twoheadleftarrow \Psi \Rightarrow \Xi$. It follows from Theorem 5.10 and Observation 5.14 that we can prove (I) by showing that $\circ$, as applied to descriptors satisfiable within its outcome set, satisfies closure, relative success, regularity, cumulativity, and confirmation. Closure follows from unitarity. Regularity and relative success follow from reflexivity. The cumulativity of $\circ$ follows from that of $\Rightarrow$. For confirmation we have:

$K \Vdash \Psi$

$\langle K, \circ \rangle \twoheadleftarrow \mathfrak{B}\top \Rightarrow \Psi$ (our definition of K)

For all Ξ, $\langle K, \circ \rangle \twoheadleftarrow \mathfrak{B}\top \Rightarrow \Xi$ iff $\langle K, \circ \rangle \twoheadleftarrow \{\mathfrak{B}\top\} \cup \Psi \Rightarrow \Xi$ (cumulativity of $\Rightarrow$)

For all Ξ, $\langle K, \circ \rangle \twoheadleftarrow \mathfrak{B}\top \Rightarrow \Xi$ iff $\langle K, \circ \rangle \twoheadleftarrow \Psi \Rightarrow \Xi$ (left logical equivalence)

For all Ξ, $K \Vdash \Xi$ iff $K \circ \Psi \Vdash \Xi$ (definition of K)

$K = K \circ \Psi$ (substitute Π_K for Ξ) □

Proof of Observation 7.5 Due to unitarity there is a belief set Y such that $\langle K, \circ \rangle \twoheadleftarrow \Psi_1 \veebar \Psi_2 \Rightarrow \Xi$ iff $Y \Vdash \Xi$. It follows from reflexivity that $Y \Vdash \Psi_1 \veebar \Psi_2$, and due to Observation 4.11, either $Y \Vdash \Psi_1$ or $Y \Vdash \Psi_2$. It is sufficient to prove the former case. In that case we have $\langle K, \circ \rangle \twoheadleftarrow \Psi_1 \veebar \Psi_2 \Rightarrow \Psi_1$, thus:

$\langle K, \circ \rangle \twoheadleftarrow \Psi_1 \Rightarrow \Xi$

$\langle K, \circ \rangle \twoheadleftarrow (\Psi_1 \veebar \Psi_2) \cup \Psi_1 \Rightarrow \Xi$ (left logical equivalence)

$\langle K, \circ \rangle \twoheadleftarrow \Psi_1 \veebar \Psi_2 \Rightarrow \Xi$ (cumulativity, since $\langle K, \circ \rangle \twoheadleftarrow \Psi_1 \veebar \Psi_2 \Rightarrow \Psi_1$) □

Proof for Section 7.5

Proof of Theorem 7.7 *General structure of the proof*: We will use the following alternative notation for the centrolinear model $\langle \mathbb{X}, \leqq \rangle$:

$\mathcal{X}_0 = X_1, X_2, X_3, X_4, \ldots$

where X_1 is the $\leqq$-minimal element of $\mathbb{X}$, X_2 the $\leqq$-minimal element of $\mathbb{X} \setminus \{X_1\}$, etc. Clearly, $\mathcal{X}_0$ contains the same information as $\langle \mathbb{X}, \leqq \rangle$, and therefore we can use $\langle \mathcal{X}_0, \ell \rangle$ as an alternative notation for $\langle \mathbb{X}, \leqq, \ell \rangle$. (This alternative notation will also be used for other centrolinear models to be constructed in the proof.)

We will use K as an alternative notation for the original belief set, i.e. $K = X_1$.

The proof will proceed by mathematical induction. In the base case we need to show that there is a series

$\mathcal{X}_1 = Z^1_1, \ldots, Z^1_{m_1}, X_2, X_3, X_4, \ldots$

and a delimiter ℓ_1 for the set consisting of its elements, such that each of $Z_1, \ldots, Z_{m_1}$ is self-limited according to ℓ and that $\langle \mathcal{X}_1, \ell_1 \rangle$ generates the same inference relation as $\langle \mathcal{X}_0, \ell \rangle$ (and thus the same as $\langle \mathbb{X}, \leqq, \ell \rangle$). We also need to show that $Z^1_1 \subseteq X_1$.

In the inductive step we will assume that we have a series

$\mathcal{X}_{n-1} = Y_1, \ldots, Y_k, X_n, X_{n+1}, X_{n+2}, \ldots$

(where $X_n, X_{n+1}, X_{n+2}, \ldots$ is the remaining part of $\mathcal{X}_0$ that has not been affected by the previous steps) and a delimiter ℓ_{n-1} according to which each of $Y_1, \ldots, Y_k$ is self-limited. We need to show that there is a series

$\mathcal{X}_n = Y_1, \ldots, Y_k, Z^n_1, \ldots, Z^n_{m_n}, X_{n+1}, X_{n+2}, \ldots$

and a delimiter ℓ_n for the set consisting of its elements, such that each of $Y_1, \ldots, Y_k$, $Z^n_1, \ldots, Z^n_{m_n}$ is self-limited according to ℓ_n and that $\langle \mathcal{X}_n, \ell_n \rangle$ generates the same inference relation as $\langle \mathcal{X}_{n-1}, \ell_{n-1} \rangle$.

Based on this, the whole of $\mathcal{X}_0$ can be replaced by the series

$\mathcal{X}_\omega = Z^1_1, \dots, Z^1_{m_1}, Z^2_1, \dots, Z^2_{m_2}, Z^3_1, \dots, Z^3_{m_3}, \dots$

with a delimiter ℓ_ω that is simply the identify function, i.e. $\ell_\omega(Z) = Z$ for all elements Z of the series. Then $\langle \mathcal{X}_\omega, \ell_\omega \rangle$ will yield the same inference relation as $\langle \mathcal{X}_0, \ell \rangle$. Furthermore, the inference relation obtainable from $\langle \mathcal{X}_\omega, \ell_\omega \rangle$ is the same as the conditional obtainable from $\mathcal{X}_\omega$ via the Ramsey test.

The proofs of the base case and the inductive step are so similar that only the latter will be given in detail.

The inductive step: Construction: We start with a series

$\mathcal{X}_{n-1} = Y_1, \dots, Y_k, X_n, X_{n+1}, X_{n+2}, \dots$

and a delimiter ℓ_{n-1} according to which each of $Y_1, \dots, Y_k$ is self-limited. Let $\mathbb{Z}$ be the set of subsets V of $\{X_n, X_{n+1}, X_{n+2}, \dots\}$ such that (i) $X_n \in V$ and (ii) there is some sentence p such that $K * p = X_n$ and $V = \{Z \mid X_n \leqq Z \leqq \ell(X_n) \text{ and } p \in Z\}$.

Case i, $\mathbb{Z} = \varnothing$:

Let $\mathcal{X}_n = Y_1, \dots, Y_k, X_{n+1}, X_{n+2}, \dots$ and let ℓ_n be the restriction of ℓ_{n-1} to the elements of the new series.

Case ii, $\mathbb{Z} \neq \varnothing$: Let $\hat{\mathbb{Z}} = \{\bigcap V \mid V \in \mathbb{Z}\}$ and let $Z^n_1, Z^n_2, \dots, Z^n_{m_n}$ be a list on which each element of $\hat{\mathbb{Z}}$ appears exactly once, and such that if $Z, Z' \in \hat{\mathbb{Z}}$ and $Z' \subset Z$ then Z' comes before Z on the list. (The existence of such a series follows from the order extension principle that follows from the axiom of choice, see [145, p. 19].)

Now let:

$\mathcal{X}_n = Y_1, \dots, Y_k, Z^n_1, \dots, Z^n_{m_n}, X_{n+1}, X_{n+2}, \dots$

and let ℓ_n be such that (i) $\ell_n(X) = X$ for all $X \in \{Y_1, \dots, Y_k, Z^n_1, \dots, Z^n_{m_n}\}$, and (ii) $\ell_n(X) = \ell_{n-1}(X)$ for all $X \in \{X_{n+1}, X_{n+2}, \dots\}$.

The inductive step: Verification: The verification is straightforward in Case i of the construction. In Case ii, let $\mid\!\sim_{n-1}$ be the inference relation derivable from $\langle \mathcal{X}_{n-1}, \ell_{n-1} \rangle$, and let $\mid\!\sim_n$ be the inference relation derivable from $\langle \mathcal{X}_n, \ell_n \rangle$. We are going to show that for all p and q: $p \mid\!\sim_n q$ if and only if $p \mid\!\sim_{n-1} q$. There are three cases.

Case 1, $p \in \bigcup\{Y_1, \dots, Y_k\}$: The desired result follows directly since ℓ_n and ℓ_{n-1} coincide in this part of the series.

Case 2, $p \notin \bigcup\{Y_1, \dots, Y_k\}$ and $p \in X_n$: Then $p \mid\!\sim_{n-1} q$ holds if and only if q holds in all elements of $\{X \mid X_n \leqq X \leqq \ell_{n-1}(X_n) \text{ and } p \in X\}$, i.e. if and only if $q \in \bigcap\{X \mid X_n \leqq X \leqq \ell_{n-1}(X_n) \text{ and } p \in X\}$. Now, $\bigcap\{X \mid X_n \leqq X \leqq \ell_{n-1}(X_n) \text{ and } p \in X\}$ is an element of $\widehat{\mathbb{Z}}$, and moreover it is the leftmost element of $\widehat{\mathbb{Z}}$ that contains p. (This is because it is a proper subset of all other elements of $\widehat{\mathbb{Z}}$ that contain p.) It follows that $p \mid\!\sim_n q$ holds if and only if $q \in \bigcap\{X \mid X_n \leqq X \leqq \ell_{n-1}(X_n) \text{ and } p \in X\}$, thus $p \mid\!\sim_n q$ holds if and only if $p \mid\!\sim_{n-1} q$ holds.

Case 3, $p \notin \bigcup\{Y_1, \dots, Y_k, X_n\}$: The desired result follows directly since ℓ_n and ℓ_{n-1} coincide in the subseries $X_{n+1}, X_{n+2}, \dots$. □

Proof for Section 7.6

Proof of Observation 7.9 Left to the reader with the reminder that if XaY then there is some Ψ with $X \circ \Psi = Y$, namely $\Psi = \Pi_Y$. □

Proofs for Section 8.1

Proof of Theorem 8.2 *From I to II*: Left to the reader.

From II to I: Let $\mathbb{X} = \{X \mid (\exists p)(K * p = X)\}$ and let $\widehat{C}$ be a monoselective choice function from descriptor-definable subsets of $\mathbb{X}$ to $\mathbb{X}$, such that for all Ψ:

(i) If there is some p such that $[\![\Psi]\!] = [\![\mathfrak{B}p]\!] \neq \varnothing$, then $\widehat{C}([\![\Psi]\!]) = K * p$ (which is possible due to uniformity)

(ii) Otherwise: (a) if $K \in [\![\Psi]\!]$, then $\widehat{C}([\![\Psi]\!]) = K$, (b) if $[\![\Psi]\!] \neq \varnothing$, then $\widehat{C}([\![\Psi]\!]) \in [\![\Psi]\!]$, and (c) if $[\![\Psi]\!] = \varnothing$, then $\widehat{C}([\![\Psi]\!])$ is undefined.

Furthermore, let $\circ$ be the descriptor revision that is based on $\widehat{C}$.

To verify the construction we need to show (1) that $\mathbb{X}$ is a set of belief sets, (2) that $K \in \mathbb{X}$, (3) that $\widehat{C}([\![\Psi]\!]) \in [\![\Psi]\!]$ whenever $[\![\Psi]\!] \neq \varnothing$, (4) that $\widehat{C}$ is K-favouring, and (5) that $*$ is based on $\widehat{C}$ in the way specified in Definitions 8.1, 5.2, and 5.6.

(1) follows from closure. (2) follows from confirmation. (3) follows from the construction. In clause (i), note that $K * p \in [\![\mathfrak{B}p]\!]$ follows from regularity when $[\![\mathfrak{B}p]\!] \neq \varnothing$. (4) follows from confirmation in clause (i) and directly from the definition in clause (ii). Likewise, (5) follows directly from the construction. □

Proof of Observation 8.3 Left to the reader. □

Lemma 8.1 *Let $*$ be a partial meet revision on the consistent belief set K. Then $*$ satisfies:*

If it holds for all q that $K * q \vdash p$ if and only if $K * q \vdash p'$, then $K * p = K * p'$. (uniformity)

Proof of Lemma 8.1 We prove the lemma in its converse form. Let $K * p \neq K * p'$. Due to the AGM postulate of extensionality, $\nvdash p \leftrightarrow p'$. It follows truthfunctionally that either $p \& \neg p'$ or $p' \& \neg p$ is consistent. We only need to treat the former case. It follows from the AGM postulate of consistency that $K * (p \& \neg p')$ is consistent. Success yields $K * (p \& \neg p') \vdash p$. Success and consistency yield $K * (p \& \neg p') \nvdash p'$. Thus it does not hold for all sentences q that $K * q \vdash p$ if and only if $K * q \vdash p'$. □

Proof of Theorem 8.4 The equivalence between (II) and (III) is known from [1].

From (I) to (III): Closure follows from the definition of monoselective descriptive revision. (Definition 5.2)

Success: It follows from ($\mathbb{X}$2) that for all $p \in \mathcal{L}$ there is some $X \in \mathbb{X}$ such that $p \in X$. It then follows from the definition of mono-selective descriptor revision that success holds.

Inclusion follows from ($\widehat{C}$*3*).

Vacuity: Let $\neg p \notin K$. Then $K \subseteq K * p$ follows from ($\widehat{C}$*4*) and $p \in K * p$ from success. Closure yields $K + p \subseteq K * p$.

Consistency follows from ($\widehat{C}2$).

Extensionality: Let $\vdash p \leftrightarrow p'$. Then $K * p = \widehat{C}([\![\mathfrak{B}p]\!]) = \widehat{C}([\![\mathfrak{B}p']\!]) = K * p'$.

From (III) to (I): Let $\mathbb{X} = \{X \mid (\exists p)(K * p = X)\}$ and let $\widehat{C}$ be a monoselective choice function on the descriptor-definable subsets of $\mathbb{X}$ such that for all Ψ:

(i) If there is some p such that $[\![\Psi]\!] = [\![\mathfrak{B}p]\!]$, then $\widehat{C}([\![\Psi]\!]) = K * p$ (which is possible due to Lemma 8.1).

(ii) Otherwise: (a) if $K \Vdash \Psi$, then $\widehat{C}([\![\Psi]\!]) = K$, (b) if $[\![\Psi]\!] \neq \varnothing$, then $\widehat{C}([\![\Psi]\!]) \in [\![\Psi]\!]$, and (c) if $[\![\Psi]\!] = \varnothing$, then $\widehat{C}([\![\Psi]\!])$ is undefined.

It follows from closure that $\mathbb{X}$ is a set of belief sets. Using inclusion and vacuity we obtain $K * \top = K$, thus $K \in \mathbb{X}$. It follows directly from the construction that $\widehat{C}$ is a monoselective choice function and that $*$ is the sentential revision based on it. We also need to verify that ($\mathbb{X}$1), ($\mathbb{X}$2), ($\widehat{C}1$), ($\widehat{C}2$), ($\widehat{C}3$), and ($\widehat{C}4$) hold.

($\mathbb{X}$1) follows from consistency.

($\mathbb{X}$2): When $K \nvdash \neg p$ then this follows from inclusion and vacuity. When $K \vdash \neg p$ then $\mathrm{Cn}(K \cup \{p\}) = \mathrm{Cn}(\{\bot\})$. It follows from success and closure that $K * \bot = \mathrm{Cn}(\{\bot\})$, thus $\mathrm{Cn}(\{\bot\}) \in \mathbb{X}$.

($\widehat{C}1$): Let $K \in [\![\mathfrak{B}p]\!]$, i.e. $p \in K$. Since K is consistent it follows from inclusion and vacuity that $\widehat{C}([\![\mathfrak{B}p]\!]) = K * p = K$.

($\widehat{C}2$) follows from consistency.

($\widehat{C}3$) follows from inclusion.

($\widehat{C}4$) Let $K \nvdash \neg p$. It follows from vacuity that $K \subseteq K * p = \widehat{C}([\![\mathfrak{B}p]\!])$. □

Proofs for Section 8.2, except Theorem 8.8

Lemma 8.2 *If $*$ satisfies closure, extensionality, relative success, confirmation, finite-based outcome, and cumulativity, then it satisfies: $K * p = K * \&(K * p)$.*[4]

Proof of Lemma 8.2 It follows from closure and finite-based outcome that $\&(K * p) \in K * p$. If $K * p = K$ then confirmation yields $K = K * \&(K * p)$. If $K * p \neq K$ we use cumulativity to obtain $K * (p \,\&\, \&(K * p)) = K * p$. Relative success yields $\&(K * p) \vdash p$ and consequently $\vdash p \,\&\, \&(K * p) \leftrightarrow \&(K * p)$, and then extensionality yields $K * \&(K * p) = K * p$. □

Definition A.2 *Let $\star$ be a sentential revision on the belief set K and let $p, q \in \mathcal{L}$. Then q* weakens *p if and only if: $p \vdash q$, $K * p \vdash p$, $K * q \vdash q$, and $K * p \neq K * q$.*

Let K' and K'' be outcomes of revisions of K. Then K'' is a weakening *of K' if and only if there are p and q such that $K' = K * p$, $K'' = K * q$, and q weakens p.*

Lemma 8.3 *Let $*$ be a sentential revision on the belief set K that satisfies finite gradation and non-circularity. Let K' be a belief set. Then there is a weakening K'' of K' such that no other weakening of K' is a weakening of K''.*

---

[4]For any finite-based set Z, $\&Z$ is the conjunction of all elements of some finite set Z' with $\mathrm{Cn}(Z) = \mathrm{Cn}(Z')$.

Proof of Lemma 8.3 It follows from finite gradation and Lemma 8.2 that K' has a finite number of weakenings and from non-circularity that the relation "is a weakening of" among its weakenings is non-circular. □

Proof of Theorem 8.5 *From (I) to (II)*: Left to the reader.

From (II) to (I): We are going to construct a set $\mathbb{X}$ and a relation $\leqq$ on $\mathbb{X}$, define $\circ$ as indicated in the theorem, and then verify the construction.

The construction: Let $\mathbb{X} = \{X \mid (\exists p)(X = K * p)\}$. We are going to construct inductively a relation $\leqq$ on $\mathbb{X}$, numbering its elements $K_0, K_1 \ldots$. This series will also inductively be shown to have the following property:

If X is a weakening of K_m, then $X \in \{K_0, K_1, \ldots K_{m-1}\}$. (the tightness condition)

We begin by setting $K_0 = K$. Clearly, since K_0 is the first element of the series the tightness condition is satisfied vacuously at this stage. For the inductive construction we use a list containing all sentences $p \in \mathcal{L}$ such that $K * p \vdash p$. In each step, we assume that we already have a series $K_0, \ldots K_m$ of belief sets and that this series satisfies the tightness condition.

Let p be the first sentence on our list such that $K * p \vdash p$ and $p \notin K_0 \cup \ldots \cup K_m$.

If it holds for all sentences q that weaken $\&(K * p)$ that $K * q \in \{K_0, \ldots K_m\}$, then let $K_{m+1} = K * p$. If not, then let K_{m+1} be a weakening of $K * p$ such that all weakenings of it are in $\{K_0, \ldots K_m\}$. The existence of such a minimal weakening follows from Lemma 8.3. Clearly there is one less weakening of $K * p$ not included in the series $K_0, \ldots K_{m+1}$ than not included in the series $K_0, \ldots K_m$. We repeat this process, finding a sentence q that weakens p and such that $K * q$ has no weakening in the series $K_0, \ldots K_{m+1}$, etc., until we arrive at some belief set K_{m+k} that is identical to $K * p$. This process is repeated with the next sentence after p on the list of sentences, etc. Due to Lemma 8.3 and the construction, the tightness condition is still satisfied after each addition to the series $K_0, K_1, \ldots$.

To finish the construction we define $\leqq$ such that $K_s \leqq K_t$ if and only if $s \leq t$. We define $\circ$ such that $K \circ \Psi$ is the unique $\leqq$-minimal element of $\mathbb{X}$ that satisfies Ψ, unless Ψ is not satisfied by any element of $\mathbb{X}$, in which case $K \circ \Psi = K$. Furthermore, we let $*$ be a sentential revision such that for for all $p \in \mathcal{L}$, $K * p = K \circ \mathfrak{B}p$.

Verification of the construction: It follows straight-forwardly that conditions (a), (b), and (c) of clause (I) of the theorem are satisfied. It remains to show that $K * r = K \circ \mathfrak{B}r$ for all r.

For that purpose, first consider the case when there is some $X \in \mathbb{X}$ with $r \in X$. We are going to show inductively that $K * r$ is equal to the $\leqq$-minimal element of $\mathbb{X}$ that contains r. Due to regularity and our construction of $\mathbb{X}$ it follows from $r \in X \in \mathbb{X}$ that $K * r \vdash r$.

It follows from $K_0 = K$ and confirmation that for all $r \in \mathcal{L}$: If $r \in K_0$ then $K * r = K_0$ iff K_0 is the $\leqq$-minimal set containing r. For the inductive step, we assume that for all r, if $r \in K_0 \cup \ldots \cup K_m$, then $K * r$ is equal to the $\leqq$-minimal r-containing element of $\{K_0, \ldots, K_m\}$. In order to show that the same holds for all sentences $r \in K_0 \cup \ldots \cup K_{m+1}$, let $r \in K_{m+1} \setminus (K_0 \cup \ldots \cup K_m)$. Due to Lemma 8.2 we have $K_{m+1} = K * \&(K_{m+1})$. Since $\&(K_{m+1}) \vdash r$ we have $\vdash r \leftrightarrow \&(K_{m+1}) \vee r$, and extensionality yields $K * r = K * (\&(K_{m+1}) \vee r)$. Therefore, if $K * r \neq$

$K * \mathcal{E}(K_{m+1})$, then $K * r$ is a weakening of $K * \mathcal{E}(K_{m+1})$. Since our construction satisfies the tightness condition, all weakenings of $K * \mathcal{E}(K_{m+1})$, i.e. K_{m+1}, are in the set $\{K_0, \dots K_m\}$. None of its elements contains r, thus $K * r$ is not a weakening of $K * \mathcal{E}(K_{m+1})$. We can conclude that $K * r = K * \mathcal{E}(K_{m+1}) = K_{m+1}$, as desired.

In the other case, $X \nvdash r$ for all $X \in \mathbb{X}$. Then $K * r \nvdash r$, and relative success yields $K * r = K$ which coincides with the outcome of the construction. □

Proof of Observation 8.6 Let p and p' be such that for all q, $K * q \vdash p$ if and only if $K * q \vdash p'$.

Case 1, there is no q with $K * q \vdash p$. Then there is no q with $K * q \vdash p'$. It follows from relative success that $K * p = K$ and $K * p' = K$.

Case 2, there is some sentence q' with $K * q' \vdash p$. Then $K * q' \vdash p'$. Due to regularity, $K * p \vdash p$ and $K * p' \vdash p'$. Due to our assumption (substituting p for q) we have $K * p \vdash p'$, and similarly $K * p' \vdash p$. Cumulativity yields $K * p = K * (p \& p')$ and $K * p' = K * (p' \& p)$. Due to extensionality, $K * (p \& p') = K * (p' \& p)$, thus $K * p = K * p'$ as desired. □

Proof of Observation 8.7 *Part 1, from cumulativity to reciprocity*: Let $q \in K * p$ and $p \in K * q$. Cumulativity yields $K * p = K * (p \& q)$ and $K * q = K * (q \& p)$. Extensionality yields $K * (p \& q) = K * (q \& p)$, and we have $K * p = K * q$.

Part 1, reciprocity to cumulativity: Let $q \in K * p$. There are two cases.

First case, $p \notin K * p$: Regularity yields $p \notin K * (p \& q)$, thus $p \& q \notin K * (p \& q)$. Relative success yields $K * p = K = K * (p \& q)$.

Second case, $p \in K * p$: Then due to closure, $p \& q \in K * p$. Regularity yields $p \& q \in K * (p \& q)$. We thus have $p \& q \in K * p$ and $p \in K * (p \& q)$. Reciprocity yields $K * p = K * (p \& q)$.

Part 2: Due to closure it follows from $p \in K * p$ and $K * p = K * (p \vee q \vee r)$ that $p \vee q \in K * (p \vee q \vee r)$. Cumulativity yields $K * (p \vee q \vee r) = K * ((p \vee q \vee r) \& (p \vee q))$, and due to extensionality $K * (p \vee q \vee r) = K * (p \vee q)$. Equivalently, $K * p = K * (p \vee q)$. □

Proof for Section 8.2: Theorem 8.8

Definition A.3 *For all* $A \subseteq \mathcal{L}$: $[A] = \{W \in \mathcal{L} \perp\!\!\perp \mid A \subseteq W\}$.

Brackets of singletons can be omitted, thus $[p] = [\{p\}]$.

Lemma 8.4 $\bigcap[A] = \text{Cn}(A)$

Proof of Lemma 8.4 See [99, p. 52]. □

Lemma 8.5 *(1)* $[X] = [Y]$ *if and only if* $\text{Cn}(X) = \text{Cn}(Y)$
(2) $[X] \subseteq [Y]$ *if and only if* $\text{Cn}(Y) \subseteq \text{Cn}(X)$.
(3) If X and Y are logically closed, then $[X \cup Y] = [X] \cap [Y]$.
(4) If X and Y are logically closed, then $[X \cap Y] = [X] \cup [Y]$.

Proof of Lemma 8.5 *Parts 1-3*: See [99, pp. 52–53].

Part 4: Right-to-left: It follows from Part 2 that $[X] \subseteq [X \cap Y]$ and $[Y] \subseteq [X \cap Y]$

Left-to-right: Suppose to the contrary that $[X \cap Y] \nsubseteq [X] \cup [Y]$. Then there is a possible world W such that $X \cap Y \subseteq W$, $X \nsubseteq W$, and $Y \nsubseteq W$, and there must be some $x \in X$ such that $x \notin W$ and some $y \in Y$ such that $y \notin W$. It follows from $x \vee y \in X \cap Y$ and $X \cap Y \subseteq W$ that $x \vee y \in W$. Furthermore, it follows from $x \notin W$ and the maximality of W that $\neg x \in W$, and $\neg y \in W$ follows in the same way. Thus W is inconsistent which it cannot be since it is a possible world. We can conclude from this contradiction that $[X \cap Y] \subseteq [X] \cup [Y]$. □

Definition A.4 ([80]) *Let $\mathcal{W} \subseteq \mathcal{L}\perp_{\perp}$. A set $\mathfrak{S}$ of subsets of $\mathcal{L}\perp_{\perp}$ is a* system of spheres centered on *$\mathcal{W}$ if and only if:*

- *($\mathfrak{S}1$) If $\mathcal{S}_1, \mathcal{S}_2 \in \mathfrak{S}$, then either $\mathcal{S}_1 \subseteq \mathcal{S}_2$ or $\mathcal{S}_2 \subseteq \mathcal{S}_1$.*
- *($\mathfrak{S}2$) $\mathcal{W} \in \mathfrak{S}$, and $\mathcal{W} \subseteq \mathcal{S}$ for all $\mathcal{S} \in \mathfrak{S}$.*
- *($\mathfrak{S}3$) $\mathcal{L}\perp_{\perp} \in \mathfrak{S}$.*
- *($\mathfrak{S}4$) For all sentences p and spheres $\mathcal{S} \in \mathfrak{S}$, if $[p] \cap \mathcal{S} \neq \varnothing$, then there is some $\mathcal{S}' \in \mathfrak{S}$ such that $[p] \cap \mathcal{S}' \neq \varnothing$ and that $[p] \cap \mathcal{S}'' = \varnothing$ for all $\mathcal{S}'' \in \mathfrak{S}$ with $\mathcal{S}'' \subset \mathcal{S}'$.*

Definition A.5 *Let $\mathfrak{S}$ be a system of spheres. A* world-ring *in $\mathfrak{S}$ is a set $\mathcal{R}$ consisting of all elements of some $\mathcal{S} \in \mathfrak{S}$ that are not elements of any $\mathcal{S}' \in \mathfrak{S}$ with $\mathcal{S}' \subset \mathcal{S}$.*

Lemma 8.6 *Let $\mathfrak{S}$ be a system of spheres. Its world-rings form a partition (set of equivalence classes) of the set of worlds.*

Proof of Lemma 8.6 Left to the reader. □

Definition A.6 *(1) Let $\mathfrak{S}$ be a system of spheres and p a sentence. Then $\mathcal{S}_p$ is the innermost sphere in $\mathfrak{S}$ that contains some p-world. If no sphere in $\mathfrak{S}$ contains any p-world, then $\mathcal{S}_p$ is undefined.*

(2) Let K be a belief set. A sentential revision $*$ on K is based on *a system $\mathfrak{S}$ of spheres centered on $[K]$ if and only if for all p, $[K * p] = [p] \cap \mathcal{S}_p$; unless there are no p-worlds, in which case $[K * p] = \varnothing$.*

Lemma 8.7 *Let $\mathbb{X}$ be the outcome set of a sphere-based revision in a sphere system $\mathfrak{S}$, and let $X \in \mathbb{X}$. Then all worlds W with $X \subseteq W$ are elements of the same ring in $\mathfrak{S}$.*

Proof of Lemma 8.7 Since $X \in \mathbb{X}$, we have $X = K * p$ for some p. Due to Definition A.6, $[K * p] = [p] \cap \mathcal{S}_p$, and due to Definition A.5, $[p] \cap \mathcal{S}_p = \mathcal{R}_p \cap [p]$ where $\mathcal{R}_p$ is the world-ring corresponding to $\mathcal{S}_p$. Thus $[K * p] \subseteq \mathcal{R}_p$, i.e. $[X] \subseteq \mathcal{R}_p$, thus for any world W if $X \subseteq W$, then $W \in \mathcal{R}_p$.[5] □

---

[5]This result is obtained since we define $*$ via $[K * p]$ as in Definition A.6, rather than through the weaker relationship $K * p = \bigcap([p] \cap \mathcal{S}_p)$. See [80, p. 163].

Lemma 8.8 *Let $\mathbb{X}$ be the outcome set of a sphere-based revision, and let $X, Y \in \mathbb{X}$. Furthermore, let $W_1, W_2 \in \mathcal{L} \perp \perp$ be such that $X \subseteq W_1$ and $Y \subseteq W_2$. Then: $X \cap Y \in \mathbb{X}$ if and only if W_1 and W_2 are elements of the same world-ring.*

Proof of Lemma 8.8 Since $X, Y \in \mathbb{X}$ there are p and q such that $X = K * p$ and $Y = K * q$.

For one direction, let W_1 and W_2 be elements of the same world-ring. It follows from Lemma 8.7 that the innermost world-ring containing some p-world coincides with the innermost world-ring containing some q-world. Let $\mathcal{R}$ be that world-ring. Since a world is a $p \vee q$-world if and only if it is either a p-world or a q-world (as can be seen from Lemma 8.5, part 4), $\mathcal{R}$ is also the innermost world-ring containing some $p \vee q$-world. Thus $[K * (p \vee q)] = \mathcal{R} \cap [p \vee q]$, $[K * p] = \mathcal{R} \cap [p]$, and $[K * q] = \mathcal{R} \cap [q]$. We then have:

$[p \vee q] = [p] \cup [q]$

$\mathcal{R} \cap [p \vee q] = (\mathcal{R} \cap [p]) \cup (\mathcal{R} \cap [q])$

$[K * (p \vee q)] = [K * p] \cup [K * q]$

$[K * (p \vee q)] = [X] \cup [Y]$

$\bigcap [K * (p \vee q)] = \bigcap ([X] \cup [Y])$

$\bigcap [K * (p \vee q)] = \bigcap [X \cap Y]$ (Lemma 8.5)

$K * (p \vee q) = X \cap Y$ (Lemma 8.4)

and we can conclude that $X \cap Y \in \mathbb{X}$.

For the other direction, let $X \cap Y \in \mathbb{X}$. Due to Lemma 8.5, $[X \cap Y] = [X] \cup [Y]$, and it follows from Lemma 8.7 that the worlds containing X and the worlds containing Y all belong to the same ring. □

Lemma 8.9 *Let $\mathbb{X}$ be the outcome set of a sphere-based revision and let $X, Y, Z \in \mathbb{X}$. If $X \cap Y \in \mathbb{X}$ and $Y \cap Z \in \mathbb{X}$, then $X \cap Z \in \mathbb{X}$.*

Proof of Lemma 8.9 Directly from Lemma 8.8. □

Lemma 8.10 *Let $\mathbb{X}$ be the outcome set of a sphere-based revision, and let $X, Y \in \mathbb{X}$. If $X \cup Y \nvdash \perp$, then $X \cap Y \in \mathbb{X}$.*

Proof of Lemma 8.10 Since $X \cup Y \nvdash \perp$ there is some world $W \in \mathcal{L} \perp \perp$ such that $X \cup Y \subseteq W$. It follows from Lemma 8.8 that $X \cap Y \in \mathbb{X}$. □

Proof of Theorem 8.8 The equivalence between (II) and (III) is known from [1]. We will make use of the well-known result from Grove [80] showing that (II) is equivalent with:

(II$^+$) $*$ is a sentential revision based on some sphere system $\mathfrak{S}$.

The proof will therefore proceed by showing the equivalence between (I) and (II$^+$).

From (I) to (II$^+$): We are first going to define an equivalence relation on the set consisting of those worlds that contain at least one element of $\mathbb{X}$, namely the relation $\sim$ such that:

$W_1 \sim W_2$ iff there are $X_1, X_2 \in \mathbb{X}$ such that $X_1 \subseteq W_1$, $X_2 \subseteq W_2$, and $X_1 \cap X_2 \in \mathbb{X}$.

This relation is obviously reflexive and symmetric. To prove that it is an equivalence relation it remains to show that it is transitive. Let $W_1 \sim W_2$ and $W_2 \sim W_3$. Then there are $X_1, X_2, Y_2, Y_3 \in \mathbb{X}$ such that $X_1 \subseteq W_1$, $X_2, Y_2 \subseteq W_2$, $Y_3 \subseteq W_3$, $X_1 \cap X_2 \in \mathbb{X}$ and $Y_2 \cap Y_3 \in \mathbb{X}$.

It follows from ($\mathbb{X}$3) that $X_2 \cap Y_2 \in \mathbb{X}$. Since $X_1 \cap X_2 \in \mathbb{X}$ another use of ($\mathbb{X}$3) yields $X_1 \cap X_2 \cap Y_2 \in \mathbb{X}$. We also have $Y_2 \cap Y_3 \in \mathbb{X}$, and a third use of ($\mathbb{X}$3) yields $X_1 \cap X_2 \cap Y_2 \cap Y_3 \in \mathbb{X}$. Combining this with $X_1 \cap X_2 \subseteq W_1$ and $Y_2 \cap Y_3 \subseteq W_3$, we obtain $W_1 \sim W_3$.

We will use the equivalence classes over worlds based on $\sim$ as world-rings, and define the following relations over these world-rings:

$\mathcal{R} \lhd \mathcal{R}'$ if and only if it holds for all $X, X' \in \mathbb{X}$ and all $W, W' \in \mathcal{L} \perp\!\!\!\perp$ that if $X \subseteq W \in \mathcal{R}$ and $X' \subseteq W' \in \mathcal{R}'$, then $X < X'$.
$\mathcal{R} \unlhd \mathcal{R}'$ if and only if either $\mathcal{R} \lhd \mathcal{R}'$ or $\mathcal{R} = \mathcal{R}'$.

($<$ is the strict part of $\leqq$.) If there are any worlds not containing any element of $\mathbb{X}$ then they are added as the lowest-ranked world-ring.

We need to show that $\unlhd$ is a linear ordering of the world-rings, i.e. that it is transitive, complete and antisymmetric. In this proof we will have use for Lemma 5.2 (p. 171) according to which $\leqq$ is transitive, complete, anti-symmetric, and descriptor-wellfounded.

To show the transitivity of $\unlhd$, let $\mathcal{R}_1 \unlhd \mathcal{R}_2$ and $\mathcal{R}_2 \unlhd \mathcal{R}_3$. Excluding trivial cases we can assume that $\mathcal{R}_1$, $\mathcal{R}_2$, and $\mathcal{R}_3$ are pairwise non-identical. Thus $\mathcal{R}_1 \lhd \mathcal{R}_2$ and $\mathcal{R}_2 \lhd \mathcal{R}_3$, and we suppose for contradiction that $\mathcal{R}_1 \not\lhd \mathcal{R}_3$. Then there are $X_1, X_3 \in \mathbb{X}$ and $W_1, W_3 \in \mathcal{L} \perp\!\!\!\perp$ such that $X_1 \subseteq W_1 \in \mathcal{R}_1$ and $X_3 \subseteq W_3 \in \mathcal{R}_3$ and $X_1 \not< X_3$. Since $\leqq$ is complete we then have $X_3 \leqq X_1$. Since $\mathcal{R}_1$ and $\mathcal{R}_3$ are disjoint, $X_3 < X_1$. Let $X_2 \subseteq W_2 \in \mathcal{R}_2$. Due to $\mathcal{R}_1 \lhd \mathcal{R}_2$ and $\mathcal{R}_2 \lhd \mathcal{R}_3$ we have $X_1 < X_2$ and $X_2 < X_3$. This makes $<$ cyclic, contrary to the properties of $\leqq$ just referred to.

Antisymmetry of $\unlhd$: Suppose to the contrary that $\mathcal{R} \unlhd \mathcal{R}'$, $\mathcal{R}' \unlhd \mathcal{R}$ and $\mathcal{R} \neq \mathcal{R}'$. Then $\mathcal{R} \lhd \mathcal{R}' \lhd \mathcal{R}$. Let $X \subseteq W \in \mathcal{R}$ and $X' \subseteq W' \in \mathcal{R}$. Then $X < X' < X$, contrary to the properties of $\leqq$.

As a preparation for the proving the completeness of $\unlhd$ we prove the following:

(X) If $\mathcal{R}_1 \neq \mathcal{R}_2$, $X_1 \subseteq W_1 \in \mathcal{R}_1$, $X_1' \subseteq W_1' \in \mathcal{R}_1$, $X_2 \subseteq W_2 \in \mathcal{R}_2$, $X_2' \subseteq W_2' \in \mathcal{R}_2$ and $X_1 < X_2$, then $X_1' < X_2'$.

Proof of (X): First suppose that $X_2' < X_1$. We then have $X_2' < X_1 < X_2$. Since $W_2, W_2' \in \mathcal{R}_2$, we have $W_2 \sim W_2'$, and consequently there are Y and Y' such that $Y \subseteq W_2$, $Y' \subseteq W_2'$, and $Y \cap Y' \in \mathbb{X}$. It follows from $X_2 \cap Y \subseteq W_2$ and ($\mathbb{X}$3) that $X_2 \cap Y \in \mathbb{X}$ and from $Y' \cap X_2' \subseteq W_2'$ that $Y' \cap X_2' \in \mathbb{X}$. Using ($\mathbb{X}$5) twice we conclude from this that $X_2 \cap X_2' \in \mathbb{X}$. Using ($\leqq$2) we can conclude from this and $X_2' < X_1 < X_2$ that $X_1 \cap X_2 \in \mathbb{X}$, contary to our assumptions that $X_1 \subseteq W_1 \in \mathcal{R}_1$ and $X_2 \subseteq W_2 \in \mathcal{R}_2$ and that $\mathcal{R}_1$ and $\mathcal{R}_2$ are different equivalence classes for $\sim$. Thus $X_2' \not< X_1$. Since $\leqq$ is complete we can conclude that $X_1 \leqq X_2'$.

Next suppose that $X_2' < X_1'$. Then due to the result we just obtained we have $X_1 \leqq X_2' < X_1'$. In the same way that we proved $X_2 \cap X_2' \in \mathbb{X}$ we can prove $X_1 \cap X_1' \in \mathbb{X}$. Combining this with $X_1 \leqq X_2' < X_1'$ we obtain from ($\leqq$2) that

$X'_1 \cap X'_2 \in \mathbb{X}$, contrary to our assumptions that $X'_1 \subseteq W'_1 \in \mathcal{R}_1$ and $X'_2 \subseteq W'_2 \in \mathcal{R}_2$ and that $\mathcal{R}_1$ and $\mathcal{R}_2$ are different equivalence classes for $\sim$. Thus not $X'_2 < X'_1$, and since $\mathcal{R}_1$ and $\mathcal{R}_2$ are disjoint, it follows from the completeness and antisymmetry of $\leqq$ that $X'_1 < X'_2$ as desired.

Completeness of $\trianglelefteq$: For the two distinct world-rings $\mathcal{R}$ and $\mathcal{R}'$, let $X \subseteq W \in \mathcal{R}$ and $X' \subseteq W' \in \mathcal{R}'$, Since $\mathcal{R}$ and $\mathcal{R}'$ are disjoint, it follows from the completeness and antisymmetry of $\leqq$ that either $X < X'$ or $X' < X$. In the former case, (X) yields $\mathcal{R} \lhd \mathcal{R}'$ and in the latter case $\mathcal{R}' \lhd \mathcal{R}$.

Next, let $\mathfrak{S}$ be the set consisting of the sets $\mathcal{S}$ such that $\mathcal{S} = \bigcup\{\mathcal{R}' \mid \mathcal{R}' \trianglelefteq \mathcal{R}\}$ for some world-ring $\mathcal{R}$. We need to show that $\mathfrak{S}$ satisfies the four conditions given in Definition A.4 (p. 201) for being a system of spheres centered on $[K]$, and that $*$ is based on $\mathfrak{S}$ in the manner described in Definition A.6.

Since $\trianglelefteq$ is a linear ordering of the world-rings, ($\mathfrak{S}$1) follows directly from the construction of $\mathfrak{S}$.

For ($\mathfrak{S}$2), note that it follows from the properties of $\leqq$ that $K < X$ for all $X \in \mathbb{X} \setminus \{K\}$. It follows from ($\leqq$1) that $\mathbb{X}$ contains no proper subset of K, from the construction of $\sim$ that the K-containing worlds form a world-ring of their own, and from our definition of $\mathfrak{S}$ that this world-ring is also the innermost sphere.

($\mathfrak{S}$3) follows from the construction of $\mathfrak{S}$, since all worlds are included in one of the world-rings.

For ($\mathfrak{S}$4), let $[p] \cap \mathcal{S} \neq \varnothing$. Due to ($\mathbb{X}$1) there is some p-containing element of $\mathbb{X}$. According to Lemma 5.2 (p. 177), $\leqq$ is descriptor-wellfounded. Thus there is some $X_p \in \mathbb{X}$ such that $p \in X_p$ and $X_p \leqq Y$ for all $Y \in \mathbb{X}$ with $p \in Y$. Let $\mathcal{R}_p$ be the world-ring containing the worlds including X_p. Then it holds for all world-rings $\mathcal{R}'$ that if $\mathcal{R}' \lhd \mathcal{R}_p$ then there is no $W \in \mathcal{R}'$ such that $p \in X \subseteq W$ for some $X \in \mathbb{X}$. Suppose that there is nevertheless some $W \in \mathcal{R}'$ such that $p \in W$. Since $\mathcal{R}'$ consists of possible worlds containing elements of $\mathbb{X}$ there is then some $Y \subseteq W$ with $Y \in \mathbb{X}$. It follows from $p \in W \nvdash \perp$ and $Y \subseteq W$ that $Y \nvdash \neg p$, and from ($\mathbb{X}$4) that $\mathrm{Cn}(Y \cup \{p\}) \in \mathbb{X}$. It follows from $Y \subseteq W$ and $p \in W$ that $\mathrm{Cn}(Y \cup \{p\}) \subseteq W$, contrary to our previous conclusion that $\mathcal{R}'$ contains no world that includes a p-containing element of $\mathbb{X}$. This contradiction shows that there is no $W \in \mathcal{R}'$ with $p \in W$. It follows that the sphere $\mathcal{S}_p = \bigcup\{\mathcal{R}' \mid \mathcal{R}' \trianglelefteq \mathcal{R}_p\}$ has the desired property, i.e. it has a non-empty intersection with $[p]$ but no sphere that is its proper subset has a non-empty intersection with $[p]$.

Finally, it follows from the construction of the world-rings that for all worlds W, if $X_p \subseteq W$ then $W \in \mathcal{R}_p$. Due to Lemma 8.4, $X_p = \bigcap\{W \in \mathcal{R}_p \mid p \in W\} = \bigcap\{W \in \mathcal{S}_p \mid p \in W\}$, i.e. X_p is the outcome of $\mathfrak{S}$-based revision of K by p according to Definition A.6. Due to (I), $K * p = X_p$, which concludes this direction of the proof.

From (II$^+$) to (I): For each world-ring $\mathcal{R}$ we define a *cluster* $\overline{\mathcal{R}}$ of elements of $\mathbb{X}$:

$$\overline{\mathcal{R}} = \{X \in \mathbb{X} \mid (\exists W \in \mathcal{R})(X \subseteq W)\}$$

It follows from Lemma 8.7 that each belief set is an element of exactly one cluster, and from Lemma 8.8 that the relation $\sqsubseteq$ on $\mathbb{X}$ such that $X \sqsubseteq Y$ iff $X \cap Y \in \mathbb{X}$ is an equivalence relation with the clusters as equivalence classes.

Next we define the relation $\sqsubseteq$ on $\mathbb{X}$ such that $X \sqsubseteq Y$ if and only if the world-ring corresponding to the cluster containing X is included in every sphere that contains the world-ring corresponding to the cluster containing Y. Clearly, $\underline{\square}$ is the symmetric part of $\sqsubseteq$. Its strict part is denoted $\sqsubset$.

Let $\breve{\subset}$ be a strict ordering[6] on $\mathbb{X}$ satisfying the condition: If $X \subset Y$ then $X \breve{\subset} Y$. (The existence of such a relation is guaranteed by the order extension principle that follows from the axiom of choice, see [145, p. 19].) We let $<$ be the relation on $\mathbb{X}$ such that for all $X, Y \in \mathbb{X}$:

$X < Y$ if and only if either (i) $X \sqsubset Y$ or (ii) both $X \underline{\square} Y$ and $X \breve{\subset} Y$.

Furthermore, let $X \leqq Y$ be the ordering such that $X \leqq Y$ iff either $X < Y$ or $X = Y$.

It follows directly that $\leqq$ is an ordering on the outcome set $\mathbb{X}$ and that K is its minimal element. It remains to show that ($\mathbb{X}$1), ($\mathbb{X}$3), ($\mathbb{X}$4), ($\mathbb{X}$5), ($\leqq$1), and ($\leqq$2) are satisfied and that for all p, $K * p$ is the $\leqq$-minimal element of $[\![\mathfrak{B}p]\!]$.

($\mathbb{X}$1) follows from the AGM postulate consistency and ($\mathbb{X}$3) from Lemma 8.10. The proofs of ($\mathbb{X}$4) and ($\mathbb{X}$5) were outlined in the main text in Section 8.2.

For ($\leqq$1), note that if $X \subseteq Y$ then it follows from Lemma 8.7 that X and Y are included only in worlds in one and the same world-ring, thus they belong to the same cluster, i.e. $X \underline{\square} Y$. Since $X \subseteq Y$ we have either $X = Y$ or $X \breve{\subset} Y$, and in both cases $X \leqq Y$ follows directly.

For ($\leqq$2), we use Lemma 8.8 to conclude that X and Z belong to the same cluster. Due to the definition of $\leqq$, Y belongs to that same cluster. Again using Lemma 8.8 we find that $X \cap Y \in \mathbb{X}$ and $Y \cap Z \in \mathbb{X}$.

Finally we have to prove that $K * p$ is the $\leqq$-minimal element of $[\![\mathfrak{B}p]\!]$: $K * p$ is the intersection of the p-worlds in the innermost world-ring that has p-worlds. Therefore it is an element of the corresponding cluster. Let X be a p-containing element of $\mathbb{X}$ such that $X \neq K * p$. If X belongs to the same cluster as $K * p$, then $K * p \,\underline{\square}\, X$ and $K * p \subset X$, and if it belongs to some other cluster then $K * p \sqsubset X$. (The well-foundedness of $\sqsubset$ follows from property ($\mathfrak{S}$4).) Thus in both cases $K * p < X$. □

Proof for Section 8.3

Proof of Observation 8.11 See Fig. A.1. *DP1*: Part (a) of the figure is compatible with $q \vdash p$. Let $X = \text{Cn}(\{\neg p, \neg q, r\})$. Then $(X * p) * q = \text{Cn}(\{p, q, r\})$ and $X * q = \text{Cn}(\{p, q, \neg r\})$.

DP2: Part (b) of the figure is compatible with $q \vdash \neg p$. Let $X = \text{Cn}(\{\neg p, q, \neg r\})$. Then $(X * p) * q = \text{Cn}(\{\neg p, q, r\})$ and $X * q = X = \text{Cn}(\{\neg p, q, \neg r\})$.

DP3: In part (c) of the figure, let $X = \text{Cn}(\{\neg p, \neg q, r\})$. Then $(X * p) * q = \text{Cn}(\{\neg p, q, r\})$ and $X * q = \text{Cn}(\{p, q, r\})$, hence $X * q \vdash p$ but $(X * p) * q \nvdash p$.

DP4: Use the same example as in part 3, and note that $X * q \nvdash \neg p$ but $(X * p) * q \vdash \neg p$. □

[6] A strict ordering is a binary relation $<$ that satisfies transitivity and trichotomy (exactly one of $X < Y$, $Y < X$, and $X = Y$).

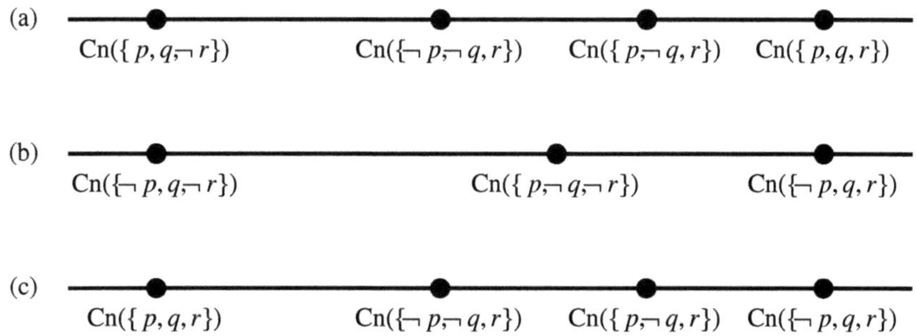

Fig. A.1 Diagram for the proof of Observation 8.11.

Proofs for Section 8.4

Proof of Observation 8.13 For one direction, let $Z \nsubseteq \bigcup\{Y \in \mathbb{X} \mid Y < Z\}$. Then there is some p such that $p \in Z$ and $p \notin Y$ for all $Y < Z$. Hence $Z = K * p$.

For the other direction, let $Z = K * p$. If $Z = K$ then the condition is vacuously satisfied. If $Z \neq K$ then we have $p \in Z$ and $p \notin Y$ for all $Y < Z$, thus $p \notin \bigcup\{Y \in \mathbb{X} \mid Y < Z\}$. □

Proof of Observation 8.15 *Part 1*: In a language with only the two atoms p and q, let $\leqq$ and $\leqq'$ be exhaustively described by:

$\mathrm{Cn}(\varnothing) < \mathrm{Cn}(\{p\&q\}) < \mathrm{Cn}(\{p\}) < \mathrm{Cn}(\{p\&\neg q\}) < \mathrm{Cn}(\{\perp\})$

respectively:

$\mathrm{Cn}(\varnothing) <' \mathrm{Cn}(\{p\&q\}) <' \mathrm{Cn}(\{p\&\neg q\}) <' \mathrm{Cn}(\{\perp\})$

Let $\circ$ be the descriptor revision generated by $\leqq$ and $\circ'$ that generated by $\leqq'$. Then $K \circ \{\mathfrak{B}p, \neg\mathfrak{B}q\} = \mathrm{Cn}(\{p\})$ but $K \circ' \{\mathfrak{B}p, \neg\mathfrak{B}q\} = \mathrm{Cn}(\{p\&\neg q\})$. However, $\leqq$ and $\leqq'$ generate the same sentential revision.

Part 2: We can use the same example as in part 1, and note that $\leqq'$ is the additive restriction of both $\leqq$ and itself.

Part 3: Let $\leqq$ and $\leqq'$ be as in part 1. They generate different proximity relations on descriptors, namely $\succeq$ and $\succeq'$, such that $\{\mathfrak{B}p, \neg\mathfrak{B}q\} \succ \{\mathfrak{B}p, \mathfrak{B}\neg q\}$ but $\{\mathfrak{B}p, \mathfrak{B}\neg q\} \succeq' \{\mathfrak{B}p, \neg\mathfrak{B}q\}$. However, $\leqq$ and $\leqq'$ generate the same believability relation. □

Proof of Observation 8.16 *Part 1*: Directly from Definitions 5.38 and 8.1.

Part 2: We need to construct two additively restricted belief set orderings that generate the same sentential revision. Let

$K_0 = K'_0 = \mathrm{Cn}(\{s \vee t\})$
$K_1 = \mathrm{Cn}(\{s\})$ and $K'_1 = \mathrm{Cn}(\{t\})$
$K_2 = \mathrm{Cn}(\{t\})$ and $K'_2 = \mathrm{Cn}(\{s\})$
$K_n = K'_n$ for all $n > 2$

Let $\leqq$ and $\leqq'$ (with the strict parts $<$ and $<'$) be the belief set orderings that are completely characterized by the series $K_0 < K_1 < K_2 \dots$ and $K'_0 <' K'_1 <' K'_2 \dots$.

Let $*$ and $*'$ be the revisions based on the two orderings. They have the same outcome set, to be denoted $\mathbb{X}$. We need to show that for all $p \in \mathcal{L}$ and all $X \in \mathbb{X}$, $K * p = X$ if and only if $K *' p = X$. This is straightforwardly shown for all $X \in \mathbb{X}$ except possibly for $\text{Cn}(\{s\})$ and $\text{Cn}(\{t\})$. The proofs in these two cases are symmetrical. One of them is:
$K * p = \text{Cn}(\{t\})$ iff $p \notin K_0$, $p \notin K_1$, and $p \in K_2$
iff $s \vee t \nvdash p$, $s \nvdash p$, and $t \vdash p$
iff $s \vee t \nvdash p$ and $t \vdash p$
iff $p \notin K'_0$ and $p \in K'_1$
iff $K *' p = \text{Cn}(\{t\})$ □

Proof of Observation 8.17 *Part 1*: In a language with only the two atoms p and q, let $K = \text{Cn}(\{p\})$ and consider the belief set orderings exhaustively characterized as follows:

$\text{Cn}(\{p\}) \leqq \text{Cn}(\{p\&q\}) \leqq \text{Cn}(\{\perp\})$
$\text{Cn}(\{p\}) \leqq' \text{Cn}(\{p\&q\})$

It follows from Definitions 5.38 and 8.12 that they give rise to the same believability relation. However, they give rise to different sentential revisions (e.g. $K * \neg q = \text{Cn}(\{\perp\})$ but $K *' \neg q = K$).

Part 2: In the example presented in the proof of Observation 8.16, part 2, the believability relation $\overset{\star}{\succeq}$ generated from $\leqq$ and the corresponding believability relation $\overset{\star}{\succeq}'$ generated from $\leqq'$ are not the same, as can be seen from $s \overset{\star}{\succ} t$ and $t \overset{\star}{\succ}' s$. However, as shown in the proof of Observation 8.16, part 2, $\overset{\star}{\succeq}$ and $\overset{\star}{\succeq}'$ are associated with the same sentential revision $*$. □

Proof of Observation 8.18 *Part 1*:
$p \overset{\star}{\succeq} q$ iff $\mathfrak{B}p \succeq \mathfrak{B}q$ (Definition 8.1)
iff either $M_{\mathfrak{B}p} \leqq M_{\mathfrak{B}q}$ or $\mathfrak{B}q$ is unsatisfiable within the domain of $\leqq$.
(Lemma 5.8, p. 186; M defined as in that lemma)

Part 2: Use the same example as in the proof of Observation 8.17, part 1. □

Proof for Section 8.5

Proof of Theorem 8.20 *From (I) to (II)*: Left to the reader.

From (II) to (I): Let $\mathbb{X}$ be the outcome set of $\overset{*}{=}$, and let $\widehat{C}$ be a function from subsets of $\mathbb{X}$ to elements of $\mathbb{X}$, such that:

(A) If there is some p with $[\![\Psi]\!] = [\![\mathfrak{B}p \vee \mathfrak{B}\neg p]\!] \neq \varnothing$, then $\widehat{C}([\![\Psi]\!]) = K \overset{*}{=} p$.
(B) For all other Ψ, $\widehat{C}([\![\Psi]\!]) \in [\![\Psi]\!]$ if $[\![\Psi]\!] \neq \varnothing$ and $\widehat{C}([\![\Psi]\!]) = K$ if $K \in [\![\Psi]\!]$.
(C) If $[\![\Psi]\!] = \varnothing$, then $\widehat{C}([\![\Psi]\!])$ is undefined.

To see that this construction is well-defined, it is sufficient to note considering (A) that due to uniformity, if $[\![\mathfrak{B}p \vee \mathfrak{B}\neg p]\!] = [\![\mathfrak{B}p' \vee \mathfrak{B}\neg p']\!]$, then $K \overset{*}{=} p = K \overset{*}{=} p'$.

We need to verify (a) that $\widehat{C}$ is a monoselective choice function, (b) that it is K-favouring, and (c) that $\overset{*}{=}$ is the operation of resolution that is derived from it according to Definition 8.19.

(a) follows from the construction. Note that due to negatable regularity, if $[\![\mathfrak{B}p \vee \mathfrak{B}\neg p]\!] \neq \varnothing$, then $K \overset{*}{=} p \in [\![\mathfrak{B}p \vee \mathfrak{B}\neg p]\!]$.

For (b) it is sufficient to show that if $K \in [\![\mathfrak{B}p \vee \mathfrak{B}\neg p]\!]$ then $\widehat{C}([\![\mathfrak{B}p \vee \mathfrak{B}\neg p]\!]) = K$. There are two cases:

First case, $p \in K$: We get $K \overset{*}{\div} p = K$ directly from confirmation.

Second case, $\neg p \in K$: It follows from uniformity (by substituting $\neg p$ for p') that $K \overset{*}{\div} p = K \overset{*}{\div} \neg p$. We have $K \overset{*}{\div} \neg p = K$ from confirmation, thus $K \overset{*}{\div} p = K$.

For (c), let $\circ$ be the descriptor revision that is based on $\widehat{C}$ and $\overset{*}{\div}'$ the operation of resolution that is based on $\circ$. We need to show that $K \overset{*}{\div}' p = K \overset{*}{\div} p$ for all p. If $[\![\mathfrak{B}p \vee \mathfrak{B}\neg p]\!] \neq \varnothing$ then this follows directly from (A). If $[\![\mathfrak{B}p \vee \mathfrak{B}\neg p]\!] = \varnothing$ then it follows from Definitions 5.2 and 8.19 that $K \overset{*}{\div}' p = K$ and from negatable relative success that $K \overset{*}{\div} p = K$. □

Proofs for Section 9.1

Proof of Theorem 9.2 *From I to II*: Left to the reader.

From II to I: Let $\mathbb{X} = \{X \mid (\exists p)(K \div p = X)\}$ and let $\widehat{C}$ be a monoselective choice function on the descriptor-definable subsets of $\mathbb{X}$, such that for all $p \in \mathcal{L}$, if $[\![\neg\mathfrak{B}p]\!] \neq \varnothing$, then $\widehat{C}([\![\neg\mathfrak{B}p]\!]) = K \div p$. (To see that this is possible, first note that due to the construction of $\mathbb{X}$, when $\neg\mathfrak{B}p$ is satisfiable within $\mathbb{X}$ then there is some $q \in \mathcal{L}$ with $K \div q \Vdash \neg\mathfrak{B}p$, i.e. $K \div q \nvdash p$. Persistence yields $K \div p \nvdash p$, i.e. $K \div p \in [\![\neg\mathfrak{B}p]\!]$. Secondly note that if $[\![\neg\mathfrak{B}p]\!] = [\![\neg\mathfrak{B}p']\!]$, then $K \div p = K \div p'$ due to uniformity.) Let $\circ$ be the descriptor revision on K that is based on $\widehat{C}$.

To verify the construction we need to prove that (1) $\mathbb{X}$ is a set of belief sets, (2) $K \in \mathbb{X}$ (as required by Definition 5.2), and (3) $K \div p = K \circ \neg\mathfrak{B}p$ for all $p \in \mathcal{L}$.

(1) follows from closure.

For (2) note that since $K \div \top \vdash \top$ it follows from relative success that $K \div \top = K$.

For (3) there are two cases. Case (A), $\neg\mathfrak{B}p$ is satisfiable within $\mathbb{X}$: Due to the construction of $\circ$ we have $K \circ \neg\mathfrak{B}p = \widehat{C}([\![\neg\mathfrak{B}p]\!])$ and due to the construction of $\widehat{C}$ we then have $\widehat{C}([\![\neg\mathfrak{B}p]\!]) = K \div p$.

Case (B), $\neg\mathfrak{B}p$ is not satisfiable within $\mathbb{X}$: Then $K \div p \vdash p$. It follows from relative success that $K \div p = K$ and from the construction of $\circ$ that $K \circ \neg\mathfrak{B}p = K$. □

Proof of Observation 9.3 Left to the reader. □

Proofs for Section 9.2

Proof of Theorem 9.6 *From (I) to (II)*: The proof for the first four postulates is left to the reader.

Decomposition: If $\neg\mathfrak{B}(p\&q)$ is unsatisfiable within $\mathbb{X}$, then so is $\neg\mathfrak{B}p$. It follows that $K \circ \neg\mathfrak{B}(p\&q) = K \circ \neg\mathfrak{B}p = K$ and consequently $K \div (p\&q) = K \div p = K$. If $\neg\mathfrak{B}(p\&q)$ is satisfiable within $\mathbb{X}$, then either $p \notin K \circ \neg\mathfrak{B}(p\&q)$ or $q \notin K \circ \neg\mathfrak{B}(p\&q)$, i.e. equivalently, either $p \notin K \div (p\&q)$ or $q \notin K \div (p\&q)$. In the former case, suppose that there is some $X \in \mathbb{X}$ with $X < K \div (p\&q)$ and $p \notin X$. Then $p\&q \notin X$, contrary to $X \in \mathbb{X}$ and the construction of $\div$. Thus $K \div (p\&q)$ is the $\leqq$-minimal element of $\mathbb{X}$ not containing p, thus $K \div p = K \div (p\&q)$. In the latter case we obtain $K \div q = K \div (p\&q)$ in the same way.

Conjunctive adjunction: Let $p \notin K \div q$. Case 1, $q \in K \div q$: Then $q \in X$ for all $X \in \mathbb{X}$. It follows that for all or all $X \in \mathbb{X}$, $p\&q \in X$ if and only if $p \in X$. Thus $K \div (p\&q) = K \div p$.

Case 2, $q \notin K \div q$: Then $p\&q \notin K \div q$, thus $K \div (p\&q) \leqq K \div q$. Since $\leqq$ is antisymmetric we have two subcases.

Subcase 2A, $K \div (p\&q) < K \div q$: Since $K \div q$ is the $\leqq$-minimal element of $\mathbb{X}$ that does not include q, it follows that $q \in K \div (p\&q)$, thus $p \notin K \div (p\&q)$. Let $X < K \div (p\&q)$ and suppose that $p \notin X$. Then $p\&q \notin X$, which is incompatible with $X < K \div (p\&q)$. We can conclude from this contradiction that $K \div (p\&q) = K \div p$.

Subcase 2B, $K \div (p\&q) = K \div q$: Then $p \notin K \div (p\&q)$. Again let $X < K \div (p\&q)$ and suppose that $p \notin X$. Then $p\&q \notin X$, which is incompatible with $X < K \div (p\&q)$. We can conclude that $K \div (p\&q) = K \div p$.

Conjunctive composition: $K \div (p\&q)$ is the $\leqq$-minimal element of $\{K \div p, K \div q\}$ and $K \div (p'\&q)$ is the $\leqq$-minimal element of $\{K \div p', K \div q\}$. It follows from $K \div p = K \div p'$ that $K \div (p\&q) = K \div (p'\&q)$.

From (II) to (I): Let $\mathbb{X} = \{X \mid (\exists x \in \mathcal{L})(X = K \div x)\}$ and let $\leqq$ be the relation on $\mathbb{X}$ such that for all $p, q \in K$:

If $K \div p \nvdash p$ and $K \div q \nvdash q$, then $K \div p \leqq K \div q$ if and only if $K \div (p\&q) = K \div p$. Let $\circ$ be the partial linear descriptor revision that is based on $\leqq$.

We need to prove that (1) $\mathbb{X}$ is a set of belief sets, (2) $K \in \mathbb{X}$, (3) $\leqq$ is a well-defined relation, (4) $\leqq$ is complete, (5) $\leqq$ is transitive, (6) $\leqq$ is anti-symmetric, (7) $\leqq$ is wellfounded for $\neg\mathfrak{B}$ descriptors, and (8) $K \div p = K \circ \neg\mathfrak{B}p$.

(1) follows from closure and (2) from vacuity.

For (3), it is sufficient to show that if $p, q \in K$, $K \div p = K \div p'$ and $K \div q = K \div q'$, then $K \div (p\&q) = K \div (p'\&q')$. It follows from $K \div p = K \div p'$ and conjunctive composition that $K \div (p\&q) = K \div (p'\&q)$. In the same way it follows from $K \div q = K \div q'$ that $K \div (p'\&q) = K \div (p'\&q')$, and we can conclude that $K \div (p\&q) = K \div (p'\&q')$.

(4) follows from decomposition. (5) can be shown as follows:

$K \div p \leqq K \div q$ and $K \div q \leqq K \div r$	
$K \div p \leqq K \div q$ and $K \div q = K \div (q\&r)$	(construction of $\leqq$)
$K \div p \leqq K \div (q\&r)$	(as just shown, $\leqq$ is well-defined)
$K \div p = K \div (p\&q\&r)$	(construction of $\leqq$)
$K \div (p\&q) = K \div (p\&q\&r)$	($K \div p \leqq K \div q$ yields $K \div p = K \div (p\&q)$)
$K \div (p\&q) \leqq K \div r$	(construction of $\leqq$)
$K \div p \leqq K \div r$	(since $K \div p = K \div (p\&q)$)

For (6), let $K \div p \leqq K \div q$ and $K \div q \leqq K \div p$. The construction of $\leqq$ yields $K \div (p\&q) = K \div p$ and $K \div (p\&q) = K \div q$, thus $K \div p = K \div q$.

For (7) it is sufficient to show that if $\neg\mathfrak{B}p$ is satisfiable within $\mathbb{X}$, then $K \div p$ is the unique $\leqq$-minimal $\neg\mathfrak{B}p$-satisfying element of $\mathbb{X}$. Suppose not. Then due to the construction of $\mathbb{X}$ there is some $q \in \mathcal{L}$ such that $K \div q \Vdash \neg\mathfrak{B}p$ and $K \div q < K \div p$. It follows from $p \notin K \div q$ and conjunctive adjunction that $K \div (p\&q) = K \div p$ and from the construction of $\leqq$ that $K \div p \leqq K \div q$. Contradiction.

For (8) we consider two cases.

Case (A), $\neg\mathfrak{B}p$ is unsatisfiable within $\mathbb{X}$: Then $K \circ \neg\mathfrak{B}p = K$ due to the construction of $\circ$. Furthermore, $p \in K \mathbin{\bar{\div}} p$, and relative success yields $K \mathbin{\bar{\div}} p = K$, thus $K \circ \neg\mathfrak{B}p = K \mathbin{\bar{\div}} p$.

Case (B), $\neg\mathfrak{B}p$ is satisfiable within $\mathbb{X}$: Then due to the construction of $\mathbb{X}$ there is some $K \mathbin{\bar{\div}} q$ such that $K \mathbin{\bar{\div}} q \nvdash p$. Persistence yields $K \mathbin{\bar{\div}} p \nvdash p$. We know from the proof of part (7) that $K \mathbin{\bar{\div}} p$ is the unique $\leqq$-minimal $\neg\mathfrak{B}p$-satisfying element of $\mathbb{X}$, thus $K \circ \neg\mathfrak{B}p = K \mathbin{\bar{\div}} p$. □

Proof of Observation 9.7 *Part 1*: Left to the reader. *Part 2*. Let it be the case that:

For all q: $K \mathbin{\bar{\div}} q \vdash p$ if and only if $K \mathbin{\bar{\div}} q \vdash p'$. (the antecedent condition)

There are two cases:

Case A, $p \in K \mathbin{\bar{\div}} p$: Due to relative success, $K \mathbin{\bar{\div}} p = K$. It follows from persistence that $K \mathbin{\bar{\div}} p' \vdash p$ and from the antecedent condition that $K \mathbin{\bar{\div}} p' \vdash p'$. thus relative success yields $K \mathbin{\bar{\div}} p' = K$ and we have $K \mathbin{\bar{\div}} p = K \mathbin{\bar{\div}} p'$.

Case B, $p \notin K \mathbin{\bar{\div}} p$: By substituting p for q in the antecedent condition and applying closure we obtain $p' \notin K \mathbin{\bar{\div}} p$. Persistence and closure yield $K \mathbin{\bar{\div}} p' \nvdash p'$. By substituting p' for q in the antecedent condition and again applying closure we obtain $p \notin K \mathbin{\bar{\div}} p'$. It remains to apply reciprocity to $p' \notin K \mathbin{\bar{\div}} p$ and $p \notin K \mathbin{\bar{\div}} p'$, and conclude that $K \mathbin{\bar{\div}} p = K \mathbin{\bar{\div}} p'$. □

Proof of Observation 9.8 Let $p \rightarrow q \in K \mathbin{\bar{\div}} q$ and $q \rightarrow p \in K \mathbin{\bar{\div}} p$. Proofs are needed for the following two cases:

Case 1, $p \in \mathrm{Cn}(\varnothing)$: Then $p \rightarrow q$ is logically equivalent with q, and we conclude from $p \rightarrow q \in K \mathbin{\bar{\div}} q$ and closure that $q \in K \mathbin{\bar{\div}} q$. It follows from success that $q \in \mathrm{Cn}(\varnothing)$. Then p and q are logically equivalent, and extensionality yields $K \mathbin{\bar{\div}} p = K \mathbin{\bar{\div}} q$.

Case 2, $p \notin \mathrm{Cn}(\varnothing)$ and $q \notin \mathrm{Cn}(\varnothing)$: Then it follows from success, closure, and $p \rightarrow q \in K \mathbin{\bar{\div}} q$ that $p \notin K \mathbin{\bar{\div}} q$. It follows in the same way that $q \notin K \mathbin{\bar{\div}} p$, and then reciprocity yields $K \mathbin{\bar{\div}} p = K \mathbin{\bar{\div}} q$. □

Proof of Observation 9.9 Let $\mathcal{L}$ consist of the two atomic sentences a and b and their truth-functional combinations. Let $K = \mathrm{Cn}(\{a, b\})$. Let $\trianglelefteq$ be a transitive and complete relation on remainders of K (with $X \trianglelefteq Y$ signifying that X is at least as highly ranked as Y). Let $\triangleleft$ be its strict part and $\boxminus$ its symmetric part. Let $\mathrm{Cn}(\{a\}) \boxminus \mathrm{Cn}(\{b\}) \triangleleft \mathrm{Cn}(\{a \leftrightarrow b\})$. Let $\div$ be the transitively relational partial meet contraction that is based on $\trianglelefteq$ via the selection function γ. We have $K \perp a = \{\mathrm{Cn}(\{b\}), \mathrm{Cn}(\{a \leftrightarrow b\})\}$, $\gamma(K \perp a) = \{\mathrm{Cn}(\{b\})\}$ and thus $K \div a = \mathrm{Cn}(\{b\})$. We also have $K \perp (a\&b) = \{\mathrm{Cn}(\{a\}), \mathrm{Cn}(\{b\}), \mathrm{Cn}(\{a \leftrightarrow b\})\}$, $\gamma(K \perp (a\&b)) = \{\mathrm{Cn}(\{a\}), \mathrm{Cn}(\{b\})\}$ and thus $K \div (a\&b) = \mathrm{Cn}(\{a\}) \cap \mathrm{Cn}(\{b\}) = \mathrm{Cn}(\{a \vee b\})$. Hence $a\&b \notin K \div a$ and $a \notin K \div (a\&b)$ but $K \div a \neq K \div (a\&b)$. □

Proofs for Section 9.3

Proof of Observation 9.10 For one direction, let $\bigcap\{Y \in \mathbb{X} \mid Y < Z\} \nsubseteq Z$. Then there is some p such that $p \in Y$ for all $Y < Z$ and $p \notin Z$. It follows that $Z = K \mathbin{\bar{\div}} p$.

For the other direction, let $Z = K \mathbin{\bar{\div}} p$. If $Z = K$ then the condition is vacuously satisfied. If $Z \neq K$ then it follows that $p \in Y$ for all $Y < Z$, thus $p \in \bigcap\{Y \in \mathbb{X} \mid Y < Z\}$ and $p \notin Z$. □

Proof of Observation 9.12 *Part 1*: In a language with only the two atoms p and q, let $\leqq$ and $\leqq'$ be completely characterized by:

$\mathrm{Cn}(\{p\&q\}) < \mathrm{Cn}(\{p \vee q\}) < \mathrm{Cn}(\{p\}) < \mathrm{Cn}(\{q \rightarrow p\})$

$\mathrm{Cn}(\{p\&q\}) <' \mathrm{Cn}(\{p \vee q\}) <' \mathrm{Cn}(\{q \rightarrow p\})$

The descriptor revisions $\circ$ and $\circ'$ that $\leqq$ respectively $\leqq'$ give rise to differ since $K \circ \{\mathfrak{B}(q \rightarrow p), \neg\mathfrak{B}q\} = \mathrm{Cn}(\{p\})$ and $K \circ' \{\mathfrak{B}(q \rightarrow p), \neg\mathfrak{B}q\} = \mathrm{Cn}(\{q \rightarrow p\})$. However, they generate the same revocation, which is a contraction.

Part 2: In the example introduced in part 1, $\leqq'$ is the subtractive restriction of both $\leqq$ and itself.

Part 3: In a language with only the two atoms p and q, let $\leqq$ and $\leqq'$ be completely characterized by:

$\mathrm{Cn}(\{p\&q\}) < \mathrm{Cn}(\{p\}) < \mathrm{Cn}(\{q\}) < \mathrm{Cn}(\varnothing)$

$\mathrm{Cn}(\{p\&q\}) <' \mathrm{Cn}(\{p\}) <' \mathrm{Cn}(\{p \vee q\}) <' \mathrm{Cn}(\varnothing)$

$\leqq$ and $\leqq'$ give rise to different proximity relations $\succeq$ and $\succeq'$, as can be seen from $\{\neg\mathfrak{B}(p \vee q)\} \succeq \{\neg\mathfrak{B}p, \neg\mathfrak{B}q\}$ and $\{\neg\mathfrak{B}p, \neg\mathfrak{B}q\} \succ' \{\neg\mathfrak{B}(p \vee q)\}$. To prove that they give rise to the same relation of entrenchment, we need to show that for all sentences z, $\mathrm{Cn}(\{q\})$ is the $\leqq$-minimal belief set in the domain of $\leqq$ not containing z if and only if $\mathrm{Cn}(\{p \vee q\})$ is the $\leqq'$-minimal belief set in the domain of $\leqq'$ not containing z. This is done as follows:

$z \in \mathrm{Cn}(\{p\&q\})$ and $z \in \mathrm{Cn}(\{p\})$ and $z \notin \mathrm{Cn}(\{q\})$
iff $z \in \mathrm{Cn}(\{p\}) \setminus \mathrm{Cn}(\{q\})$
iff $z \in \mathrm{Cn}(\{p\}) \setminus (\mathrm{Cn}(\{p\}) \cap \mathrm{Cn}(\{q\}))$
iff $z \in \mathrm{Cn}(\{p\}) \setminus \mathrm{Cn}(\{p \vee q\})$
iff $z \in \mathrm{Cn}(\{p\&q\})$ and $z \in \mathrm{Cn}(\{p\})$ and $z \notin \mathrm{Cn}(\{p \vee q\})$ □

Proof of Observation 9.13 *Part 1*: Due to Definition 9.1, $K \div p = K \circ \neg\mathfrak{B}p$. According to t^{bo}, if $\neg\mathfrak{B}p$ is satisfiable within the domain of $\leqq$, then $K \div p$ is the $\leqq$-minimal set in the domain of $\leqq$ not containing p. According to Definition 9.11, this is also the $\bar{\leqq}$-minimal set not containing p. The case when $\neg\mathfrak{B}p$ is not satisfiable within the domain of $\leqq$ follows directly.

Part 2: It follows from Definition 9.11 that the domain of $\bar{\leqq}$ consists of the sets $K \circ \neg\mathfrak{B}p$ such that $K \circ \neg\mathfrak{B}p \Vdash \neg\mathfrak{B}p$, i.e. equivalently the sets $K \div p$ such that $K \div p \nvdash p$. Furthermore:

$K \div p \,\bar{\leqq}\, K \div q$ iff $K \circ \neg\mathfrak{B}p \leqq K \circ \neg\mathfrak{B}q$ (Definition 9.11)
iff $K \circ \neg\mathfrak{B}p = K \circ (\neg\mathfrak{B}p \vee \neg\mathfrak{B}q)$ (t^{ob}, Part 4 of Theorem 5.39)
iff $K \circ \neg\mathfrak{B}p = K \circ \neg\mathfrak{B}(p\&q)$ (uniformity; Lemma 5.1, p. 176)
iff $K \div p = K \div (p\&q)$ (Definition 9.1) □

Proof of Observation 9.14 *Part 1*:
$p \leq q$ iff $\neg\mathfrak{B}p \succeq \neg\mathfrak{B}q$ (Definition 5.36)
iff $K \circ \neg\mathfrak{B}p \leqq K \circ \neg\mathfrak{B}q$ or $\neg\mathfrak{B}q$ is unsatisfiable within the domain of $\leqq$ (Lemma 5.8, p. 186)
iff $K \div p \,\bar{\leqq}\, K \div q$ or q is included in all elements of the domain of $\bar{\leqq}$ (Definition 9.1)

iff $K \div p = K \div (p\&q)$ or q is included in all elements of the domain of $\leqq$
iff $K \div p = K \div (p\&q)$

Part 2: In the example presented in the proof of Observation 9.12, part 3, $\leqq$ and $\leqq'$ give rise to the same entrenchment relation $\lesssim$. However, they give rise to different revocations (that are both contractions), as can be seen from revoking p (contracting by p). □

Proof of Observation 9.15 *Part 1*:
$p \lesssim q$ iff $\neg\mathfrak{B}p \succeq \neg\mathfrak{B}q$ (Definition 5.36)
iff $M_{\neg\mathfrak{B}p} \leqq M_{\neg\mathfrak{B}q}$ or $\neg\mathfrak{B}q$ is unsatisfiable within the domain of $\leqq$
(Lemma 5.8, p. 186; M defined as in the lemma)
iff $M_{\neg\mathfrak{B}p} \bar{\leqq} M_{\neg\mathfrak{B}q}$ or $\neg\mathfrak{B}q$ is unsatisfiable within the domain of $\bar{\leqq}$
(Definition 9.11)

Part 2: In the example given in the proof of Observation 9.12, part 3, we have $\bar{\leqq} = \leqq$ and $\bar{\leqq}' = \leqq'$. Thus $\leqq$ and $\leqq'$ are different. However, they give rise to the same entrenchment relation $\lesssim$. □

Proof of Observation 9.16 Let the language consist of the two atoms p and q and their truth-functional combinations. Let the original belief set be $K = \text{Cn}(\{p, q\})$.

We are first going to construct an entrenchment relation that satisfies the properties. Let $\lesssim$ be the entrenchment relation that is derivable from the belief set ordering $\leqq$ (with the strict part $<$) that is completely characterized by:

$$\text{Cn}(\{p, q\}) < \text{Cn}(\{p \vee q\}) < \text{Cn}(\varnothing).$$

It follows from Observation 5.37 that $\lesssim$ has the desired properties.

Next we construct the operation $\div$ that is based on $\lesssim$ in the manner stated in the observation. We will show that it has the following three properties:

(a) $q \in K \div p$
(b) $p \in K \div q$
(c) $K \div (p\&q) = \text{Cn}(\{p \vee q\})$

(a) follows from $p \lessdot p \vee q$ and (b) from $q \lessdot p \vee q$. To prove (c) we first note that $p\&q \lessdot ((p\&q) \vee (p \vee q))$, thus $p \vee q \in K \div (p\&q)$, i.e. $\text{Cn}(\{p \vee q\}) \subseteq K \div (p\&q)$. Suppose that $\text{Cn}(\{p \vee q\}) \subset K \div (p\&q)$. Then, due to the atomic structure of the language, it must be the case that either $p \rightarrow q \in K \div (p\&q)$ or $q \rightarrow p \in K \div (p\&q)$. However, since $p\&q \not\lessdot ((p\&q) \vee (p \rightarrow q))$ and $p\&q \not\lessdot ((p\&q) \vee (q \rightarrow p))$, neither $p \rightarrow q$ nor $q \rightarrow p$ is in $K \div (p\&q)$, thus (c) holds.

Since (a)–(c) all hold, $\div$ does not satisfy decomposition, and therefore it is not a centrolinear revocation. □

Lemma 9.1 *Let X be a logically closed set and let $\dot{\simeq}$ be a symmetric relation on sentences such that if $\vdash x \leftrightarrow x'$ then $x \dot{\simeq} y$ if and only if $x' \dot{\simeq} y$. Then:*

$X \subseteq \{s \mid (\forall t)(s \vee t \not\dot{\simeq} p)\}$ if and only if $X \subseteq \{s \mid p \not\dot{\simeq} s\}$.

Proof of Lemma 9.1 For one direction, let $X \nsubseteq \{s \mid p \not\dot{\simeq} s\}$. Then there is some $x \in X$ such that $x \dot{\simeq} p$. Then $x \vee x \dot{\simeq} p$, and consequently $X \nsubseteq \{s \mid (\forall t)(s \vee t \not\dot{\simeq} p)\}$.

For the other direction, let $X \not\subseteq \{s \mid (\forall t)(s \vee t \not\doteq p)\}$. Then there is some $x \in X$ and some t such that $x \vee t \doteq p$, thus $x \vee t \notin \{s \mid p \not\doteq s\}$. It follows from the logical closure of X that $x \vee t \in X$, thus $X \not\subseteq \{s \mid p \not\doteq s\}$. □

Proof of Theorem 9.17 *First step*: We begin by constructing $\leqq$ using the lower limit of each E_p, i.e. setting it equal to $\{s \mid p < s\}$. For all sentences p that are included in some element of the domain of $\leqq$, let X_p be the $\leqq$-minimal element of the domain of $\leqq$ that does not contain p, if there is such an element.

We are going to show that with this construction, $\leq$ is the entrenchment relation derivable from $\leqq$. Since we have constructed $\leqq$ to be shrinking (cf. Definition 9.19), $\{s \mid p < s\} = X_p$ and we have:
$X_p \leqq X_q$ or q holds in all elements in the domain of $\leqq$
iff $\{s \mid p < s\} \leqq \{s \mid q < s\}$ or q holds in all elements in the domain of $\leqq$
iff $E_p \leqq E_q$ or q holds in all elements in the domain of $\leqq$
iff $p \leq q$. (Lemma 5.8, p. 186)
(The logical closure of $\{s \mid p < s\}$ is shown in Observation 9.18.)

Second step: Next we are going to identify the belief sets E_p that can replace $\{s \mid p<s\}$ without changing the derived entrenchment relation. Since the construction using $\{s \mid p < s\}$ yields a shrinking belief set ordering, the $\leqq$-minimal belief set not containing p is $\{s \mid p < s\}$, and the intersection of all preceding belief sets is equal to the immediately preceding one, i.e. $\{s \mid p \leq s\}$. Therefore our criterion for E_p is that $\{s \mid p \leq s\} \setminus E_p = \{s \mid p \leq s\} \setminus \{s \mid p < s\}$. We have:
$\{s \mid p \leq s\} \setminus E_p = \{s \mid p \leq s\} \setminus \{s \mid p < s\}$
iff $E_p \cap (\{s \mid p \leq s\} \setminus \{s \mid p < s\}) = \varnothing$ and $\{s \mid p < s\} \subseteq E_p$
(since $\{s \mid p < s\} \subseteq \{s \mid p \leq s\}$)
iff $E_p \cap \{s \mid p \doteq s\} = \varnothing$ and $\{s \mid p < s\} \subseteq E_p$
iff $\{s \mid p < s\} \subseteq E_p \subseteq \{s \mid p \not\doteq s\}$
iff $\{s \mid p < s\} \subseteq E_p \subseteq \{s \mid (\forall t)(s \vee t \not\doteq p)\}$ (Lemma 9.1, E_p is logically closed) □

Proof of Observation 9.18 For the first set, let $\{s \mid p < s\} \vdash r$. We are going to show that $r \in \{s \mid p < s\}$. Due to compactness there is a finite subset $\{s_1, \ldots, s_n\}$ of $\{s \mid p < s\}$ such that $\{s_1, \ldots, s_n\} \vdash r$, equivalently $s_1 \& \ldots \& s_n \vdash r$. It follows by repeated use of conjunctiveness that there is some $s_k \in \{s_1, \ldots, s_n\}$ with $s_k \leq (s_1 \& \ldots \& s_n)$, and dominance yields $(s_1 \& \ldots \& s_n) \leq r$. Thus $p < s_k \leq (s_1 \& \ldots \& s_n) \leq r$, thus by transitivity $p < r$, i.e. $r \in \{s \mid p < s\}$.

For the second set, let $\{s \mid (\forall t)(s \vee t \not\doteq p)\} \vdash u$ and suppose that $u \notin \{s \mid (\forall t)(s \vee t \not\doteq p)\}$. Then there is some v with $(u \vee v) \doteq p$. Due to $\{s \mid (\forall t)(s \vee t \not\doteq p)\} \vdash u \vee v$ and compactness there is a finite set $\{s_1, \ldots, s_n\} \subseteq \{s \mid (\forall t)(s \vee t \not\doteq p)\}$ such that $\{s_1, \ldots, s_n\} \vdash u \vee v$. It follows truth-functionally that $\vdash u \vee v \leftrightarrow (s_1 \vee u \vee v) \& \ldots \& (s_n \vee u \vee v)$, and dominance yields $u \vee v \doteq (s_1 \vee u \vee v) \& \ldots \& (s_n \vee u \vee v)$. Due to conjunctiveness there is some s_k such that $(s_k \vee u \vee v) \leq (s_1 \vee u \vee v) \& \ldots \& (s_n \vee u \vee v)$. Dominance yields $(s_k \vee u \vee v) \doteq (s_1 \vee u \vee v) \& \ldots \& (s_n \vee u \vee v)$. We already have $(u \vee v) \doteq p$ and $u \vee v \doteq (s_1 \vee u \vee v) \& \ldots \& (s_n \vee u \vee v)$. Transitivity yields $(s_k \vee u \vee v) \doteq p$, contrary to $s_k \in \{s \mid (\forall t)(s \vee t \not\doteq p)\}$. We can conclude from this contradiction that $u \in \{s \mid (\forall t)(s \vee t \not\doteq p)\}$. □

Proof of Theorem 9.20 The proofs from (1) to (2), from (2) to (1) and from (1) to (3) are straightforward.

From (3) to (2): We have $K \div p = \{s \mid p \lessdot s\}$ and $K \div q = \{s \mid q \lessdot s\}$. Due to Observation 5.37 and Lemma 5.7 (p. 179), $\leqslant$ is transitive and complete. Thus either $p \leqslant q$ or $q \leqslant p$. Due to transitivity, $\{s \mid q \lessdot s\} \subseteq \{s \mid p \lessdot s\}$ in the former case and $\{s \mid p \lessdot s\} \subseteq \{s \mid q \lessdot s\}$ in the latter.

From (1) to (4): Due to Observation 9.10, since $\leqq$ is subtractively restricted, the domain of $\leqq$ is the outcome set of $\div$, which according to (3) is the set of sets $\{s \mid p \lessdot s\}$. Since $\leqq$ is shrinking and subtractively restricted, $X \leqq Y$ if and only if $Y \subseteq X$.

From (4) to (2): Due to the completeness of $\leqq$, it follows from (4) that either $K \div p \subseteq K \div q$ or $K \div q \subseteq K \div p$. □

Proofs for Section 9.4

Proof of Observation 9.23 For one direction, let $Y \not\subseteq Z$ for all $Y < Z$ and let

$A = \{p \mid (\exists Y)(Y < Z \text{ and } p \in Y \setminus Z)\}$

Then $Z = K \div_{\forall} A$.

For the other direction, suppose to the contrary that there is some Y with $Y < Z$ and $Y \subseteq Z$, and some A with $Z = K \div_{\forall} A$. Since K is the minimal element of $\leqq$ it follows that $Z \nvdash_{\exists} A$. Since $Y \subseteq Z$ we then also have $Y \nvdash_{\exists} A$, but that is incompatible with $Y < Z$ and $Z = K \div_{\forall} A$. This contradiction concludes the proof. □

Proof of Observation 9.25 Let $K = \text{Cn}(\{p\&q\})$ and let $\leqq_1$ and $\leqq_2$ be the belief set orderings that are completely characterized as follows:

$\text{Cn}(\{p\&q\}) <_1 \text{Cn}(\{p \vee q\}) <_1 \text{Cn}(\varnothing)$

$\text{Cn}(\{p\&q\}) <_2 \text{Cn}(\{p \vee q\}) <_2 \text{Cn}(\{p\}) <_2 \text{Cn}(\varnothing)$

It follows from Definition 9.22 and Observation 9.23 that $\overset{=}{\leqq}_1 = \overset{=}{\leqq}_2 = \leqq_1$ and consequently $\div_{\forall 1} = \div_{\forall 2}$ and $\overset{=}{\leqslant}_1 = \overset{=}{\leqslant}_2$. However, $K \circ_1 \{\mathfrak{B}p, \neg\mathfrak{B}q\} \neq K \circ_2 \{\mathfrak{B}p, \neg\mathfrak{B}q\}$, and $\neg\mathfrak{B}(p \vee q) \succeq_1 ((\neg\mathfrak{B}(p \vee q)) \vee (\mathfrak{B}p\&\neg\mathfrak{B}q))$, whereas $((\neg\mathfrak{B}(p \vee q)) \vee (\mathfrak{B}p\&\neg\mathfrak{B}q)) \succ_2 \neg\mathfrak{B}(p \vee q)$. □

Proof of Observation 9.27 *Part 1*: Left to the reader.

Part 2: Let $K \div_{\forall} A \nvdash_{\exists} A$ and $K \div_{\forall} B \nvdash_{\exists} B$. Then $K \div_{\forall} A \overset{=}{\leqq} K \div_{\forall} B$ holds if and only if $K \div_{\forall} A$ is $\overset{=}{\leqq}$-minimal among the elements X of the outcome set of $\div_{\forall}$ such that either $X \nvdash_{\exists} A$ or $X \nvdash_{\exists} B$. We have:

$X \nvdash_{\exists} A$ or $X \nvdash_{\exists} B$
iff not: $X \vdash_{\exists} A$ and $X \vdash_{\exists} B$
iff not: $(\exists p \in A)(X \vdash p)$ and $(\exists q \in B)(X \vdash q)$
iff not: $(\exists p \in A)(\exists q \in B)(X \vdash p\&q)$
iff not: $X \vdash_{\exists} \{p\&q \mid p \in A \text{ and } q \in B\}$
iff: $X \nvdash_{\exists} (A \overset{\&}{\times} B)$ □

Proof of Observation 9.28 Left to the reader. In Part 1, proceed as in Observation 9.14, using Observation 9.27, part 2. □

Proof of Observation 9.29 *Part 1* can be proved in the same way as part 1 of Observation 9.15.

Part 2: The feasibility of the construction follows from Observation 9.23. The rest of this proof is left to the reader. □

Proof of Observation 9.30 *Parts 1–3* are left to the reader. For *part 4*, let $K = \mathrm{Cn}(\{p\&q\})$ and let $\leqq_1$ and $\leqq_2$ be the belief set orderings that are completely specified as follows:

$\mathrm{Cn}(\{p\&q\}) <_1 \mathrm{Cn}(\{p\}) <_1 \mathrm{Cn}(\{q\}) <_1 \mathrm{Cn}(\{p \vee q\}) <_1 \mathrm{Cn}(\varnothing)$

$\mathrm{Cn}(\{p\&q\}) <_2 \mathrm{Cn}(\{p\}) <_2 \mathrm{Cn}(\{q\}) <_2 \mathrm{Cn}(\varnothing)$

It follows from Definitions 9.11 and 9.24 that $\bar{\bar{\leqq}}_1 \neq \bar{\bar{\leqq}}_2$ but $\bar{\leqq}_1 = \bar{\leqq}_2$.

Part 5: Use the same example as in part 4 and consider multiple revocation by $\{p, q\}$.

Part 6: Use the same example as in part 4 and note that $\{p, q\}\bar{\bar{\lessdot}}_1\{p \vee q\}$ but $\{p \vee q\}\bar{\bar{\leqslant\!\cdot}}_2\{p, q\}$. □

Proofs for Section 10.2

Proof of Theorem 10.2 Obtainable with small additions to the proof of Theorem 9.2.

Proof of Theorem 10.3 Obtainable with small additions to the proof of Theorem 9.6.

Proof of Observation 10.6 Let $\div$ have the outcome set $\mathbb{X}$, and let it be based via Definition 10.1 on a belief set ordering $\leqq$ such that $K \leqq X$ for all $X \in \mathbb{X}$. Suppose that $\div$ does not satisfy maximality. Then there are p and q such that $K \div p \subset K \div q$ and $p \notin K \div q$. Due to the antisymmetry of $\leqq$ (Lemma 5.2, p. 177) it follows from $K \div p \neq K \div q$ that either $K \div p < K \div q$ or $K \div q < K \div p$. The latter case is impossible since $K \div q \nvdash p$ and due to the construction, $K \div p$ cannot be preceded in $\leqq$ by a belief set not containing p. It follows that $K \div p < K \div q$, thus $K \div q \neq K$, thus $q \notin K \div q$. Since $K \div p \subset K \div q$ we then have $q \notin K \div p < K \div q$ which is impossible. This contradiction concludes the proof. □

Proofs for Section 10.3

Proof of Theorem 10.7 *From I to II*: Left to the reader.

From II to I: In order to show that $\div$ is maxichoice, let $q \in K \setminus (K \div p)$ and $r \in K$. We need to show that $(K \div p) \cup \{q\} \vdash r$.

It follows from recovery that $(K \div (p \vee q)) \cup \{p \vee q\} \vdash r$, thus $(K \div (p \vee q)) \cup \{q\} \vdash r$.

It also follows from recovery that $K \div (q \rightarrow p) \cup \{q \rightarrow p\} \vdash q$, thus due to closure, $q \in K \div (q \rightarrow p)$. Since $q \notin K \div p$ we can conclude that $K \div p \neq K \div (q \rightarrow p)$, and extensionality yields $K \div ((p \vee q)\&(q \rightarrow p)) \neq K \div (q \rightarrow p)$. Due to decomposition (shown in Theorem 10.3), either $K \div ((p \vee q)\&(q \rightarrow p)) = K \div (p \vee q)$ or $K \div ((p \vee q)\&(q \rightarrow p)) = K \div (q \rightarrow p)$. We have just seen that the latter is not the case, thus $K \div ((p \vee q)\&(q \rightarrow p)) = K \div (p \vee q)$, or equivalently $K \div p = K \div (p \vee q)$.

Combining this with $K \div (p \vee q) \cup \{q\} \vdash r$ we obtain $(K \div p) \cup \{q\} \vdash r$, as desired. Thus $\div$ is maxichoice.

It follows from decomposition that $\div$ also satisfies the postulate of conjunctive factoring (either $K \div (p\&q) = K \div p$, $K \div (p\&q) = K \div q$, or $K \div (p\&q) = (K \div p) \cap (K \div q)$). As shown in the AGM paper, in the presence of the basic postulates, conjunctive factoring implies that $\div$ is transitively relational. ([1], cf. [99, pp. 119–120]) This concludes the proof. □

Proof of Theorem 10.8 Suppose to the contrary that $\div$ is reconstructible as a centrolinear contraction. It then follows from Theorem 10.7 that it is maxichoice. However, it follows from Observation 2.4, part 3, that a maxichoice contraction cannot satisfy finite-based outcome, contrary to the conditions. □

Proofs for Section 10.4

Proof of Observation 10.10 Left to the reader.

Proof of Theorem 10.12 *From construction to postulates*: It follows from Theorem 10.2 that the first five postulates hold. For unique maximum, let it hold for all r that $K \div r \nvdash p$ if and only if $K \div r \subseteq K \div q$. Since all elements of $\mathbb{X}$ are contraction outcomes it follows that $\bigcup [\![\neg\mathfrak{B}p]\!] \subseteq K \div q$ and $K \div q \in [\![\neg\mathfrak{B}p]\!]$. Since $\widehat{C}$ is monomaximizing it follows that $\widehat{C}([\![\neg\mathfrak{B}p]\!]) = K \div q$, i.e. $K \div p = K \div q$.

From postulates to construction: We can construct $\widehat{C}$ as in the proof of Theorem 9.2, with the additional condition that if $\varnothing \neq [\![\Psi]\!] \neq [\![\neg\mathfrak{B}p]\!]$ for all p, then $\widehat{C}([\![\Psi]\!])$ is an inclusion-maximal element of $[\![\Psi]\!]$ (i.e. $\widehat{C}([\![\Psi]\!]) \in [\![\Psi]\!]$ and there is no X with $\widehat{C}([\![\Psi]\!]) \subset X \in [\![\Psi]\!]$). To verify the construction we can use the proofs of Theorems 9.2 and 10.2. It only remains to verify that $\widehat{C}$ is monomaximizing. If $\mathbb{Y} \neq [\![\neg\mathfrak{B}p]\!]$ for all p then this follows from the additional condition on $\widehat{C}$ given above (using Observation 10.10). In the main case, let $\mathbb{Y} = [\![\neg\mathfrak{B}p]\!]$ and $\bigcup \mathbb{Y} \subseteq Y \in \mathbb{Y}$. Since $\mathbb{X}$ has by construction only contraction outcomes as its elements, there is some q with $Y = K \div q$ and we then have:

$\bigcup [\![\neg\mathfrak{B}p]\!] \subseteq K \div q \in [\![\neg\mathfrak{B}p]\!]$
$\bigcup\{K \div r \mid K \div r \nvdash p\} \subseteq K \div q \in \{K \div r \mid K \div r \nvdash p\}$
$K \div r \subseteq K \div q$ for all $K \div r$ with $K \div r \nvdash p$, and $K \div q \nvdash p$
For all r: $K \div r \nvdash p$ iff $K \div r \subseteq K \div q$
$K \div p = K \div q$ (unique maximum)
$\widehat{C}([\![\neg\mathfrak{B}p]\!]) = K \div q$
$\widehat{C}(\mathbb{Y}) = Y$ □

Proof for Section 10.5

Proof of Theorem 10.14 *From construction to postulates*: This is left to the reader with the exception of preservativity. For preservativity, let $K \div q \nsubseteq K \div p$ and $K \div q \nvdash p$. First suppose that $[\![\neg\mathfrak{B}p]\!]$ has only one inclusion-maximal element. It then follows from $K \div q \nvdash p$ and Definition 10.13 that $K \div q \subseteq K \div p$, contrary to the conditions. Thus $[\![\neg\mathfrak{B}p]\!]$ has more than one inclusion-maximal element. Suppose that $K \div q$ is a subset of all of them. Then it is a subset of all elements of $C([\![\neg\mathfrak{B}p]\!])$, and Definition 10.13 yields $K \div q \subseteq K \div p$, contrary to the assumption. Thus there is some $K \div r$ that is an inclusion-maximal element of $[\![\neg\mathfrak{B}p]\!]$ and such that

$K \div q \not\subseteq K \div r$. Let $(K \div q) \cup (K \div r) \subseteq K \div s$. Then $K \div r \subset K \div s$, and since $K \div r$ is an inclusion-maximal element of $[\![\neg\mathfrak{B}p]\!]$ it follows that $K \div s \vdash p$.

From postulates to construction: It follows from the first five postulates and Theorem 10.2 that there is a monoselective choice function $\widehat{C}'$ such that $\div$ is the monoselective contraction based on $\widehat{C}'$. Now define C as a function on subsets of $\mathbb{X}$ such that for all p, $C([\![\neg\mathfrak{B}p]\!])$ is equal to the set of inclusion-maximal elements of $[\![\neg\mathfrak{B}p]\!]$ that contain $\widehat{C}'([\![\neg\mathfrak{B}p]\!])$. To show that C is a choice function, it is sufficient to note that if $[\![\neg\mathfrak{B}p]\!] \neq \varnothing$, then $\widehat{C}'([\![\neg\mathfrak{B}p]\!])$ is an element of $\mathbb{X}$ that does not imply p and consequently, $C([\![\neg\mathfrak{B}p]\!]) \neq \varnothing$. That C is maximizing follows directly. It remains to show that for each p, the perimaximal contraction based on C yields $K \div p$ as its outcome. This follows directly if all elements of $\mathbb{X}$ contain p, so we only need to prove the case when this is not so.

We need to prove the following intermediate result:

(X) For all $Z \in \mathbb{X}$: $Z \subseteq \bigcap C([\![\neg\mathfrak{B}p]\!])$ if and only if $Z \subseteq \widehat{C}'([\![\neg\mathfrak{B}p]\!])$.

The right-to-left direction follows directly from the definition of C. For the left-to-right direction, suppose to the contrary that there is some $K \div q$ such that $K \div q \subseteq \bigcap C([\![\neg\mathfrak{B}p]\!])$ and $K \div q \not\subseteq \widehat{C}'([\![\neg\mathfrak{B}p]\!])$. Since $\widehat{C}'([\![\neg\mathfrak{B}p]\!]) = K \div p$ by definition, we then have $K \div q \not\subseteq K \div p$. It follows by preservativity that there is some r such that $K \div p \subseteq K \div r \nvdash p$ and that for all s, if $(K \div q) \cup (K \div r) \subseteq K \div s$ then $K \div s \vdash p$. Let r' be such that $K \div r \subseteq K \div r'$ and that $K \div r'$ is inclusion-maximal in $[\![\neg\mathfrak{B}p]\!]$. Since $K \div p \subseteq K \div r'$ we then have $K \div r' \in C([\![\neg\mathfrak{B}p]\!])$, thus $\bigcap C([\![\neg\mathfrak{B}p]\!]) \subseteq K \div r'$. It follows from preservativity that $K \div q \not\subseteq K \div r'$. Consequently, $K \div q \not\subseteq \bigcap C([\![\neg\mathfrak{B}p]\!])$, contrary to the assumption. This concludes the proof of (X).

It follows from Definition 10.13 that the outcome of the C-based perimaximal contraction is:

$$\bigcup\{Z \in \mathbb{X} \mid Z \subseteq \bigcap C([\![\neg\mathfrak{B}p]\!])\}$$
$$= \bigcup\{Z \in \mathbb{X} \mid Z \subseteq \widehat{C}'([\![\neg\mathfrak{B}p]\!])\} \qquad \text{(result X above)}$$
$$= \bigcup\{Z \in \mathbb{X} \mid Z \subseteq K \div p\}$$
$$= K \div p$$

This concludes the proof. □

Proof of Observation 10.15 *Part 1*: Let $K \div q \nvdash p$ and $K \div q \not\subseteq K \div p$. It follows from persistence that $K \div p \nvdash p$. Furthermore, it follows from $K \div q \not\subseteq K \div p$ that if $(K \div q) \cup (K \div p) \subseteq K \div s$, then $K \div p \subset K \div s$, and maximality yields $K \div s \vdash p$. We can therefore prove preservativity by setting $r = p$.

Part 2: Let preservativity and persistence hold. In order to prove that unique maximum holds, we make the following assumption:

$$\text{It holds for all } t \text{ that } K \div t \nvdash p \text{ if and only if } K \div t \subseteq K \div q. \qquad \text{(A)}$$

It follows from (A) that $K \div q \nvdash p$. Persistence yields $K \div p \nvdash p$, and with one more use of (A) we obtain $K \div p \subseteq K \div q$.

Suppose that $K \div p \subset K \div q$. We then have $K \div q \nvdash p$ and $K \div q \not\subseteq K \div p$. It follows from preservativity that there is some r such that $K \div p \subseteq K \div r \nvdash p$ and:

It holds for all s that if $(K \div q) \cup (K \div r) \subseteq K \div s$, then $K \div s \vdash p$. (S)

It follows from $K \div r \nvdash p$ and (A) that $K \div r \subseteq K \div q$. By setting $s = q$ we can show that (S) does not hold. It follows from this contradiction that $K \div p \subset K \div q$ does not hold. Since $K \div p \subseteq K \div q$ we can conclude that $K \div p = K \div q$. □

Proofs for Section 10.6

Proof of Observation 10.16 Let $B = \mathcal{L} \setminus X$. □

Proof of Observation 10.19 Left to the reader. □

Bibliography

References

1 Alchourrón, C. E., Gärdenfors, P., & Makinson, D. (1985). On the logic of theory change: Partial meet contraction and revision functions. *Journal of Symbolic Logic*, *50*(2), 510–530.

2 Alchourrón, C. E., & Makinson, D. (1981). Hierarchies of regulations and their logic. In R. Hilpinen (Ed.), *New Studies in Deontic Logic* (pp. 125–148). Dordrecht: D. Reidel Publishing Company.

3 Alchourrón, C. E., & Makinson, D. (1982). On the logic of theory change: Contraction functions and their associated revision functions. *Theoria*, *48*(1), 14–37.

4 Alchourrón, C. E., & Makinson, D. (1985). On the logic of theory change: Safe contraction. *Studia Logica*, *44*(4), 405–422.

5 Areces, C., & Becher, V. (2001). Iterable AGM functions. In H. Rott & M.-A. Williams (Eds.), *Frontiers in Belief Revision* (pp. 261–277). Dordrecht: Kluwer Academic Publishers.

6 Arlo-Costa, H. (2014). The logic of conditionals. In E. N. Zalta (Ed.), *Stanford Encyclopedia of Philosophy*. Center for the Study of Language and Information (CSLI), Stanford University. http://plato.stanford.edu/archives/sum2014/entries/logic-conditionals.

7 Arrow, K. J. (1948). The possibility of a universal social welfare function. Mimeographed. http://130.154.3.14/content/dam/rand/pubs/papers/2013/P41.pdf.

8 Audi, R. (2001). Doxastic voluntarism and the ethics of belief. In M. Steup (Ed.), *Knowledge, Truth, and Duty. Essays on Epistemic Justification, Responsibility, and Virtue* (pp. 93–111). Oxford: Oxford University Press.

9 Ayers, M. R. (1965). Counterfactuals and subjunctive conditionals. *Mind*, *74*, 347–364.

10 Bellot, D., Godefroid, C., Han, P., Prost, J. P., Schlechta, K., & Wurbel, E. (1997). A semantical approach to the concept of screened revision. *Theoria*, *63*(1–2), 24–33.

11 Bennett, J. (1974). Counterfactuals and possible worlds. *Canadian Journal of Philosophy*, *4*, 381–402.

12 Bezzazi, H., Makinson, D., & Pino Pérez, R. (1997). Beyond rational monotony: Some strong non-Horn rules for nonmonotonic inference relations. *Journal of Logic and Computation*, *7*(5), 605–631.

13 Billington, D., Antoniou, G., Governatori, G., & Maher, M. (1999). Revising nonmonotonic theories: The case of defeasible logic. In *KI-99: Advances in Artificial Intelligence* (pp. 101–112). Berlin: Springer.

14 Bonnano, G. (2009). Revealed preference, iterated belief revision and dynamic games. In G. Bonanno, J. Delgrande, & H. Rott (Eds.), *Information Processing, Rational Belief Change and Social Interaction*. Schloss Dagstuhl – Leibniz Center for Informatics.

S.O. Hansson, *Descriptor Revision*, Trends in Logic 46,
DOI 10.1007/978-3-319-53061-1

15 Booth, R. (2001). A negotiation-style framework for non-prioritised revision. In *Proceedings of the 8th Conference on Theoretical Aspects of Rationality and Knowledge* (pp. 137–150). Morgan Kaufmann Publishers Inc.
16 Booth, R., Chopra, S., Ghose, A., & Meyer, T. (2005). Belief liberation (and retraction). *Studia Logica*, *79*(1), 47–72.
17 Booth, R., & Meyer, T. A. (2006). Admissible and restrained revision. *Journal of Artificial Intelligence Research (JAIR)*, *26*, 127–151.
18 Booth, R., Meyer, T. A., & Wong, K.-S. (2006). A bad day surfing is better than a good day working: How to revise a total preorder. In P. Doherty, J. Mylopoulos, & C. Welty (Eds.), *Proceedings of KR2006, Tenth International Conference on the Principles of Knowledge Representation and Reasoning* (pp. 230–238). AAAI Press.
19 Boutilier, C. (1990). Conditional logics of normality as modal systems. In *AAAI-90: Proceedings. National Conference on Artificiell Intelligence, American Association for Artificial Intelligence* (pp. 594–599). AAAI Press.
20 Boutilier, C. (1993). Revision sequences and nested conditionals. In *Proceedings of the Thirteenth International Joint Conference on Artificial Intelligence (IJCAI'93)* (pp. 519–531). AAAI Press.
21 Boutilier, C. (1996). Iterated revision and minimal change of conditional beliefs. *Journal of Philosophical Logic*, *25*(3), 263–305.
22 Boyle, M. (2009). Active belief. *Canadian Journal of Philosophy*, *39*(supl), 119–147.
23 Bradley, R. (2007). A defence of the Ramsey test. *Mind*, *116*(461), 1–21.
24 Bradley, R. (2012). Restricting preservation: A response to Hill. *Mind*, *121*(481), 147–159.
25 Brewka, G. (1991). Belief revision in a framework for default reasoning. In A. Fuhrmann & M. Morreau (Eds.), *The Logic of Theory Change* (pp. 206–222). Berlin: Springer.
26 Burgess, J. P. (1981). Quick completeness proofs for some logics of conditionals. *Notre Dame Journal of Formal Logic*, *22*(1), 76–84.
27 Burgess, J. P. (1999). Which modal logic is the right one? *Notre Dame Journal of Formal Logic*, *40*, 81–93.
28 Cantwell, J. (1997). On the logic of small changes in hypertheories. *Theoria*, *63*(1–2), 54–89.
29 Cantwell, J. (1999). Some logics of iterated belief change. *Studia Logica*, *63*(1), 49–84.
30 Chandler, J. (2013). Transmission failure, AGM style. *Erkenntnis*, *78*(2), 383–398.
31 Chisholm, R. M. (1946). The contrary-to-fact conditional. *Mind*, *55*, 289–307.
32 Chopra, S., Ghose, A., & Meyer, T. (2003). Non-prioritized ranked belief change. *Journal of Philosophical Logic*, *32*(4), 417–443.
33 Chopra, S., Ghose, A., Meyer, T., & Wong, K.-S. (2008). Iterated belief change and the recovery axiom. *Journal of Philosophical Logic*, *37*(5), 501–520.
34 Cross, C. B., & Thomason, R. H. (1992). Conditionals and knowledge-base update. In P. Gärdenfors (Ed.), *Belief Revision* (pp. 247–275). Cambridge: Cambridge University Press.
35 Dalal, M. (1988). Investigations into a theory of knowledge base revision: Preliminary report. In *Proceedings of the Seventh National Conference on Artificial Intelligence (AAAI-88)* (Vol. 2, pp. 475–479), Menlo Park, CA: AAAI Press.
36 Darwiche, A., & Pearl, J. (1997). On the logic of iterated belief revision. *Artificial Intelligence*, *89*, 1–29.
37 de Rijke, M. (1994). Meeting some neighbours. In J. van Eijck & A. Visser (Eds.), *Logic and Information Flow* (pp. 170–195). Cambridge: MIT Press.
38 Dechter, R., & Pearl, J. (1992). Structure identification in relational data. *Artificial Intelligence*, *58*, 237–270.
39 Del Val, A. (1997). Non monotonic reasoning and belief revision: Syntactic, semantic, foundational and coherence approaches. *Journal of Applied Non-Classical Logics*, *7*(1–2), 213–240.
40 Delgrande, J. P. (1988). An approach to default reasoning based on a first-order conditional logic: Revised report. *Artificial Intelligence*, *36*, 63–90.
41 Delgrande, J. P., Dubois, D., & Lang, J. (2006). Iterated revision as prioritized merging. In P. Doherty, J. Mylopoulos, & C. Welty (Eds.), *Tenth International Conference on Principles of Knowledge Representation and Reasoning* (pp. 210–220). Menlo Park, California: AAAI Press.

42 Douven, I. (2002). A new solution to the paradoxes of rational acceptability. *British Journal for the Philosophy of Science*, *53*(3), 391–410.
43 Douven, I. (2003). The preface paradox revisited. *Erkenntnis*, *59*(3), 389–420.
44 Doyle, J. (1979). A truth maintenance system. *Artificial Intelligence*, *12*(3), 231–272.
45 Edgington, D. (2007). On conditionals. In D. M. Gabbay & F. Guenthner (Eds.), *Handbook of philosophical logic* (2nd ed., Vol. 14, pp. 127–221). Dordrecht: Springer.
46 Fagin, R., Ullman, J. D., & Vardi, M. Y. (1983). On the semantics of updates in databases. In *Proceedings of the 2nd ACM SIGACT-SIGMOD Symposium on Principles of Database Systems* (pp. 352–365). ACM.
47 Falappa, M. A., Kern-Isberner, G., Reis, M. D. L., & Simari, G. R. (2012). Prioritized and non-prioritized multiple change on belief bases. *Journal of Philosophical Logic*, *41*(1), 77–113.
48 Falappa, M. A., García, A. J., Kern-Isberner, G., & Simari, G. R. (2013). Stratified belief bases revision with argumentative inference. *Journal of Philosophical Logic*, *42*(1), 161–193.
49 Fermé, E. (2000). Irrevocable belief revision and epistemic entrenchment. *Logic Journal of the IGPL*, *8*(5), 645–652.
50 Fermé, E., & Hansson, S. O. (1999). Selective revision. *Studia Logica*, *63*(3), 331–342.
51 Fermé, E., & Hansson, S. O. (2001). Shielded contraction. In H. Rott & M.-A. Williams (Eds.), *Frontiers in Belief Revision* (pp. 85–107). Dordrecht: Kluwer Academic Publishers.
52 Fermé, E., & Hansson, S. O. (2011). AGM 25 years. *Journal of Philosophical Logic*, *40*(2), 295–331.
53 Fermé, E., & Rodriguez, R. (1998). Semi-contraction: Axioms and construction. *Notre Dame Journal of Formal Logic*, *39*(3), 332–345.
54 Fermé, E., & Rott, H. (2004). Revision by comparison. *Artificial Intelligence*, *157*(1), 5–47.
55 Fermé, E., Saez, K., & Sanz, P. (2003). Multiple kernel contraction. *Studia Logica*, *73*(2), 183–195.
56 Fishburn, P. C. (1970). *Utility Theory for Decision Making*. New York: Wiley.
57 Freund, M., & Lehmann, D. (1994). Nonmonotonic reasoning: From finitary relations to infinitary inference operations. *Studia Logica*, *53*(2), 161–201.
58 Friedman, N., & Halpern, J. Y. (1999). Belief revision: A critique. *Journal of Logic, Language and Information*, *8*(4), 401–420.
59 Fritz, P. (2014). What is the correct logic of necessity, actuality and apriority? *Review of Symbolic Logic*, *7*, 385–414.
60 Fuhrmann, A. (1988). *Relevant Logics, Modal Logics and Theory Change*. PhD thesis, The Australian National University, Canberra.
61 Fuhrmann, A. (1989). Reflective modalities and theory change. *Synthese*, *81*(1), 115–134.
62 Fuhrmann, A. (1991). On the modal logic of theory change. In A. Fuhrmann & M. Morreau (Eds.), *The Logic of Theory Change* (pp. 259–281). Berlin: Springer.
63 Fuhrmann, A. (1991). Theory contraction through base contraction. *Journal of Philosophical Logic*, *20*(2), 175–203.
64 Fuhrmann, A., & Hansson, S. O. (1994). A survey of multiple contractions. *Journal of Logic, Language and Information*, *3*(1), 39–75.
65 Fuhrmann, A., & Levi, I. (1994). Undercutting and the Ramsey test for conditionals. *Synthese*, *101*(2), 157–169.
66 Galliers, J. R. (1992). Autonomous belief revision and communication. In P. Gärdenfors (Ed.), *Belief Revision* (pp. 220–246). Cambridge: Cambridge University Press.
67 Gärdenfors, P. (1982). Rules for rational changes of belief. In T. Pauli (Ed.), *Philosophical Essays Dedicated to Lennart Åqvist on His Fiftieth Birthday* (pp. 88–101). Department of Philosophy, Uppsala University.
68 Gärdenfors, P. (1986). Belief revisions and the Ramsey test for conditionals. *Philosophical Review*, *95*, 81–93.
69 Gärdenfors, P. (1988). *Knowledge in Flux: Modeling the Dynamics of Epistemic States*. Cambridge: MIT Press.

70 Gärdenfors, P. (2011). Notes on the history of ideas behind AGM. *Journal of Philosophical Logic*, *40*(2), 115–120.
71 Gärdenfors, P., & Makinson, D. (1988). Revisions of knowledge systems using epistemic entrenchment. In *Proceedings of the 2nd Conference on Theoretical Aspects of Reasoning about Knowledge* (pp. 83–95). Los Angeles: Morgan Kaufmann Publishers.
72 Gärdenfors, P., & Makinson, D. (1994). Nonmonotonic inference based on expectations. *Artificial Intelligence*, *65*, 197–245.
73 Gärdenfors, P., & Rott, H. (1995). Belief revision. In D. M. Gabbay, C. J. Hogger, & J. A. Robinson (Eds.), *Handbook of Logic in Artificial Intelligence and Logic Programming, Vol. 4, Epistemic and Temporal Reasoning* (pp. 35–132). Oxford: Oxford University Press.
74 Garson, J. W. (2013). *Modal Logic for Philosophers* (2nd ed.). Cambridge: Cambridge University Press.
75 Ghose, A., & Goebel, R. (1998). Belief states as default theories: Studies in non-prioritized belief change. In *Proceedings of the 13th European Conference on Artificial Intelligence, Brighton, U.K. (ECAI'98)* (pp. 8–12).
76 Ginsberg, M. L. (1986). Counterfactuals. *Artificial Intelligence*, *30*, 35–79.
77 Glaister, S. M. (1998). Symmetry and belief revision. *Erkenntnis*, *49*(1), 21–56.
78 Glaister, S. M. (2000). Recovery recovered. *Journal of Philosophical Logic*, *29*(2), 171–206.
79 Goldstick, D. (1978). The truth-conditions of counterfactual conditional sentences. *Mind*, *87*, 1–21.
80 Grove, A. (1988). Two modellings for theory change. *Journal of Philosophical Logic*, *17*(2), 157–170.
81 Hansson, S. O. (1989). New operators for theory change. *Theoria*, *55*(2), 114–132.
82 Hansson, S. O. (1991). Belief contraction without recovery. *Studia Logica*, *50*(2), 251–260.
83 Hansson, S. O. (1992). A dyadic representation of belief. In P. Gärdenfors (Ed.), *Belief Revision* (pp. 89–121). Cambridge: Cambridge University Press.
84 Hansson, S. O. (1992). In defense of base contraction. *Synthese*, *91*(3), 239–245.
85 Hansson, S. O. (1992). In defense of the Ramsey test. *The Journal of Philosophy*, *89*, 522–540.
86 Hansson, S. O. (1992). Similarity semantics and minimal changes of belief. *Erkenntnis*, *37*(3), 401–429.
87 Hansson, S. O. (1993). Changes of disjunctively closed bases. *Journal of Logic, Language, and Information*, *2*, 255–284.
88 Hansson, S. O. (1993). Reversing the Levi identity. *Journal of Philosophical Logic*, *22*(6), 637–669.
89 Hansson, S. O. (1993). Theory contraction and base contraction unified. *The Journal of Symbolic Logic*, *58*(2), 602–625.
90 Hansson, S. O. (1994). Kernel contraction. *The Journal of Symbolic Logic*, *59*(3), 845–859.
91 Hansson, S. O. (1994). Taking belief bases seriously. In D. Prawitz & D. Westerståhl (Eds.), *Logic and Philosophy of Science in Uppsala* (pp. 13–28). Berlin: Springer.
92 Hansson, S. O. (1995). The emperor's new clothes: Some recurring problems in the formal analysis of counterfactuals. In G. Crocco, L. Fariñas del Cerro, & A. Herzig (Eds.), *Conditionals: From Philosophy to Computer Science* (pp. 13–31). Oxford: Clarendon Press.
93 Hansson, S. O. (1996). Hidden structures of belief. In A. Fuhrmann & H. Rott (Eds.), *Logic, Actions and Information* (pp. 79–100). Berlin: de Gruyter.
94 Hansson, S. O. (1996). Knowledge-level analysis of belief base operations. *Artificial Intelligence*, *82*(1), 215–235.
95 Hansson, S. O. (1997). Semi-revision. *Journal of Applied Non-Classical Logics*, *7*(1–2), 151–175.
96 Hansson, S. O. (1997). What's new isn't always best. *Theoria*, *63*(1–2), 1–13.
97 Hansson, S. O. (1999). Recovery and epistemic residue. *Journal of Logic, Language and Information*, *8*(4), 421–428.
98 Hansson, S. O. (1999). A survey of non-prioritized belief revision. *Erkenntnis*, *50*(2–3), 413–427.

99 Hansson, S. O. (1999). *A Textbook of Belief Dynamics: Theory Change and Database Updating*. Dordrecht: Kluwer.
100 Hansson, S. O. (2000). Formalization in philosophy. *Bulletin of Symbolic Logic*, *6*, 162–175.
101 Hansson, S. O. (2002). Preference logic. In D. Gabbay & F. Guenthner (Eds.), *Handbook of Philosophical Logic* (2nd ed., Vol. 4, pp. 319–393). Dordrecht: Kluwer.
102 Hansson, S. O. (2002). The rôle of language in belief revision. *Studia Logica*, *70*, 5–21.
103 Hansson, S. O. (2003). Ten philosophical problems in belief revision. *Journal of Logic and Computation*, *13*, 37–49.
104 Hansson, S. O. (2004). Belief revision from an epistemological point of view. In I. Niiniluoto, M. Sintonen, & J. Wolenski (Eds.), *Handbook of Epistemology* (pp. 255–279). Dordrecht: Kluwer.
105 Hansson, S. O. (2006). Coherence in epistemology and belief revision. *Philosophical Studies*, *128*, 93–108.
106 Hansson, S. O. (2006). In praise of full meet contraction. *Análysis Filosófico*, *26*, 134–146.
107 Hansson, S. O. (2006). Logic of belief revision. In E. N. Zalta (Ed.), *The Stanford Encyclopedia of Philosophy*. Center for the Study of Language and Information (CSLI), Stanford University. http://plato.stanford.edu/entries/logic-belief-revision.
108 Hansson, S. O. (2007). Contraction based on sentential selection. *Journal of Logic and Computation*, *17*(3), 479–498.
109 Hansson, S. O. (2008). Specified meet contraction. *Erkenntnis*, *69*(1), 31–54.
110 Hansson, S. O. (2009). Replacement—a Sheffer stroke for belief change. *Journal of Philosophical Logic*, *38*(2), 127–149.
111 Hansson, S. O. (2010). Multiple and iterated contraction reduced to single-step single-sentence contraction. *Synthese*, *173*(2), 153–177.
112 Hansson, S. O. (2011). Changing the scientific corpus. In E. J. Olsson & S. Enqvist (Eds.), *Belief Revision Meets Philosophy of Science* (pp. 43–58). Dordrecht: Springer.
113 Hansson, S. O. (2012). Eradication. *Journal of Applied Logic*, *10*, 75–84.
114 Hansson, S. O. (2012). Finite contractions on infinite belief sets. *Studia Logica*, *100*(5), 907–920.
115 Hansson, S. O. (2012). Global and iterated contraction and revision: An exploration of uniform and semi-uniform approaches. *Journal of Philosophical Logic*, *41*(1), 143–172.
116 Hansson, S. O. (2013). Blockage contraction. *Journal of Philosophical Logic*, *42*(2), 415–442.
117 Hansson, S. O. (2013). Bootstrap contraction. *Studia Logica*, *101*(5), 1013–1029.
118 Hansson, S. O. (2013). Cognitive realism in belief revision. In E. L. Fermé, D. M. Gabbay, & G. R. Simari (Eds.), *Trends in Belief Revision and Argumentation Dynamics* (pp. 57–74). Norcross: College Publications.
119 Hansson, S. O. (2013). Contraction, revision, expansion: Representing belief change operations. In R. Trypuz (Ed.), *Krister Segerberg on Logic of Actions* (pp. 135–151). Berlin: Springer.
120 Hansson, S. O. (2013). Maximal and perimaximal contraction. *Synthese*, *190*(16), 3325–3348.
121 Hansson, S. O. (2013). Outcome level analysis of belief contraction. *Review of Symbolic Logic*, *6*(2), 183–204.
122 Hansson, S. O. (2013). Repertoire contraction. *Journal of Logic, Language and Information*, *22*(1), 1–21.
123 Hansson, S. O. (2014). Decomposition of multiple contraction: Possibility and impossibility results. *Logic Journal of the IGPL*, *22*, 696–710.
124 Hansson, S. O. (2014). Descriptor revision. *Studia Logica*, *102*(5), 955–980.
125 Hansson, S. O. (2014). How context dependent is scientific knowledge? In F. Lihoreau & M. Rebuschi (Eds.), *Epistemology, Context, and Formalism* (pp. 127–140). Berlin: Springer.
126 Hansson, S. O. (2014). Relations of epistemic proximity for belief change. *Artificial Intelligence*, *217*, 76–91.
127 Hansson, S. O. (2015). Choice and change in an epistemic context. *South American Journal of Logic*, *1*, 365–385.

128 Hansson, S. O. (2015). A monoselective presentation of AGM revision. *Studia Logica, 103*, 1019–1033.
129 Hansson, S. O. (2016). Alternatives to the Ramsey test. In C. Beierle, G. Brewka, & M. Thimm (Eds.), *Computational Models of Rationality. Essays Dedicated to Gabriele Kern-Isberner on the Occasion of Her 60th Birthday* (pp. 84–97). Norcross: College Publications.
130 Hansson, S. O. (2016). Blockage revision. *Journal of Logic, Language, and Information, 25*, 37–50.
131 Hansson, S. O. (2016). The co-occurrence test for non-monotonic inference. *Artificial Intelligence, 234*, 190–195.
132 Hansson, S. O. (2016). Iterated descriptor revision and the logic of Ramsey test conditionals. *Journal of Philosophical Logic, 45*, 429–450.
133 Hansson, S. O. (2016). Reconstruction of contraction operators. *Erkenntnis, 81*, 185–199.
134 Hansson, S. O. (2017). AGM contraction is not reconstructible as a descriptor operation. *Journal of Logic and Computation, 27*, 1133–1141.
135 Hansson, S. O. (2017, in press). Back to basics: Belief revision through direct selection. *Studia Logica.*
136 Hansson, S. O. (2017). How values can influence science without threatening its integrity. In H. Leitgeb, I. Niiniluoto, P. Seppälä, & E. Sober (Eds.), *Logic, Methodology and Philosophy of Science – Proceedings of the 15th International Congress (CLMPS 2015)* (pp. 207–221). College Publications.
137 Hansson, S. O., Fermé, E., Cantwell, J., & Falappa, M. A. (2001). Credibility-limited revision. *Journal of Symbolic Logic, 66*(4), 1581–1596.
138 Hansson, S. O., & Gärdenfors, P. (2014). David Makinson and the extension of classical logic. In S. O. Hansson (Ed.), *David Makinson on Classical Methods for Non-Classical Problems* (pp. 11–18). Berlin: Springer.
139 Hansson, S. O., & Wassermann, R. (2002). Local change. *Studia Logica, 70*(1), 49–76.
140 Harsanyi, J. C. (1977). On the rationale of the Bayesian approach: Comments on professor Watkins's paper. In R. E. Butts & J. Hintikka (Eds.), *Proceedings of the Fifth International Congress of Logic, Methodology and Philosophy of Science, London, Ontario, Canada 1975. Part 2, Foundational Problems in the Special Sciences*. Dordrecht: Reidel.
141 Hieronymi, P. (2006). Controlling attitudes. *Pacific Philosophical Quarterly, 87*(1), 45–74.
142 Hieronymi, P. (2009). Believing at will. *Canadian Journal of Philosophy, 39*(supl), 149–187.
143 Hild, M., & Spohn, W. (2008). The measurement of ranks and the laws of iterated contraction. *Artificial Intelligence, 172*(10), 1195–1218.
144 Hill, B. (2012). Defending the Ramsey test: What is wrong with preservation? *Mind, 121*(481), 131–146.
145 Jech, T. J. (2008 [1973]). *The Axiom of Choice*. Mineola, NY: Dover Publications.
146 Jin, Y., & Thielscher, M. (2007). Iterated belief revision, revised. *Artificial Intelligence, 171*(1), 1–18.
147 Kálmar, L., (1948). Andrzej Mostowski. Axiom of choice for finite sets. *Journal of Symbolic Logic, 13*(1), 45–46.
148 Kautz, H. A., Kearns, M. J., & Selman, B. (2007). Reasoning with characteristic models. In *Proceedings of the Eleventh National Conference on Artificial Intelligence* (pp. 34–39). AAAI Press.
149 Kern-Isberner, G. (2008). Linking iterated belief change operations to nonmonotonic reasoning. In G. Brewka & J. Lang (Eds.), *Proceedings of the 11th International Conference on Knowledge Representation and Reasoning (KR2008)* (pp. 166–176).
150 Kim, B. (2015). This paper surely contains some errors. *Philosophical Studies, 172*, 1013–1029.
151 Konieczny, S., & Pino Pérez, R. (2000). A framework for iterated revision. *Journal of Applied Non-Classical Logics, 10*(3–4), 339–367.
152 Kourousias, G., & Makinson, D. (2007). Parallel interpolation, splitting, and relevance in belief change. *Journal of Symbolic Logic, 72*, 994–1002.

153 Kraus, S., Lehmann, D., & Magidor, M. (1990). Nonmonotonic reasoning, preferential models and cumulative logics. *Artificial Intelligence*, *44*(1), 167–207.
154 Lehmann, D. (1995). Belief revision, revised. In *Proceedings of the 14th International Joint Conference on Artificial intelligence (IJCAI-95)* (Vol. 2, pp. 1534–1540). Morgan Kaufmann Publishers Inc.
155 Lehmann, D., Magidor, M., & Schlechta, K. (2001). Distance semantics for belief revision. *Journal of Symbolic Logic*, *66*(1), 295–317.
156 Leitgeb, H. (2014). A way out of the preface paradox? *Analysis*, *74*(1), 11–15.
157 Leitgeb, H., & Segerberg, K. (2007). Dynamic doxastic logic: Why, how, and where to? *Synthese*, *155*, 167–190.
158 Levesque, H. J. (1990). All I know: A study in autoepistemic logic. *Artificial Intelligence*, *42*, 263–309.
159 Levi, I. (1977). Four types of ignorance. *Social Research*, *44*, 745–756.
160 Levi, I. (1977). Subjunctives, dispositions and chances. *Synthese*, *34*, 423–455.
161 Levi, I. (1979). Serious possibility. In E. Saarinen, R. Hilpinen, I. Niiniluoto, & M. P. Hintikka (Eds.), *Essays in Honour of Jaakko Hintikka* (pp. 219–236). Dordrecht: Reidel.
162 Levi, I. (1980). *The Enterprise of Knowledge: An Essay on Knowledge, Credal Probability, and Chance*. Cambridge, MA: MIT Press.
163 Levi, I. (1988). Iteration of conditionals and the Ramsey test. *Synthese*, *76*(1), 49–81.
164 Levi, I. (1991). *The Fixation of Belief and its Undoing: Changing Beliefs Through Inquiry*. Cambridge: Cambridge University Press.
165 Levi, I. (2003). Counterexamples to recovery and the filtering condition. *Studia Logica*, *73*(2), 209–218.
166 Lewis, D. (1979). Counterfactual dependence and time's arrow. *Noûs*, *13*, 455–476.
167 Lewis, D. K. (1973). *Counterfactuals*. Oxford: Blackwell.
168 Li, J. (1998). A note on partial meet package contraction. *Journal of Logic, Language and Information*, *7*(2), 139–142.
169 Lindström, S., & Rabinowicz, W. (1991). Epistemic entrenchment with incomparabilities and relational belief revision. In A. Fuhrmann & M. Morreau (Eds.), *The Logic of Theory Change* (pp. 93–126). Berlin: Springer.
170 Lindström, S., & Rabinowicz, W. (1992). Belief revision, epistemic conditionals and the Ramsey test. *Synthese*, *91*(3), 195–237.
171 Lindström, S., & Rabinowicz, W. (1999). Belief change for introspective agents. In B. Hansson & N.-E. Sahlin (Eds.), *Spinning Ideas, Electronic Essays Dedicated to Peter Gärdenfors on His Fiftieth Birthday*. Department of Philosophy, Lund University. http://www.lucs.lu.se/spinning.
172 Lindström, S., & Rabinowicz, W. (1999). DDL unlimited: Dynamic doxastic logic for introspective agents. *Erkenntnis*, *50*(2), 353–385.
173 Makinson, D. (1965). The paradox of the preface. *Analysis*, *25*, 205–207.
174 Makinson, D. (1985). How to give it up: A survey of some formal aspects of the logic of theory change. *Synthese*, *62*(3), 347–363.
175 Makinson, D. (1987). On the status of the postulate of recovery in the logic of theory change. *Journal of Philosophical Logic*, *16*(4), 383–394.
176 Makinson, D. (1990). The Gärdenfors impossibility theorem in non-monotonic contexts. *Studia Logica*, *49*(1), 1–6.
177 Makinson, D. (1994). General patterns in nonmonotonic reasoning. In D. M. Gabbay, C. J. Hogger, & J. A. Robinson (Eds.), *Handbook of Logic in Artificial Intelligence and Logic Programming. Vol. 3. Nonmonotonic Reasoning and Uncertain Reasoning* (pp. 35–110). Oxford: Clarendon.
178 Makinson, D. (1997). On the force of some apparent counterexamples to recovery. In E. Garzón Valdés (Ed.), *Normative Systems in Legal and Moral Theory, Festschrift for Carlos Alchourrón and Eugenio Bulygin* (pp. 475–481). Berlin: Duncker and Humblot.
179 Makinson, D. (1997). Screened revision. *Theoria*, *63*(1–2), 14–23.
180 Makinson, D. (2003). Ways of doing logic: What was different about AGM 1985? *Journal of Logic and Computation*, *13*(1), 3–13.

181 Makinson, D. (2005). *Bridges From Classical to Nonmonotonic Logic*. Norcross: College Publications.
182 Makinson, D. (2008). *Sets, Logic and Maths for Computing*. Berlin: Springer.
183 Makinson, D. (2012). Logical questions behind the lottery and preface paradoxes: Lossy rules for uncertain inference. *Synthese*, *186*(2), 511–529.
184 Makinson, D., & Gärdenfors, P. (1991). Relations between the logic of theory change and nonmonotonic logic. In A. Fuhrmann & M. Morreau (Eds.), *The Logic of Theory Change* (pp. 183–205). Berlin: Springer.
185 Mares, E. D. (2002). A paraconsistent theory of belief revision. *Erkenntnis*, *56*(2), 229–246.
186 Mazzieri, M., & Dragoni, A. F. (2007). Ontology revision as non-prioritized belief revision. In P. Haase, A. Hotho, L. Chen, E. Ong, & P. C. Mauroux (Eds.), *Proceedings of the International Workshop on Emergent Semantics and Ontology Evolution (ESOE2007) at ISWC/ASWC2007, Busan, South Korea* (pp. 58–69). CEUR Workshop Proceedings (CEUR-WS.org).
187 McKinsey, J. C. C. (1943). The decision problem for some classes of sentences without quantifiers. *Journal of Symbolic Logic*, *8*, 61–76.
188 Meyer, T., Heidema, J., Labuschagne, W., & Leenen, L. (2002). Systematic withdrawal. *Journal of Philosophical Logic*, *31*(5), 415–443.
189 Montmarquet, J. (2008). Virtue and voluntarism. *Synthese*, *161*(3), 393–402.
190 Moore, R. C. (1985). Semantical considerations on nonmonotonic logic. *Artificial Intelligence*, *25*, 75–94.
191 Moulin, H. (1985). Choice functions over a finite set: A summary. *Social Choice and Welfare*, *2*(2), 147–160.
192 Mourad, R. (2008). Choosing to believe. *International Journal for Philosophy of Religion*, *63*, 55–69.
193 Nayak, A., Goebel, R., Orgun, M., & Pham, T. (2006). Taking Levi identity seriously: A plea for iterated belief contraction. In J. Lang, F. Lin, & J. Wang (Eds.), *Proceedings of the First International Conference on Knowledge Science, Engineering and Management, (KSEM'06), LNAI 4092* (pp. 305–317). Berlin: Springer.
194 Nayak, A. C. (1994). Iterated belief change based on epistemic entrenchment. *Erkenntnis*, *41*(3), 353–390.
195 Nayak, A. C., Goebel, R., & Orgun, M. A. (2007). Iterated belief contraction from first principles. In *Proceedings of the Twentieth International Joint Conference of Artificial Intelligence (IJCAI-07)* (pp. 2568–2573).
196 Nayak, A. C., Pagnucco, M., & Peppas, P. (2003). Dynamic belief revision operators. *Artificial Intelligence*, *146*(2), 193–228.
197 Nickel, P. J. (2010). Voluntary belief on a reasonable basis. *Philosophy and Phenomenological Research*, *81*(2), 312–334.
198 Niederée, R. (1991). Multiple contraction. A further case against Gärdenfors' principle of recovery. In A. Fuhrmann & M. Morreau (Eds.), *The Logic of Theory Change* (pp. 322–334). Berlin: Springer.
199 Nottelmann, N. (2006). The analogy argument for doxastic voluntarism. *Philosophical Studies*, *131*(3), 559–582.
200 Nozick, R. (1981). *Philosophical Explanations*. Oxford: Clarendon.
201 Nute, D. (1975). Counterfactuals and the similarity of worlds. *Journal of Philosophy*, *12*, 773–778.
202 Nute, D. (1981). Introduction. *Journal of Philosophical Logic*, *10*, 127–147.
203 Nute, D. (1984). Conditional logic. In D. M. Gabbay & F. Guenthner (Eds.), *Handbook of Philosophical Logic* (Vol. 2, pp. 387–439). Dordrecht: Reidel.
204 Olsson, E. J. (1997). A coherence interpretation of semi-revision. *Theoria*, *63*(1–2), 105–134.
205 Olsson, E. J. (2003). Belief revision, rational choice and the unity of reason. *Studia Logica*, *73*(2), 219–240.
206 Perlis, D. (1997). Sources of, and exploiting, inconsistency: Preliminary report. *Journal of Applied Non-Classical Logics*, *7*(1–2), 13–24.

207 Pollock, J. L. (1981). A refined theory of counterfactuals. *Journal of Philosophical Logic, 10*, 239–266.
208 Pollock, J. L. (1986). The paradox of the preface. *Philosophy of Science, 53*, 246–258.
209 Pollock, J. L. (1994). Justification and defeat. *Artificial Intelligence, 67*(2), 377–407.
210 Ramsey, F. P. (1931). *Foundations of Mathematics and Other Logical Essays*. New York: Routledge.
211 Roberts, F. S. (1979). *Measurement Theory with Applications to Decisionmaking, Utility and the Social Sciences*. Reading, Massachusetts: Addison-Wesley.
212 Rott, H. (1989). Conditionals and theory change: Revisions, expansions, and additions. *Synthese, 81*(1), 91–113.
213 Rott, H. (1992). Preferential belief change using generalized epistemic entrenchment. *Journal of Logic, Language and Information, 1*(1), 45–78.
214 Rott, H. (1993). Belief contraction in the context of the general theory of rational choice. *Journal of Symbolic Logic, 58*(4), 1426–1450.
215 Rott, H. (1997). Vom Primat der praktischen Vernunft: Logische Regeln als Regeln rationaler Wahl. In G. Meggle (Ed.), *Analyomen 2. Proceedings of the 2nd Conference 'Perspectives in Analytical Philosophy'*, (Vol. 1, pp. 138–147). Berlin: Walter de Gruyter.
216 Rott, H. (1999). Moody conditionals: Hamburgers, switches, and the tragic death of an American president. In J. Gerbrandy, M. Marx, M. de Rijke, & Y. Venema (Eds.), *Essays Dedicated to Johan van Benthem on the Occasion of His 50th Birthday* (pp. 98–112). Amsterdam: Amsterdam University Press. http://www.illc.uva.nl/j50/contribs/rott.
217 Rott, H. (2001). *Change, Choice and Inference: A Study of Belief Revision and Nonmonotonic Reasoning*. Oxford: Oxford University Press.
218 Rott, H. (2003). Basic entrenchment. *Studia Logica, 73*(2), 257–280.
219 Rott, H. (2003). Coherence and conservatism in the dynamics of belief II: Iterated belief change without dispositional coherence. *Journal of Logic and Computation, 13*(1), 111–145.
220 Rott, H. (2009). Shifting priorities: Simple representations for twenty-seven iterated theory change operators. In D. Makinson, J. Malinowski, & H. Wansing (Eds.), *Towards Mathematical Philosophy* (pp. 269–296). Dordrecht: Springer.
221 Rott, H. (2011). Reapproaching Ramsey: Conditionals and iterated belief change in the spirit of AGM. *Journal of Philosophical Logic, 40*(2), 155–191.
222 Rott, H. (2014). Four floors for the theory of theory change: The case of imperfect discrimination. In E. Fermé & J. Leite (Eds.), *Logics in Artificial Intelligence. Proceedings of the 14th European Conference on Logics in Artificial Intelligence (JELIA 2014)* (pp. 368–382). Berlin: Springer.
223 Rott, H. (2017). Negative doxastic voluntarism and the concept of belief. *Synthese, 194*, 2695–2720.
224 Rott, H., & Hansson, S. O. (2014). Safe contraction revisited. In S. O. Hansson (Ed.), *David Makinson on Classical Methods for Non-Classical Problems* (pp. 35–70). Dordrecht: Springer.
225 Rott, H., & Pagnucco, M. (1999). Severe withdrawal (and recovery). *Journal of Philosophical Logic, 28*(5), 501–547.
226 Ryan, S. (1991). The preface paradox. *Philosophical Studies, 64*(3), 293–307.
227 Sandqvist, T. (2000). On why the best should always meet. *Economics and Philosophy, 16*(2), 287–313.
228 Satoh, K. (1988). Nonmonotonic reasoning by minimal belief revision. In *Proceedings of the International Conference on Fifth Generation Computer Systems, Tokyo 1988* (pp. 455–462). Institute for New Generation Computer Technology.
229 Schlechta, K. (1997). Non-prioritized belief revision based on distances between models. *Theoria, 63*(1–2), 34–53.
230 Schlossberger, E. (1978). Similarity and counterfactuals. *Analysis, 38*, 80–82.
231 Schwitzgebel, E. (2009). Knowing your own beliefs. *Canadian Journal of Philosophy, 39*(supl), 41–62.
232 Scott-Kakures, D. (1994). On belief and the captivity of the will. *Philosophy and Phenomenological Research, 54*, 77–103.

233 Segerberg, K. (1989). A note on an impossibility theorem of Gärdenfors. *Noûs*, *23*, 351–354.
234 Segerberg, K. (1995). Belief revision from the point of view of doxastic logic. *Bulletin of the IGPL*, *3*, 535–553.
235 Segerberg, K. (1999). Two traditions in the logic of belief: Bringing them together. In H. J. Ohlbach & U. Reyle (Eds.), *Logic, Language and Reasoning: Essays in Honour of Dov Gabbay* (pp. 135–147). Dordrecht: Kluwer Academic Publishers.
236 Segerberg, K. (2010). Some completeness theorems in the dynamic doxastic logic of iterated belief revision. *Review of Symbolic Logic*, *3*, 228–246.
237 Sen, A. K. (1970). *Collective Choice and Social Welfare*. San Francisco: Holden-Day.
238 Sen, A. K. (1971). Choice functions and revealed preference. *Review of Economic Studies*, *38*, 307–317.
239 Spohn, W. (2010). Multiple contraction revisited. In M. Suárez, M. Dorato, & M. Rédei (Eds.), *EPSA Epistemology and Methodology of Science: Launch of the European Philosophy of Science Association* (pp. 279–288). Dordrecht: Springer.
240 Stalnaker, R. (1968). A theory of conditionals. In N. Rescher (Ed.), *Studies in Logical Theory, American Philosophical Quarterly Monograph Series, I* (pp. 99–112). Cambridge: Blackwell.
241 Stalnaker, R. (1993). A note on non-monotonic modal logic. *Artificial Intelligence*, *64*, 183–196.
242 Stalnaker, R. (2009). Iterated belief revision. *Erkenntnis*, *70*(2), 189–209.
243 Stebbins, W. C. (1973). Hearing of old world monkeys (cercopithecinae). *American Journal of Physical Anthropology*, *38*, 357–364.
244 Tamminga, A. (2004). Expansion and contraction of finite states. *Studia Logica*, *76*(3), 427–442.
245 Ullmann-Margalit, E., & Morgenbesser, S. (1977). Picking and choosing. *Social Research*, *44*, 757–785.
246 van Benthem, J. (1989). Semantic parallels in natural language and computation. In H.-D. Ebbinghaus, J. Fernandez-Prida, M. Garrido, D. Lascar, & M. Rodríguez Artalejo (Eds.), *Logic Colloquium '87* (pp. 331–375). Amsterdam: North-Holland.
247 van Benthem, J. (1994). Logic and the flow of information. In D. Prawitz, B. Skyrms, & D. Westerståhl (Eds.), *Logic, Methodology and Philosophy of Science IX: Proceedings of the Ninth International Congress of Logic, Methodology, and Philosophy of Science, Uppsala, Sweden, August 7–14, 1991*. Studies in Logic and the Foundations of Mathematics (Vol. 134, pp. 693–724). Elsevier.
248 van Benthem, J. (2011). *Logical Dynamics of Information and Interaction*. Cambridge: Cambridge University Press.
249 van Ditmarsch, H., & Kooi, B. (2006). The secret of my success. *Synthese*, *151*(2), 201–232.
250 van Ditmarsch, H., van der Hoek, W., & Kooi, B. P. (2007). *Dynamic Epistemic Logic*. Berlin: Springer.
251 Wansing, H. (2000). A reduction of doxastic logic to action logic. *Erkenntnis*, *53*(1), 267–283.
252 Wansing, H. (2006). Doxastic decisions, epistemic justification, and the logic of agency. *Philosophical Studies*, *128*(1), 201–227.
253 Wassermann, R. (2001). Local diagnosis. *Journal of Applied Non-Classical Logics*, *11*(1–2), 107–129.
254 Wheeler, G. (2006). Rational acceptance and conjunctive/disjunctive absorption. *Journal of Logic, Language and Information*, *15*(1–2), 49–63.
255 Williams, J. N. (1987). The preface paradox dissolved. *Theoria*, *53*(2–3), 121–140.
256 Williamson, C. (1969). Analysing counterfactuals. *Dialogue*, *8*, 310–314.
257 Williamson, T. (2008). *The Philosophy of Philosophy*. New York: Wiley.
258 Wobcke, W. (1995). Belief revision, conditional logic and nonmonotonic reasoning. *Notre Dame Journal of Formal Logic*, *36*, 55–102.
259 Yap, A. (2014). Idealization, epistemic logic, and epistemology. *Synthese*, *191*, 3351–3366.
260 Zhang, D. (2004). Properties of iterated multiple belief revision. In V. Lifschitz & I. Niemelä (Eds.), *Logic Programming and Nonmonotonic Reasoning. Proceedings of the 7th International Conference (LPNMR 2004)* (pp. 314–325). Springer.

261 Zhang, D., Chen, S., Zhu, W., & Chen, Z. (1997). Representation theorems for multiple belief changes. In *Proceedings of the Fifteenth International Joint Conference on Artificial Intelligence, IJCAI 97, Nagoya, Japan, August 23–29, 1997* (Vol. 1, pp. 89–94). San Francisco: Morgan Kaufmann.

262 Zhang, D., & Foo, N. (2001). Infinitary belief revision. *Journal of Philosophical Logic*, *30*(6), 525–570.

263 Zhang, L. (2017). Believability relations for select-direct sentential revision. *Studia Logica*, *105*, 37–63.

264 Zhang, L., & Hansson, S. O. (2015). How to make up one's mind. *Logic Journal of the IGPL*, *23*, 705–717.

Index

A
absorption, left logical, 103
absurdity avoidance, 79–80
access
  direct, 48
  successive, 48
access set
  direct, 49
  finite, 49
  global, 66n
  local, 66
  successive, 49
accessibility function, 60, 86, 90
additive restriction, 128
adjunction, conjunctive, 136–137, 153
AGM model, 3–42, 105, 112n, 114, 119, 122–125, 134, 139n, 141, 151, 154–155, 157n
Alchourrón, Carlos, 6, 24, 28, 136
amplification, 79–80
And (postulate), 104, 106
Arrow, Kenneth (1921–2017), 18n
asymmetric distance, 88–89
asymmetry, 91
atom, linguistic, 24, 28n, 144n
atomic descriptor, 53–54
attitude
  epistemic, 49–50
  propositional, 100
augmented belief set, 96
autoepistemic belief, 37, 42, 52n, 54, 96, 166, 167
  dynamic, 96, 98
  static, 96
autoepistemic ignorance, 97
autoepistemic incompleteness, 98
autoepistemic omniscience, 97–98, 104
axiom of choice, 172, 193
axiomatic approach, 5, 17
Ayers, Michael R., 99

B
belief
  autoepistemic, 37, 42, 52n, 54, 96, 98, 166, 167
  dynamic, 96, 98
  false, 98, 167
  non-sentential, 45–49, 52
  perceptually based, 45–46
  scientific, 46
  static, 96
  strengthening, 52n
  weakening, 52n
belief base, 16, 24n, 50n, 65, 97n, 113n
belief change
  global, 16, 37–42, 52n
  indeterministic, 20n, 47, 58, 70–73, 74n, 113, 167
  internally generated, 47
  local, 37
  minimal, 58
  multiple, 16, 23, 47, 54, 130, 143–146, 160–161, 167
  non-prioritized, 30, 65–66, 102n, 118
belief descriptor, *see* descriptor
belief predicate, 37, 53
belief set, 4, 112n
  augmented, 96
  descriptor-definable, 57
  directly reachable, 48
  finite-based, 24
  inaccessible, 88, 90
  indirectly reachable, 48

S.O. Hansson, *Descriptor Revision*, Trends in Logic 46,
DOI 10.1007/978-3-319-53061-1

self-limited, 110
belief set ordering, 71
shrinking, 142–143
belief state, 46–52
intermediate, 24
believability, 127–129, 167
direct, 51
successive, 51
Bennett, Jonathan, 111n
Betz, Gregor, vi
blockage contraction, 73n, 157n
blockage relation, 73–74
blockage revision, 73–78, 107, 167
global, 91–93
Booth, Richard, 133n
bootstrap contraction, 159–162
bootstrapping selector, 161
bunch revision, 54n

C

C, system, 104
canonical contraction, 152
Cantor, Georg (1845–1918), 25, 190, 193
Cantwell, John, vi, 104n
cautious monotony, 69n, 103n
centrolinear descriptor revision, 70, 87, 101
contraction, 109–110, 152–154
dilated model, 109–110
global, 86, 92
revocation, 135–138, 167
sentential, 120, 167
change, *see* belief change
changeability, 48
characteristic set, 56
Chisholm, Roderick M., 34
choice
axiom, 172, 193
direct, 59
epistemic, 18–26
choice contraction, 54
choice function, 18–26, 154, 165
maximizing, 154, 156
monomaximizing, 155–156
monoselective, 59
selection function (AGM), 6, 42, 165
choice revision, 23, 54, 130, 144, 167
choice success, 130
classical logic, 4n, 34, 37, 50
closeness, 12, 58, 70
see also distance, epistemic proximity, *and* similarity
closure
logical, 4–5
belief state, 50
contraction, 8, 135–136, 138, 153
descriptor revision, 65, 68, 71
sentential revision, 9, 118–120, 123
coextensive descriptor revision, 86–87
commitment, epistemic, 4, 97, 108
compactness, 172
compartment of belief state, 162
completeness, 12n, 184
complexity, 162
composite descriptor, 53–54
composition, conjunctive, 136–137, 153
conditional sentence, 12, 34–36, 42, 52n, 98–111, 166–167
doxastic, 98
elicited, 106
epistemic, 98–99
indicative, 99
nested, 96
ontic, 98–99
subjunctive, 99
confirmation
descriptor revision, 67, 71, 101
sentential revision, 118, 120
pseudo-confirmation, 69
confuting information, 148
conjunctive adjunction, 136–137, 153
conjunctive composition, 136–137, 153
conjunctive inclusion, 8–9, 29n
conjunctive overlap, 8–9, 29n
conjunctive product, 145
conjunctiveness, 15, 81, 139
connectedness (completeness), 12n, 184
peripheral weak, 76
weak, 91
consequence relation, 4
consistency, 9, 50, 55, 67, 119, 123
context dependence, 106
contraction, 4, 151–162
blockage, 73n, 157n
bootstrap, 159–162
canonical, 152
centrolinear, 109–110, 152–154
choice, 54
entrenchment-based, 15–16
finite-based, 28, 29n
full meet, 6
global, 37–39, 143n
hypothetical, 32
kernel, 16
maxichoice, 6, 24n, 137, 154–155
maximal, 154

monomaximal, 155–156
monoselective, 152–153
multiple, 47, 54, 143, 160–161
package, 54, 160
partial meet, 6–7, 38, 122, 124, 154–155
perimaximal, 156–159
possible world model, 13–14
pure, 31–32
safe, 16
shielded, 66, 102n
contractive outcome set, 152
contractiveness, 152
co-occurrence test, 108–110
core, 66n
countable set, 190, 193
counter-dominance, 79–80
counterfactual, *see* conditional sentence
coupling, 79–80
credible sentence, 66n
CS, 111
cumulative monotony, 103
cumulative reasoning, 104
cumulativity
conditional, 103
descriptor revision, 68, 71
Ramsey descriptor, 103
sentential revision, 69n, 120–121
peripheral, 75
cut
conditional, 103
revocation, 152
sentential revision, 69n
cut primacy, 125
cycle, 91, 121

D

Darwiche-Pearl postulates, 39–41, 126–127
database, 3, 46
DDL, 113
de Rijke, Maarten, 113
decision theory, 3, 18
decomposition
postulate, 136–137, 147–148, 153
principle, 7, 23, 33–34, 137n
deductivism, 33
del Val, Alvaro, 97
delimiter, 109
density of outcome set, 65
depletability
direct, 51
successive, 51
descriptor, 53–57
atomic, 53–54
composite, 53–54
dynamic, 95–114
finite, 53
inconsistent, 55
maximally specified, 56, 80n, 81
maxispecified, 56, 80n, 81
molecular, 53–54
negatable, 113
negated, 56
non-negatable, 56
satisfiable, 54
static, 95, 101
descriptor-definable belief set, 57
descriptor disjunction, 56
descriptor falsum, 56
descriptor revision, 45–93
centrolinear, 70, 87, 101, 109–110, 120, 167
coextensive, 86–87
dilated centrolinear, 109–110
distance-based, 87–90, 126–127
exhaustive centrolinear, 109–110
global, 85–93, 112, 114, 166
linear, 67–70
local, 63
monoselective, 64–67, 86
proximity-based, 82
relational, 67–70
sentential centrolinear, 120, 167
sentential monoselective, 118–120
descriptor-wellfoundedness, 71
deterministic model, 47
dilated centrolinear model, 109–110
direct access, 48–49
direct believability, 51
direct choice, 59
direct depletability, 51
direct necessity, 112
direct possibility, 112
direct removability, 51
directly reachable belief set, 48
disjunction
descriptor, 56
infinite, 130n, 144n
disjunctive factoring, 75n, 123
disjunctive identity
descriptor revision, 75
sentential revision, 75n
peripheral, 75
disjunctive implication, 176
disjunctive inclusion, 123
disjunctive interpolation, 122

disjunctive overlap, 123
disjunctiveness, 184
distance, 88–89, 91n, 98, 126
  asymmetric, 88–89
  Euclidean, 89n
  non-negative, 89
  pseudodistance, 89
  symmetric, 88, 92
  see also closeness *and* epistemic proximity
distance-based descriptor revision, 87–90, 126–127
dominance, 15, 81, 139
doxastic conditional sentence, 98
doxastic logic, 113–114
dynamic autoepistemic belief, 96, 98
dynamic descriptor, 95–114
dynamic doxastic logic, 113
dynamic equivalence, 52
dynamic information, 95
dynamic logic, 113

E
Edgington, Dorothy, 100
elicited conditional sentence, 106
entrenchment, 15–26, 78–82, 139–146, 167
  multiple, 144, 167
entrenchment-based contraction, 15–16
epistemic attitude, 49–50
epistemic choice, 18–26
epistemic commitment, 4, 97, 108
epistemic conditional sentence, 98–99
epistemic entrenchment, 15–26, 78–82, 139–146, 167
  multiple, 144, 167
epistemic proximity, 78–84, 101, 127, 144, 167
epistemic voluntarism, 18
equivalence
  dynamic, 52
  left logical, 103
  negation, 131
  static, 52
equivalence relation, 12n
Euclidean distance, 89n
ex falso quodlibet, 102
exhaustive centrolinear model, 109–110
expansion
  operation, 4
  property of sentential revision, 33, 42, 166
expressive power, 24, 45, 52, 79, 111
extensionality
  contraction, 8, 138
  descriptor revision, 65
  sentential revision, 9, 118–121, 123

F
factoring, disjunctive, 75n, 123
false belief, 98, 167
falsum, 56
  descriptor, 56
favouring, 67, 118, 119n
Fermé, Eduardo, vi
finite descriptor, 53
finite direct access set, 49
finite gradation, 121
finite outcome set, 49
finite representability, 51
finite successive access set, 49
finite-based belief set, 24
finite-based contraction, 28, 29n
finite-based outcome, 27–29, 42, 51, 121, 166
finite-based revision, 28
finiteness
  logical, 24n, 25
  syntactical, 24n
first-order logic, 34n
Fuhrmann, André, 36–37
Fuhrmann's test, 36–37
full meet contraction, 6
function
  accessibility, 60, 86, 90
  choice, 18–26, 154, 165
  input assimilation, 47, 50
  location, 126
  selection, 6, 42, 165
  support, 49–52
  transformation, 82–84
  transition, 47n

G
Gärdenfors, Peter, 6, 15, 28, 105, 108
Gärdenfors' impossibility theorem, 36, 105
generic belief state model, 47
Glaister, Stephen Murray, 31
global access set, 66n
global blockage revision, 91–93
global centrolinear revision, 86, 92
global change, 16, 37–42, 52n, 85–93, 112, 114, 126–127, 143n, 166
global contraction, 37–39, 143n
  multiple, 143n

global descriptor revision, 85–93, 112, 114, 166
global regularity, 87
global sentential revision, 39–42, 126–127
Goranko, Valentin, vi
gradation, finite, 121
Grove, Adam, 10, 12, 202
Grove model, 9–16, 25, 112, 124, 137n, 201
  contraction, 13–14
  revision, 11–13
Grove's bijection, 14n–15n

H

Harper identity, 7, 151
Hintikka, Jaakko (1929–2015), 114
Horn clause, 57n
hypothetical contraction, 32
hypothetical input, 106n, 107n

I

idealization, 4, 28, 65, 148–149, 162
identity, disjunctive, 75
ignorance, autoepistemic, 97
implication, disjunctive, 176
implicative reciprocity, 138
impossibility theorem
  AGM contraction, 154–155
  Gärdenfors, 36, 105
impression, sensory, 45–46
inaccessible belief set, 88, 90
inclusion
  contraction, 8, 31–32, 42, 133–134, 153, 166
  sentential revision, 9, 119, 123
  conjunctive, 8–9, 29n
  disjunctive, 123
incompleteness, autoepistemic, 98
inconsistency, 9, 50, 55, 67, 119, 123
  avoidance, 67
  local, 50n
inconsistent descriptor, 55
indeterministic belief change, 20n, 47, 58, 70–73, 74n, 113, 167
indicative conditional, 99
indirectly reachable belief set, 48
inference
  non-monotonic, 69n, 107–111
  single-premiss, 66, 67n
inferribility, 108–109
infinite disjunction, 130n, 144n
information
  dynamic, 95
  minimal confuting, 148
  static, 95
injectivity, 52
input, 4, 45–47
  hypothetical, 106n, 107n
input assimilation function, 47, 50
intermediate belief state, 24
internally generated change, 47
interpolation, disjunctive, 122
intersection
  preservation of optimality, 20–22
  preservation of success, 20, 22–23, 42, 57, 165
intricacy, 162
irreflexivity, 91
iterated blockage revision, 91–93
iterated centrolinear revision, 86, 92
iterated change, 16, 37–42, 52n, 85–93, 112, 114, 126–127, 143n, 166
iterated contraction, 37–39, 143n
  multiple, 143n
iterated descriptor revision, 85–93, 112, 114, 166
iterated sentential revision, 39–42, 126–127
iterative necessity, 112
iterative possibility, 112

J

justificatory structure, 97

K

kernel contraction, 16
Konieczny, Sébastien, 40
Kraus, Sarit, 104, 107

L

left absorption, 103
left logical absorption, 103
left logical equivalence, 103
lefthand uniqueness, 89
Lehmann, Daniel, 91n, 104, 107
Levi, Isaac, 3, 4, 7, 35, 97, 111
Levi identity, 7, 33, 39
Lewis, David (1941–2001), 12, 106
limited recovery, 29n
Lindström, Sten, 98
linear descriptor revision, 67–70
linear ordering, 12n
local access set, 66
local belief change, 37
local descriptor revision, 63

local inconsistency, 50n
local regularity, 87
location function, 126
logic
  classical, 4n, 34, 37, 50
  doxastic, 113–114
  dynamic, 113
  first-order, 34n
  modal, 36, 42, 97, 111–114, 166–167
  paraconsistent, 50n
  predicate, 34n
  truth-functional, 4n, 34, 37, 50
  update, 113, 167
logical closure, 4–5
logical finiteness, 24n, 25
logical necessity, 110n
logical omniscience, 4n
loop, 91

M
Magidor, Menachem, 104, 107
making up one's mind (resolution), 55, 130–131, 165
Makinson, David, vi, 6, 15, 24, 28–29, 31, 108, 133n, 136, 162n
maxichoice contraction, 6, 24n, 137, 154–155
maximal contraction, 154
maximal preservativity, 153n
maximality
  contraction, 153–154
  entrenchment, 15, 82, 139
maximally specified descriptor, 56, 80n, 81
maximizing choice function, 154, 156
maximizing relation, 154n
maximizingness, 124–125
maximum, unique, 156, 159–161
maxispecified descriptor, 56, 80n, 81
mental state, 47–49, 167
  dynamics, 49
minimal change, 58
minimal confuting information, 148
minimality, 15, 81–82, 139
modal logic, 36, 42, 97, 111–114, 166–167
molecular descriptor, 53–54
monomaximal contraction, 155–156
monomaximizing choice function, 155–156
monoselective choice function, 59
monoselective contraction, 152–153
monoselective descriptor revision, 64–67, 86
monoselective revocation, 134–135
monoselective sentential revision, 118–120
monotony
  cautious, 69n, 103n
  cumulative, 103
  revision, 35–36
multiple change, 16, 23, 47, 54, 130, 143–146, 160–161, 167
  contraction, 47, 54, 143, 160–161
  global contraction, 143n
  revocation, 143–146
  sentential revision, 23, 54, 130, 144, 167
multiple epistemic entrenchment, 144, 167

N
near-superiority, 76–77
necessity, 111–114
  direct, 112
  iterative, 112
  logical, 110n
  physical, 110n
negatable descriptor, 113
negatable relative success, 131
negated descriptor, 56
negation equivalence, 131
negative self-knowledge, 37
negative transmission, 91–92
nested conditional sentence, 96
non-circularity, 91, 121
non-convergence, 48
non-inclusion, 37
non-inferiority, 77
non-monotonic inference, 69n, 107–111
non-negatable descriptor, 56
non-negativity, of distance, 89
non-occlusion, peripheral, 75
non-prioritized belief change, 30, 65–66, 102n, 118
non-reversion, 48
non-sentential belief, 45–49, 52
Nozick, Robert (1938–2002), 111n
Nute, Donald, 109n

O
omniscience, 24
  autoepistemic, 97–98, 104
  logical, 4n
ontic conditional sentence, 98–99
optimality, preservation under intersection, 20–22
Or (postulate), 104
order extension principle, 193
ordering, 12n
  belief set, 71

linear, 12n
partial, 12n
shrinking, 142–143
strict, 12n
weak, 12n
outcome, finite-based, 27–29, 42, 51, 121, 166
outcome set, 64, 86
contractive, 152
density, 65
finite, 49
overlap
conjunctive, 8–9, 29n
disjunctive, 123

P

P, system, 104
package contraction, 54, 160
package revision, 54, 130
package revocation, 144, 146
package success, 130
paraconsistent logic, 50n
partial meet contraction, 6
full meet, 6
global, 38
maxichoice, 6, 24n, 137, 154–155
transitively relational, 7, 122, 125, 154–155
partial meet revision, 7
transitively relational, 7
partial ordering, 12n
perceptually based belief, 45–46
perimaximal contraction, 156–159
peripheral cumulativity, 75
peripheral disjunctive identity, 75
peripheral non-occlusion, 75
peripheral weak connectedness, 76
persistence, 66n, 135–136, 153
physical necessity, 110n
pick revision, 54n
picking, 19
Pino Pérez, Ramón, 40
positive self-knowledge, 37
possibility, 36, 111–114
direct, 112
iterative, 112
serious, 111
possible theories model, 112, 114
possible world model, 9–16, 25, 112, 124, 137n, 201
contraction, 13–14
revision, 11–13
predicate, belief, 37, 53
predicate logic, 34n
preface paradox, 162n
preferential reasoning, 104
preservation under intersection
optimality, 20–22
success, 20, 22–23, 42, 57, 165
preservativity, 159
maximal, 153n
product, conjunctive, 145
property CS, 111
proposition, 46n
propositional attitude, 100
proximity, epistemic, 78–84, 101, 127, 144, 167
proximity-based descriptor revision, 82
pseudo-confirmation, 69
pseudodistance, 89
pure contraction, 31–32

R

Rabinowicz, Wlodek, 98
Ramsey, Frank, 34, 53n, 99
Ramsey descriptor, 98–105, 166
Ramsey test
for conditionals, 34–36, 42, 96, 99–100, 105, 166–167
for non-monotonic inference, 108–111
alternatives to, 105–111
rationality, 3, 5, 17–18, 25–26, 32–33, 37, 65, 67, 107
reachability, 48, 113
reasoning
cumulative, 104
non-monotonic, 69n, 107–111
preferential, 104
reciprocity
contraction, 137–138
descriptor revision, 69
sentential revision, 69, 122–123
implicative, 138
recovery, 8, 30–31, 42, 153, 166
limited, 29n
reflexivity, 103
regularity
descriptor revision, 65–66, 68, 71
sentential revision, 66n, 107, 118, 121
global, 87
local, 87
relational descriptor revision, 67–70
relative success
contraction, 135–136, 153

descriptor revision, 65–66, 68, 71
sentential revision, 66n, 107, 118, 120–121
negatable, 131
remainder, 6, 25, 157n
removability
direct, 51
successive, 51
repertoire, 64
replacement, 54
representability
finite, 51
sentential, 45–46, 49, 52
representation theorem, 5
resolution (making up one's mind), 55, 130–131, 165
restriction
additive, 128
subtractive, 140
retainability, 48
retractable sentence, 66n
retraction, 133n
revision, *see* descriptor revision *and* sentential revision
revision monotonicity, 35–36
revisionary revocation, 146–149, 167
revocation, 133–149
centrolinear, 135–138, 167
monoselective, 134–135
multiple, 143–146
package, 144, 146
revisionary, 146–149, 167
revocation cut, 152
right weakening, 103
righthand uniqueness, 89
ring, 201
Rott, Hans, 29, 50n, 54n, 75n, 97n, 99, 138, 167

S

safe contraction, 16
Sandqvist, Tor, 21
satisfiable descriptor, 54
Schlossberger, Eugene, 109n
scientific beliefs, 46
Segerberg, Krister, 113–114
select-and-intersect, 19–24, 27, 59
select-and-pick, 19
selection function, 6, 42, 165
see also choice function
selector
sentential, 28n, 160–161
bootstrapping, 161
self access, 91
self-closeness, 89
self-knowledge
negative, 37
positive, 37
self-limited belief set, 110
semi-revision, 65–66, 118
sensory impression, 45–46
sentential representation, 45–46, 49, 52
sentential revision, 4–5, 59n, 117–131
blockage, 73–78, 107, 167
bunch, 54n
centrolinear, 120, 167
choice, 23, 54, 130, 144, 167
finite-based, 28
global, 39–42, 126–127
monoselective, 118–120
multiple, 54, 130
package, 54, 130
partial meet, 7
pick, 54n
possible world model, 11–13
semi-revision, 65–66, 118
sentential selector, 28n, 160–161
sententiality, 52
serious possibility, 111
shielded contraction, 66, 102n
shiftability, 106n
shrinking belief set ordering, 142–143
similarity, 12, 58
single-premiss inference, 66, 67n
social choice theory, 18–19, 21, 110n–111n
specification, 121
sphere model, 9–16, 25, 112, 124, 137n, 201
contraction, 13–14
revision, 11–13
stalemate (tie) 72, 148, 157
Stalnaker, Robert, 34, 40, 99
Stalnaker property, 69n
state of mind, 47–48, 167
static descriptor, 95, 101
static equivalence, 52
static information, 95
strengthening of belief, 52n
strict ordering, 12n
subexpansion, 9, 123
subjunctive conditional sentence, 99
subtractive restriction, 140, 144
success, 22–23, 54, 106
contraction, 5, 8, 29–30, 42, 133, 135, 162, 165
descriptor revision, 65

package, 130
sentential revision, 5, 9, 29–30, 42, 109, 112n, 119, 123, 165
choice, 130
negatable relative, 131
relative, 65–66, 68, 71, 107, 118, 120–121, 135–136, 153
preservation under intersection, 20, 22–23, 42, 57, 165
success set, 66
successive access, 48–49
successive believability, 51
successive depletability, 51
successive removability, 51
superexpansion, 9, 123
superiority
blockage relation, 91
descriptor revision, 77
support function, 49–52
supported sentence, 50
symmetry, of distance, 88, 92
syntactical finiteness, 24n
system C, 104
system P, 104

T
ternary relation, 91
theory, 112n, 114
three-place relation, 91
tie (stalemate), 72, 148, 157
time, representation of, 47
top adjacency, 76–77
transformation function, 82–84
transition function, 47n
transitively relational partial meet contraction, 7, 122, 125, 154–155
transitively relational partial meet revision, 7
transitivity, 6
belief set ordering, 68–69
blockage relation, 91
epistemic entrenchment, 15, 81, 139
epistemic proximity, 79
selection function, 6–7
transmission, negative, 91–92
trichotomy, 12n
truth maintenance system, 3
truthfulness, 66n, 98
truth-functional logic, 4n, 34, 37, 50

U
uniformity
contraction, 135, 137, 153
descriptor revision, 59, 65, 69, 71, 176
sentential revision, 118, 121
unique maximum, 156, 159–161
uniqueness
lefthand, 89
of distance, 89
righthand, 89
unitarity, 103
update logic, 113, 167
upper bound property, 172

V
vacuity
contraction, 8, 33, 135–136, 138, 153
sentential revision, 9, 37, 119, 123
vagueness, 107n
van Benthem, Johan, vi, 113
van Ditmarsch, Hans, vi
voluntarism, epistemic, 18

W
weak connectedness, 91
peripheral, 76
weak ordering, 12n
weakening
of belief, 52n
right, 103
wellfoundedness, 110n, 135
well-ordering, 110n
withdrawal, 133n
world-ring, 201

Z
Zhang, Li, vi, 121n
Zorn's lemma, 172

MIX
Papier aus verantwortungsvollen Quellen
Paper from responsible sources
FSC® C105338

If you have any concerns about our products,
you can contact us on
ProductSafety@springernature.com

In case Publisher is established outside the EU,
the EU authorized representative is:
Springer Nature Customer Service Center GmbH
Europaplatz 3, 69115 Heidelberg, Germany

Printed by Libri Plureos GmbH
in Hamburg, Germany